Global Communications since 1844

Published in cooperation with the Center for American Places, Santa Fe, New Mexico, and Harrisonburg, Virginia

Global Communications since 1844

Geopolitics and Technology

PETER J. HUGILL

The Johns Hopkins University Press
Baltimore and London

The Johns Hopkins University Press
2715 North Charles Street
Baltimore, Maryland 21218-4363
www.press.jhu.edu

Library of Congress Cataloging-in-Publication Data will be found
at the end of this book.
A catalog record for this book is available from the British Library.

Frontispiece: Britain the World Centre (Cable and Wireless Great
Circle Map, 1945). Courtesy Cable and Wireless Archives.

ISBN 0-8018-6039-3
ISBN 0-8018-6074-1 (pbk.)

To Don Meinig, from whom I should have learned more, earlier!

Contents

List of Figures and Tables   ix

Preface   xi

Acknowledgments   xv

ONE Information Technology, Geopolitics, and the
World-System   1

TWO Telegraphy and the First Global Telecommunications
Hegemony   25

THREE "The Whole World Kin": Telephony and the Development
of the Continental Polity to 1956   53

FOUR Radio Telegraphy, Radio Telephony, and Interstate
Competition, 1896–1917   83

FIVE Challenges to British Telecommunications Hegemony:
Continuous Wave Wireless   109

SIX Military Uses of Radio Communication: The Development
of Communications, Command, and Control   139

SEVEN Communications, Command, and Control in the War
in the Air: Radar, World War II, and the Slow Transition
to American Power   159

EIGHT Telecommunications and World-System Theory   223

Glossary   253

References   257

Name Index   269

Subject Index   272

Figures and Tables

FIGURES

Frontispiece. Britain the World Centre (Cable and Wireless Great Circle Map, 1945)

1.1. The Three German *Ostmarks*   6

1.2. Afro-Eurasia United by Railways and Airplane Routes as Mackinder Saw It in 1919   13

1.3. German Cartoon, 1908, Showing a Zeppelin Penetrating Britain via the Thames Estuary   14

2.1. French Visual Telegraph System, c. 1846   26

2.2. Proposed Routes of Western Union's Russian-American Telegraph and the British Atlantic Cable, 1865   40

2.3. European Telegraphs in the 1870s   42

2.4. American, British, and German Submarine Cables, 1911   44

2.5. The Cable and Wireless Network, 1934   50

3.1. Long Lines Network of the American Telephone and Telegraph Company, 1925   58

3.2. International Wireless Telephony, 1937   59

3.3. Projected European Trunk Cable System, 1924   60

3.4. Capacity and Range of Telephone Systems, 1880–1985   64

3.5. Long-Distance Lines of the American Telephone and Telegraph Company, 1906   68

3.6. L1 Long-Distance Coaxial Cable Routes of the Bell System, 1939   72

3.7. L3 Long-Distance Coaxial Cable Routes of the Bell System, 1953   73

3.8. TD2 Microwave System Routes, 1980   78

3.9. Coaxial Cable Routes in the Third Reich, 1945   81

4.1. The Electromagnetic Spectrum   85

4.2. The Three Major Layers of the Ionosphere and Their Effects on Wireless Transmission   87

4.3. Telefunken's "Quenched Spark" Wireless Transmission   90

4.4. Marconi's Proposed "Imperial Chain," 1912   98

4.5. Marconi's Proposed "Wireless Girdle Round the Earth," 1912   100

4.6. Poulsen Arc Wireless Services of the United States Navy, 1919   106

5.1. Damped Oscillations from Spark Transmission and Uniform Waves from Continuous Waves   110

5.2. Federal Telegraph's Continuous Wave Wireless Telegraphy Network, 1912   112

5.3. International Radio Communication, 1922   126

5.4. Diurnal Variations in Quality of Transatlantic Wireless Voice Transmission, 1923   129

5.5. American Air Mail Beacons, 1931   136

7.1. The Chain Home Radar System, 1940   180

7.2. The German Himmelbett Night-Fighting System, Summer 1943   205

8.1. The Transatlantic Submarine Telephone Cables, TAT-1 through TAT-7   232

8.2. Global Submarine Cables, c. 1970   234

8.3. Global Fiberoptic Submarine Cables, 1989   238

8.4. Global Megalopolises: Present, Emergent, and Possible   248

TABLES

8.1. Long Cycles and Geopolitical World Orders since 1790   224

8.2. Submarine Telephone Cable Capacity, 1956–1983   231

8.3. Intelsat Telephone Circuits, 1965–1981   235

Preface

Historians get what my twelve-year-old daughter, Laura, rather charm- ingly calls "confussed" over dates. As a historical geographer I am less "fussed" over exact dates since area is more important to me than epoch. I am thus less confused about what dates imply for systems, which is clo- sure. I am also more of a social scientist than a humanist, and thus my ap- proach differs from historians' in that my training leads me to search for theory. I prefer a history informed by social science—as well as a social science informed by history—in which closure takes second place to the- ory formulation. But I am not unsympathetic to the historian's need for dates. I merely prefer to follow one of my mentors, Donald W. Meinig, and say that I am less concerned with boundaries, where everything is fuzzy, and more concerned with cores, where everything is sharp. I have set 1844 as a starting date for this book since the boundary there is rea- sonably clear. In that year the world's first user-friendly public telegraph line opened, reaching all the way from Washington, D.C., to Baltimore, Maryland. Please note the adjectival qualifiers. There had been telegraphs before, even electric ones. But Morse's 1844 line transmitted signals using Morse's user-friendly code rather than immensely com- plex—and slow—indicator needles. The line was also public. Previous telegraphs had been military- or company-owned, such as those built for the new railways in Britain. Starting in 1844, however, any company, agency, or person could pay to have messages transmitted on a public sys- tem.

My fuzzy ending boundary may leave historians, but, I hope, not social scientists, much more "confussed." I could plump for 1945, when Amer- ica became the undisputed military as well as economic hegemon in the capitalist world-system, because it gives me near symmetry with my start- ing date. But America did not enjoy telecommunications hegemony in 1945, as this book demonstrates. Or I could plump for 1956, when the opening of Bell's transatlantic telephone cable ushered in a new era of American telecommunications dominance. But Bell's cable was built in British cable yards. Or I could plump for 1971, when Intelsat IV opened the "fully mature phase of satellite communications" (Pritchard 1977, 295). But 1971 saw the formation of the European Space Agency, which would successfully combat American dominance in the development of satellite launch vehicles. Put simply, American hegemony in what I argue is the key technology of the past 130 years—telecommunications—has been brief and insecure compared with the remarkable global domi- nance America has enjoyed economically, starting in 1917, and militari- ly, starting in 1945. I therefore end my main text, the first seven chapters, in 1945, with America replacing Britain as military hegemon. Chapter 8 carries the narrative through the development of the transatlantic telephone and the success of Intelsat, but it also attempts to use the theoretical premises of the first seven chapters to look into the future. A huge amount of technological development occurred in telecommunications from 1945 to 1971, some of which I cover very briefly in chapter 8. From 1971 to the present there have been perhaps even more developments. To cover this period in the same detail as I cover the period up to 1945 would require a book twice as long as the present one, yet it would not change my conclusions.

In the past few years there has been an explosion of academic interest in the emergence of the information economy, some of it by geographers. Much of this research is fascinating, some of it is original, but it all suffers from the same basic problems, the most serious being that it assumes that the information economy is a recent development and best construed at the level of an academic concentration such as city planning or electrical engineering. Such a focus illustrates what is at once the great strength and the great weakness of our bureaucratized academic society: that we have compartmentalized things to such a degree. Occasionally a historian enters the fray, usually one interested in the history of technology, and reminds us that a particular technology has been around a lot longer than anyone seems to realize. This is fun, and it is a useful corrective to the technospeak beloved of much of the literature, but it hardly gets us much "forrader." The problem is how to put the elephant back together based on the various descriptions by the usual blindfolded gropers after truth: a leg here, a trunk there, and a tail floating somewhere off in left field.

Although I willingly grant that my own discipline suffers from much of the same nitpicking bureaucratic subcompartmentalization as do many other disciplines, at least some geographers seem to try to override such compartmentalization. To be fair, I have good friends in many disciplines who share the same goal, but I think geography is special. This may result in part from one of our (many) intellectual roots, the subdiscipline geopolitics, first brought into prominence by Halford Mackinder in late Victorian and Edwardian Britain. At its best, geography has been obsessed with explaining the behavior of human beings and human societies in space. Sometimes this obsession has generated its own species of mathematized technospeak, as in the more theoretical end of regional science. Sometimes it has merely emphasized mastery of a useful technology that can be marketed to students as a commodified skill, such as the current obsession with geographic information systems (GIS). Sometimes geography has drifted off into humanistic and recondite fields, such as the ethnographic studies of vanishing cultures beloved of the old Berkeley School of Cultural Geography. Sometimes it has concerned itself with human interactions with the physical world. Like all modern disciplines that claim to belong to the social sciences or the humanities, geography is currently awash with isms, some of which will likely prove useful in the long run; most, however, will prove to be intellectual dead ends. Geography is not, in short, a narrow academic discipline. This is at once both worrisome, since academics increasingly seem to prefer the confines of their compartments, and wonderful, since it allows geographers more freedom than their colleagues in most disciplines to explore the ideas of others. For me, at least, geography represents what academe as a whole ought to be but usually is not because of turf wars and predatory administrations. Readers of all stripes may, of course, wish to take me behind their respective woodsheds for not including their specialization in what follows.

Mackinder is the central focus of much of this book, not because I intended him to be when I started the book, but because he snuck in and took over. Since the book is about hegemony in the capitalist world-system it is necessarily also about hegemonic transitions. In 1919 Mackinder resoundingly forecast both the failure of World War I to solve the German question and the military resurgence of Germany within a generation. He was roundly disbelieved—until 1939! He proposed that the only solution to the German question was to prevent the emergence of any powerful, centralized, bureaucratic state in western Eurasia. This would require breaking down not only Imperial Germany but also the newly formed Soviet Union into its ethnic and religious components. As a radical devolutionist, he would have greatly enjoyed the recent failure of the Soviet experiment. But he also supported radical devolution at home. He wanted independent northern and southern political regions within England itself, not to mention home rule for the Celtic fringe. (Today, with Tony Blair as prime minister and "New Labour" in power, at least the latter may occur.) Mackinder reasoned that because of the growth of state bureaucracies, most governments were too distant from the people they served and that the needs of the people would best be served by radical devolution. It is strange to hear his thought echoed, though usually less capably expressed, in what passes for modern political discourse.

Mackinder's great strength was that he understood the impact of the technical changes occurring in his world upon the geopolitics of that world. He did not invent geopolitics—Friedrich Ratzel did—but he made it his own. Better overland communications, in the form of railways and telegraphs, and better political control, in the form of national bureaucracies, worried him when they were coupled with the rise of a new, territorial nationalism based on the example of the French Revolution. They offered the possibility of the unification, by force if necessary, of a large continental polity that could come to dominate the huge resource base of the Eurasian landmass. At that point, he noted, the "Empire of the World" would begin and the days of the maritime Atlantic trading states would be numbered.

It seems to me that Mackinder's geopolitics have much to say to world-system theory, which since 1974 has developed out of the pioneering efforts of Immanuel Wallerstein into a small and pleasingly interdisciplinary academic growth industry. This book is partly an attempt to make that link, albeit with more reference to the work of historical and political geographers, notably Donald W. Meinig and Peter J. Taylor, than to Wallerstein directly. Mackinder's obvious concern with military activities led me to the excellent recent interpretations of the relationship between coercion and capital in the work of Charles Tilly. His concern with bureaucracies led me to Michael Mann's fine updating of the work of Max Weber.

Finding a theoretical underpinning for communications media proved difficult. Compartmentalization has divided studies of communication between areas of the humanities and of engineering, which makes for no discourse. There is no modern theoretical synthesis of the role of communications to match that of Meinig, Taylor, Wallerstein, Tilly, or Mann. There is always Marshall McLuhan, whose work was trendy back in the Dark Ages, when I was in graduate school. It seemed wiser to go back to McLuhan's own mentor, the Canadian economic historian Harold A. Innis, of the University of Toronto (the real UT, although the University of Tennessee is much older), who also had an impact on one major and admirable school of historical geography—that represented by the work of Andrew Hill Clark and his students at the University of Wisconsin–Madison. It was in Innis' work published as *Empire and Communications* (1950) that I found the bare bones of a theoretical structure that I could apply to the current problem. I cannot imagine how this work, delivered as the Beit lectures on imperial economic history at Oxford University in England, was received at the time. I hope it was well received, but Innis' work seems to have faded into the obscurity of Canadian studies. He was too broad a thinker for his times, in a world where the triumph of Keynesianism meant that most economists found his brand of macro-thinking obsolete. If the work of Wallerstein and the world-system theorists has kept macrotheory from being dumped into the dustbin of history by the supporters of Marx, I hope this work helps to rescue Innis.

Acknowledgments

As always, more people and organizations deserve thanks than I can include here. I apologize to them in advance. George F. Thompson, president of the Center for American Places, encouraged me to write this extended footnote to *World Trade since 1431*. It took me a lot longer than I (and, probably, he) supposed it would. As the book developed, I found myself drawn into the rapidly expanding area of the history of defense electronics. I also had to seek out people who could introduce me to certain areas of engineering. I thank George for tolerating these delays and for handling matters with the Johns Hopkins University Press so I could get on with writing. I thank Joanne Allen, of Johns Hopkins, for doing a fine job of copy-editing. Likewise, I thank the outside readers for the Press, notably John Guilmartin, who made numerous exceptionally useful suggestions concerning my discussion of military technologies. Donald W. Meinig and Brian Blouet both read an earlier version of the manuscript. I have been greatly influenced by the work of both men and by extended periods of contact with them—Don when he served on my dissertation committee at Syracuse University, which he effectively cochaired with David Sopher, and Brian when he served as chair of the geography department at Texas A&M University. I came to Mackinder with an appreciation born of my English academic upbringing, but Don Meinig's love of Mackinder's work deepened that appreciation and Brian Blouet made me read and reread Mackinder's work so that I could understand his insightful biography of the man.

John Trenouth, curator of the television section of the British National Museum of Photography, Film, and Television; Dr. John Bryant, of the Department of Electrical Engineering and Computer Science at the University of Michigan; and Arthur Bauer, of the Centre for German Communication and Related Technology, helped me with chapter 7. It is better for their input, but any residual errors are mine. I particularly value Arthur Bauer's attempts to help me understand the development of German radar. Parts of chapters 1 and 7 were presented in a paper delivered at the conference on the history of radar organized by the Committee for the History of Defence Electronics (CHIDE) at Bournemouth University in England in December 1995. I thank Dr. Brian James for inviting me and allowing me to meet so many people who contributed to the rapid development of my ideas at that time. There I met John Bryant and Arthur Bauer and had useful discussions with G. W. Dummer and E. B. Callick.

I also had brief but useful interchanges regarding the history of the development of radio with Frederick Nebeker, of Rutgers University, and Hartmut Petzold and Oskar Blumtritt, of the Deutsches Museum, at the annual conference of the Society for the History of Technology (SHOT) in London in August 1996.

I would also like to thank the staffs of the Public Records Office, the Library of the Institution of Electrical Engineers (IEE), and the Cable and Wireless Archives, all in London, for their help. I thank Dr. Tom Going, of Southend-on-Sea, for talking to me out of the blue and at length on the phone one day in the early development of this manuscript. I also thank Dick Gargan, an old friend of my parents' and formerly at Marconi's, for making me aware of the links between the developing television technology of the 1930s and the early implementation of radar. Growing up in Southend, I was always aware of the presence of Marconi as a major employer in the region. I thank an old high-school friend and electrical engineer also formerly at Marconi's, Mike Sampson, for persuading me that the Library of the IEE was well worth a visit. It turned out to be one of the most pleasant I have been lucky enough to work in. Much of this book exists because I was able to spend some four weeks there in the winter of 1995. At the Cable and Wireless Archives, Mary Godwin, the curator, helped me make sense of material in what had to be a lightning visit in the summer of 1996. I also thank the curators of the Imperial War Museum in London, the Imperial War Museum Aviation Collection at Duxford, and the Royal Air Force Museum at Hendon for preserving at least a few bits of World War I airborne radio gear (sometimes in the appropriate airplane) and a German Würzburg ground radar from World War II. The Yorkshire Air Museum has a de Havilland Mosquito with British airborne intercept Mk IV radar. Arthur Bauer is restoring a German Lichtenstein SN-2 (FuG 220), but I have no idea where he will get a Messerschmitt Bf 110 night fighter to stick it in. A native of Yorkshire, I do not think that the British National Museum of Photography, Film, and Television in Bradford is off the beaten track, though others might. I thank my cousin Graham Hugill, who works there dealing with the vagaries of vintage electronic equipment, for his help. Bradford is worth a visit for other reasons as well: as anyone from Yorkshire will modestly tell you, Yorkshire beer is the best in the world.

At Texas A&M I thank Jonathan Coopersmith, a colleague in history, for steering me to the CHIDE and SHOT conferences and for being interested in some of the same things I am (buy his book on the fax machine when it come out). My colleague in geography Bob Bednarz was kind enough to help me solve many of the map problems. Another colleague in geography, Jonathan Smith, is both supportive of some of my wilder ideas and good at making me come down to earth. His wife, Uli Marschnigg Smith, tolerates with good grace my occasional requests for recondite translations from German. A colleague in electrical engineering, Pierce Cantrell, whom I came to know through service on the faculty senate, helped a great deal with the section on satellites. A colleague in oceanography, Tim Francis, formerly with the Ocean Drilling Program here at Texas A&M, was kind enough to introduce me to Barry Peck, of Sealine Marine Services in the United Kingdom, who helped out with the base map for figure 8.2. I thank several graduate students who produced papers for my seminar on the role of telecommunications in the world-system that helped me clarify my thoughts on some of the topics in this book: David Butler, Susan Gilbertz, Scott Hickenbottom, Dan Overton, and Jack Ramage. I also thank Jacob Sulzbach for pointing me to the Cabinet diaries of Josephus Daniels. One of my graduate students, Peter Stulting, has been generous to a fault with his personal library of military books, many of which are absent from the university library. I would, however, like to say that our library has hidden strengths in its collection of electrical-engineering journals. Several of the front-desk staff had plentiful opportunities to develop their own strength hauling bound copies of the *Electrician* up from the basement storage area.

I could not have developed this book without the hidden assistance of the NAR (so called because its founder jokingly refused to belong to any semi-NAR). This is a faculty reading and discussion group that meets weekly and to which I have "belonged" for almost the whole of my nineteen years at Texas A&M (you can check in, but you can never leave). Through this group I encountered the work of Tilly and Mann and a raft of other authors. The NAR's discussions have improved of late, since many local hostelries have upgraded their beer selections to meet its demands. For example, I have been able to move up the gravity scale from Shiner Bock to Guinness and Old Peculiar, a fine Yorkshire porter.

Finally, let me thank my wife, Judith L. Warren, and our two children, Lacey and Laura, for their help. Without them I might have finished this book a lot sooner, but I would have had a lot less fun!

Global Communications since 1844

CHAPTER ONE

Information Technology, Geopolitics, and the World-System

Who rules East Europe commands the Heartland: Who rules the Heartland commands the World-Island: Who rules the World-Island commands the World.

—Halford J. Mackinder, *Democratic Ideals and Reality*

In the present stage of aircraft development . . . England . . . is thrown back into her medieval position—in semi-isolation on the edge of the continental transport system. . . . It must still be some time before trans-oceanic flying becomes a normal service. Then . . . will the axis of air communications again be shifted to the British Isles, as was that of sea transport by the original discovery of America.

—Basil Liddell Hart, *Paris, or the Future of War*

Although much has been written about the information economy in recent years, it has tended to be narrowly compartmentalized. I draw on some of this literature, but much of it is self-serving and to review it here would be pointless and time-consuming. To support my argument I rely heavily on the technical literature, both trade magazines and professional journals, especially those contemporary to the technology under discussion. The primary idea informing this book is that human history, at least for the last 500 years, has been subject to some interesting regularities and that these regularities have been somewhat controlled by the technologies that involve the movement of ideas, goods, people, and information. I have dealt with at least some aspects of the first three of these in two previous books: *The Transfer and Transformation of Ideas and Material Culture*, with D. Bruce Dickson (1988) and, more thoroughly, *World Trade since 1431: Geography, Technology, and Capitalism* (1993). The present book stands as a rather elaborate footnote to *World Trade* and deals with the capitalist system only as it has developed over the past 150 years. The principal subject is the commodification of information in the service of society and the state, hence the subtitle "Geopolitics and Technology." In the purest way information is power.

I elaborated in *World Trade* on three technical epochs based on the work of Lewis Mumford: the Eotechnic, the Paleotechnic, and the Neotechnic. This work deals only with the last. The Neotechnic I dealt with in *World Trade* was primarily the new transportation technics made possible by electricity generation, the ideal of individual mobility, and the internal combustion engine: the streetcar, the bicycle, and the automobile. These technologies flowered in the late 1880s and 1890s. Electric communication, which needed much less energy than electric transportation, was possible much earlier and marks the real beginning of the Neotechnic.

My text starts in 1844, with the completion of the first user-friendly, public-use electric telegraph line in the world, from Baltimore, Maryland, to Washington, D.C. There were slightly earlier electric lines, mostly owned by British railroad companies, but they were not public-access and they did not use Samuel F. B. Morse's brilliant new software to make sending messages easy. Throughout this book I use the term *software* according to the definition I developed in *World Trade:* "the technologies of utilization of human labor, usually in interaction with hardware" (11). In most of this volume *hardware* thus refers to the technological infrastructure used to transmit information and *software* refers to the technologies by which humans generate and process that information.

The main text ends with the (rather delayed) American assumption of hegemony after 1945. America did not lack for effort. Valiant attempts by the American republic to expand a continental telegraphic system into a global one in the immediate aftermath of the American Civil War failed. Attempts by the United States Navy to create a dominant global radio network during World War I also failed. It was Britain that first constructed a global submarine telegraph system and Britain that kept (just) one step ahead of its competitors in all forms of civil and military telecommunications until 1945. I provide a limited commentary on events since 1945 in part because American hegemony has since been established, in part because the rate of change of both software and hardware technologies has accelerated, and in part because of the tendency to surround recent technological breakthroughs with high levels of both commercial and military security.

If information is power, whoever rules the world's telecommunications system commands the world. The old concerns of political geography, exemplified by the opening epigraph to this chapter, must thus be expanded. Command of the world economy is not necessarily about ruling territory. More properly, it is about appropriating a disproportionate share of the surplus production of economic systems. When Mackinder was writing, Lenin's insights on this process, exemplified in his essay *Imperialism, the Highest Stage of Capitalism* ([1917] 1939), had hardly penetrated political thought in the democracies, despite the brilliant pioneering insights of Hobson (1938) and of Mackinder himself (Mackinder 1900; see also Hugill 1993, ch. 7).

By 1919 Mackinder had ample evidence of how a major shift in the geopolitical balance of power could upset the overall balance of power in the world-system. He recognized clearly that "Germany under the Kaiserdom aimed at a world-empire" and that the Germans had failed to reach this goal because they had failed to subdue the Slavs (Mackinder [1919] 1942, 147). The consequences were clear: "Unless you would lay up trouble for the future," he wrote, "you cannot now accept any outcome of the war which does not finally dispose of the issue between German and Slav in East Europe. . . . If we accept anything less than a complete solution of the Eastern Question in its largest sense we shall merely have gained a respite, and our descendants will find themselves under the necessity of marshalling their power afresh for the siege of the Heartland" (149, 154).

The military theorist Basil Liddell Hart clearly understood what Mackinder's insight meant for Britain from a geopolitical perspective: Britain went, at least until the balance was redressed by transatlantic-range airplanes, from being at the geographic hub of the global economy to being on the border of the continental economies of Eurasia and North America. Liddell Hart saw America as well placed by geography to integrate the traditional Atlantic and the rising new Pacific economies, "whereas Japan suffers . . . the disadvantages of England's insular and border situation" (1925, 52). Prophetically, he did not see this as giving America a military advantage, noting that the Japanese had read their Clausewitz "and that one of his axioms reads: 'A small state which is involved with a superior power, and foresees that each year its position will become worse,' should, if it considers war inevitable, 'seize the time when the situation is furthest from the worst,' and attack. It was on this principle that Japan declared war on Russia [in 1904] and *for the United States the next decade is the danger period*" (53, emphasis added).

INNIS' MODEL OF COMMUNICATIONS

In 1950 Harold Innis, the great Canadian economic historian, wrote a little-read and often misunderstood book, *Empire and Communications*. Innis had a considerable and well-known impact on the work of Marshall McLuhan. His influence on the historical geographer Andrew Hill Clark is less well understood. The primary focus of both Clark and his numerous doctoral students was the process of expansion of Europe overseas, although the vast majority of his students dealt with expansion into North America. Innis noted that "the subject of communications . . . occupies a crucial position in the organization and administration of government and in turn of empires and western civilization" (1950, 3). I echoed the sentiment in the concluding chapter of *World Trade*. Beneath a rather rambling discourse on various world empires, Innis concealed the basics of a model showing how different communications technologies have affected cultures. Because the model was not stated explicitly, because his examples were from world empires and not from what we have come to call the world-system of capitalism, and because communication was not a topic of great academic interest in the 1950s, Innis' model has been largely ignored.

Innis' technology-driven model generated two cultural archetypes (Innis 1950, 7):

Type 1—Durable (heavy and not very portable) communications media (e.g., stone, clay, and parchment) allow cultures to control time. Such cultures express themselves best in architecture and sculpture. They tend to be decentralized geographically but to have hierarchical social systems and systems of government.

Type 2—Ephemeral (light and easily portable) communications systems (e.g., papyrus and paper) allow cultures to control
space. Such cultures express themselves best in administration and trade. They tend to be centralized geographically
but to have relatively less hierarchical social systems and
systems of government than cultures using Type 1 systems.

Innis subjected these two archetypes to three modifiers:

a. No culture is exclusively Type 1 or Type 2. Durable and ephemeral communications media frequently coexist, especially in more
recent cultures (7). Excessive concentration on one type of communications media usually elicits competition from the other
(140, 216).
b. A crucial element of the interaction between cultures is their
adoption and use of different communications systems to control
their space (216).
c. Different media have different geographies of both production
and use. Simply put, media may be ubiquitous or highly concentrated in nodes or along fixed routes. These media will interact
with the cultural archetypes, reinforcing or reducing their ten-
dencies to centralization and hierarchy (140–41, 156–57, 216).
Innis fails to distinguish between explicitly two-way media (in
which a response is called *for*) and one-way media (such as novels,
films, radio, and television, in which a response is called *forth* but
does not necessarily result in communication with the person or
persons calling it forth). Because his focus was on written information, he also did not distinguish between the different *qualities*
of information that can flow along networks constructed with different technologies. Nor did he address the problem of codes
and ciphers, simply assuming that all material is equally easy to
read. In addition, he did not address the cost of the system, which
is partly a function of its capacity. A low-capacity, high-cost system
of even ephemeral communications would be closer to his Type 1
than to his Type 2.

Innis' model is thus only a starting point, however much one modifies it.
Yet it allows us to see that communications systems are embedded in all
histories, albeit in different forms that affect humankind's ability to control space and time militarily, economically, and culturally. The technology of communication is thus the technology of much geopolitical power. My main concern here is thus the two-way flows of information that
predominate as mechanisms of military and economic control. The centralization of decision making such systems offer is a tremendous advantage, but they are subject to both disruption and interception. Polities
that can read information flowing through the communications systems
of other polities while keeping their own information flows secure have
considerable economic and military advantages. Although I recognize
the importance of the one-way flows of information that characterize cultural systems, I am less concerned with them. Mobilized as commercial

and political propaganda, such one-way flows outside the societies within which they were developed amount to a form of control as the term *cultural imperialism* attests. My concern, however, is with the older, more direct, more geopolitical forms of control.

THE TRADING AND TERRITORIAL STATES
OF MACKINDER, FOX, AND TILLY

Historians such as Fox and Tilly have recently discovered something that geographers have long known but not so clearly stated: control over space differs with types of politico-economic organization. Mackinder had inklings of it in 1919. Fox, working on France, noted that for the past several hundred years there have been two Frances; Tilly extended this division to Europe in general (Fox 1991, 1–2; Tilly 1989). France 1 is the trading state represented at the time of the French Revolution by the Girondins, whose leaders came from the department of Gironde and the trading city of Bordeaux. Theirs was a mercantile network of French coastal and riverine cities, notably Bordeaux, Toulon, Nantes, Lyon, and Marseilles, connected to other cities around the world by a fleet of trading ships and protected by naval power (Fox 1991, 56, 60). Such power could be applied even at the global level, but only to spatially concentrated targets (usually other trading cities). France 2 is the territorial nation-state represented at the time of the Revolution by the Jacobins, with a national capital at Paris, national bureaucracies, national communications along a road network with Paris as its center (the *routes nationales*), and national military power concerned partly with controlling unruly provinces. In the aftermath of the Revolution this pattern was exceptionally clear as the territorial French state had to take the port cities back from the trading interests, who sought a federal solution to France's problems rather than the "enlightened despotism" that would emerge in the person of Napoleon Bonaparte (61). At a macrotheoretical (and geopolitical) level this argument is most clearly encapsulated by Rosecrance (1986), who defined two types of states within the European system: trading states, which are mercantile in character and generate wealth through trade; and territorial states, which generate wealth through the occupation and exploitation of territory and are thus disposed to militarism.

This argument, however, is originally that of Mackinder, who, forced to account for the peculiar German behavior that led to World War I's being fought on two fronts simultaneously, distinguished between a trading Germany of Hamburg merchants pushing west into the Atlantic and a territorial Germany of mainly Prussian Junkers pushing east into the three *Ostmarks* of Poland, Silesia, and Austria (fig. 1.1).

> Berlin had not decided between her political objectives—Hamburg and overseas dominion, or Bagdad and the Heartland—and therefore her strategical aim was also uncertain. (Mackinder [1919] 1942, 154)

> Had Germany elected to stand on the defensive on her short frontier towards France and had she thrown her main strength against Russia, it is

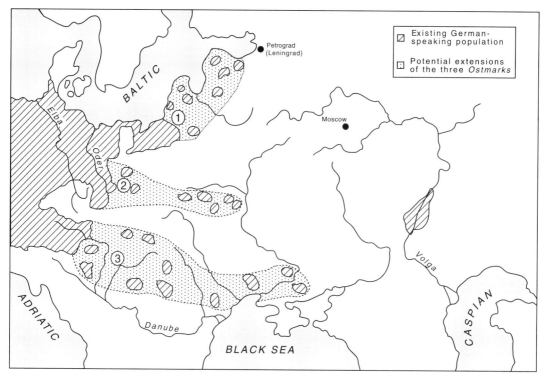

Fig. 1.1. The Three German *Ostmarks:* Tongues of German Speech Pushing into (1) Prussia, (2) Silesia, and (3) Austria and Their Potential Extensions (redrawn from Mackinder [1919] 1942, 129)

not improbable that the world would be nominally at peace to-day, but overshadowed by a German East Europe in command of all the Heartland. The British and American insular peoples would not have appreciated the danger until it was too late. (149–50)

Mackinder thus read the history of World War I as that of a Germany forced, against all the teachings of the great architect of German unity, Bismarck, and against all geopolitical sense, into a two-front war because of the tension between these two Germanys. The Hamburg trading interests simply forced the pace in the west, whereas Germany's geopolitical interests would have been better served by settling its eastern territorial question before it took on the western trading states. Here Mackinder is simply rehearsing what any intelligent observer could see, that the loss of World War I was a result of military overextension in a two-front war. In the next war German territorial aims would be better achieved by attacking on only one front at a time. This line of geopolitical reasoning is embedded deeply in Hitler's *Mein Kampf.*

Tilly has considerably extended this Mackinderian line of reasoning to include the complex relationship between the types of state interests and the need to both make war and control it. He notes that the dynamic tension between trading and territorial interests is central to the history of all of Europe, not just to that of France or Germany, and that the "pursuit of war and military capacity, after having created national states as a sort of by-product, led to a civilianization of government and domestic politics" (Tilly 1990, 206–7).

The powerful mercantile elites of Europe's great trading cities were connected primarily to the elites of other cities, both within Europe and around the world. These cities frequently became the nuclei of core regions of the world-system. The powerful landlords of the great rural peripheries controlled large amounts of agricultural surplus produced by controlled-cost labor systems. Such labor systems could involve physical coercion, as in slavery and serfdom, or financial coercion, as in sharecropping, or they could be driven by market forces, such as the family farm and mechanization. The surplus production of the last two rural labor systems has usually been appropriated by owners of financial agencies rather than owners of land. Because these controlled-cost labor systems were widely diffused by Europeans in the colonial era, the division of labor between urban core and rural periphery also became one of the great organizing principles of the historical geography of capitalism. It is, for example, embedded deep in the geographical behavior of the most successful "Europe Overseas," the United States of America. It is also found in the historical conflict between North and South that led to the Civil War and, in the aftermath of that war, in the conflict between the northern trading capital, New York, and the northern territorial capital, Chicago.

In the terms developed in Innis' pioneering work, all European cultures, since they are at least in part trading states, have concentrated on Type 2 communications systems, although the more territorial states have been willing to embrace some aspects of Type 1 systems as well. The most significant differences have been in the level of state monopolization. Territorial states tend to want state bureaucracies to control communications. In eighteenth-century France this led to the construction of the finest network of post roads in the world, the *routes nationales* radiating outward from Paris; a superb but geopolitically structured canal system; and control over the postal service. This system was foreshadowed by the royal roads of the Inca state. Trading states are more willing to accept various mixes of state and commercial control, usually depending on the extent to which these states have become embedded in territorial states. Britain, for example, controlled the mail but left the provision of roads and canals to private enterprise.

The geopolitical interests of trading states are also very different from those of territorial states. Since trading states are interested in exerting weak control over long distances, in contrast to territorial states, which wish to exert strong control over short distances, they prefer to invest in long-range military and communications systems.

THE GEOPOLITICAL STANCES OF TRADING AND TERRITORIAL STATES: MAHAN VERSUS MACKINDER

In the French example cited above the two Frances coexisted uneasily until the rise of Bonaparte, when France 1, despite the delaying action of the Girondins, was overwhelmed by a France 2 that went on to attempt to extend its territorial power over all of Europe and European Russia. The struggle in France was part of a much larger hegemonic struggle between France and Britain. France was perfecting the territorial state in order to gain hegemony at the same time that Britain was extending and refining the nature of the trading state.

The first great trading state, Holland, rose to power in the late 1500s on the basis of such software innovations as centralized banking, cheap money, and risk sharing through multiple ownership of trading vessels. To defend the immense commercial power generated by its trading vessels, Holland invested heavily in its navy. England replicated and refined Dutch software innovations, notably centralized banking, better risk sharing by insurance, and the joint-stock company, to succeed Holland as hegemon in the late 1600s. The Navigation Acts were a notable English innovation, designed to reduce the trading wealth of Holland by denying Dutch ships free access to English ports. The principle of mercantilism whereby trade with English colonies was given preference over free trade was a further successful innovation. Following the Dutch example, England invested heavily in a powerful navy beginning in the mid-1600s. Political union with Scotland created the British polity in 1707 and, despite the 1715 and 1745 rebellions, largely removed the threat of a French-backed Scottish invasion of England.

After 1707 the French challenge strengthened. After the third Dutch War the two Protestant polities, Holland and England, made common cause against their Catholic challenger. The French challenge to British power of the late 1700s, in particular its central role in the detachment of thirteen of the North American colonies from the first British Empire, helped push Britain toward a commercial-industrial rather than merely a commercial version of the trading state (Hugill 1993, 20–31). In the terms of the economic theories developed at the time of this transition, Britain moved from being a state organized for competitive advantage along the lines of agricultural production suggested by David Ricardo to one that took advantage of the industrial division of labor suggested by Adam Smith (16, 25, 324). The tremendous advantage early industrialization gave to Britain resulted in an increasing stress on free trade by British politicians representing industrial interests against the mercantilist interests of the older landed and commercial economy. The victory of the principle of free trade over mercantilism with the repeal of the Corn Laws in 1846 resulted in a veritable obsession with global telecommunications on Britain's part. The advantage of superiority in industrial production could only be realized through free trade, and the advantage of free trade could only be maximized through global telecommunications.

As a trading state, Britain took a geopolitical stance that was very different from that of a territorial state, of which France was the obvious exemplar. To maximize capital accumulation, Britain had to trade its commercial and industrial skills for the surplus agricultural production of others, which required a global network of ships and, eventually, communications designed to fill such ships with profitable cargoes. The stance of a trading state was brilliantly summarized by the American geopolitician Admiral Alfred Thayer Mahan as a navalist, insular geopolitics ([1890] 1957a, [1892] 1957b). Such insularity hinged on the use of the navy to accomplish four aims, listed here in historical order (Mahan put no. 4 first):

1. Protect the "moat defensive" (the term is from *Richard II* 2.2, the deathbed speech of John of Gaunt) of the surrounding seas that would have to be crossed by any would-be invader, where incoming merchant ships would also be at their most vulnerable.
2. Pursue a war of blockade against enemy ports to deny the enemy the profits from its merchant ships.
3. Pursue a *guerre de course* against enemy merchant ships on the high seas.
4. Seek out and destroy the enemy battle fleet in a climactic battle.

Obviously, Holland and then Britain pursued a navalist, insular geopolitics from the Dutch Wars of the 1600s to Mahan's writing in the late 1800s. Two events stand out in Mahan's interpretation of Britain's situation: the loss of the thirteen American colonies and victory against Napoleonic France. In the American Revolutionary War a British blockade of the American coast was ineffective and a *guerre de course* was useless. Blockades work only against trading states, and although to outside observers it appeared to be a trading state within the British Empire, the new American republic was also part of a self-sufficient territorial economy. For their part in the American Revolutionary War, the French avoided direct involvement outside the Colonies and never challenged Britain's "moat defensive." There was no climactic battle, nor was one ever really possible. The naval loss by the British to the French fleet at Chesapeake Bay sealed Cornwallis' fate at Yorktown. The subsequent rout of the French fleet in the Caribbean did not restore British control over the rebellious colonies, however much it may have restored British naval mastery over the French.

Some twenty years later, in the Anglo-French and Napoleonic Wars, the French got almost everything wrong and the British got it all right, as Mahan obviously recognized. France was enough of a trading state to be ultimately susceptible to the British blockade. Orthodox Anglo-centric interpretations to the contrary, the French won a substantial initial success at the Glorious First of June, when a French grain fleet broke through the British blockade to land at Brest. But from then on the blockade held. Winning no follow-up victories, Bonaparte was forced back onto his "continental system" in an attempt to become self-sufficient. French *guerre de course* tactics never came to much, although some Caribbean islands suffered severe damage. The British maintained the "moat defensive" against all Bonaparte's dreams of an invasion fleet. Trafalgar was the climactic battle to end all climactic battles. The implications of Mahanian geopolitics for the geostrategy of hegemonic trading states were clear:

1. Maintain the world's most powerful battle fleet, in quality and fighting spirit, if not in numbers.
2. Maintain an offensive naval strategy, striking constantly at enemy shipping, ports, and fleet.
3. Avoid entanglement in land wars, which in Britain's case meant allying with the weaker against the stronger continental power(s) to fight land powers in Europe, such as with Prussia against Napoleonic France.

Despite Britain's stunning success against the first great territorial nation-state in the Napoleonic Wars, by the early twentieth century Mahanian geopolitics was being challenged in three ways. Two challenges, the nation-state version of the territorial state governed by efficient professional bureaucracies and better land communications by railroads and telegraphy, combined to make possible the emergence of geographically more extensive territorial states. The third challenge was the emergence, in the first decade of the twentieth century, of the possibility of air power. For Britain this was particularly crucial. Air power made Britain "no longer an island," to quote the title of Gollin's book, by destroying the traditional "moat defensive," which worked only against sea power.

From the geographer's perspective, the first true nation-state was Revolutionary France (Taylor 1993, 195–96). Revolutionary France defined itself as a state by its territory, not by its monarch ("L'état, c'est moi"), its parliament (the British notion of "the King in Parliament"), or its people ("We, the People"). Yet France before the Revolution was always the two Frances of Fox, a trading state as well as a territorial one. Achieving its revolutionary ambitions, which were entirely those of the newly invented territorial nation-state, required forcing subject groups to adopt French systems of government and, ultimately, French culture even within what we now regard as France itself. A "post-aristocratic" form of French was "imposed on the dialect-rich provinces as a means of breaking down local loyalties" (196). Outside the traditional boundaries of the French state, at least to begin with, the French were often welcomed as liberators come to release the inhabitants from oppressive *anciens régimes,* but opposition grew as it became clear that so-called enlightened despotism was often little better than repressive royal and noble power. The expansion of French ambitions under Bonaparte to cover much of Europe meant that it was no longer possible to administer all from Paris because communications depended for the most part on the speed of human movement, and the French road, canal, and mechanical telegraph systems were national systems. For this reason, if for no other, Bonaparte had little choice but to impose his relatives and marshals as rulers in the lands he had initially so easily conquered. Although irredeemably altered by French expansion, Europe could thus not become Greater France without far better communications than were possible in the early years of the nineteenth century.

By the late nineteenth century, however, the nature of the capitalist state had changed considerably as capitalism had permeated the territorial as well as the trading states. Trading states had raised money for their navies by state-guaranteed loans from the central bank, also known as the national debt. Taxes on trade repaid the loans, and the tax-collecting bureaucracy was small because trade had to pass through a small number of easily policed ports. Although navies were expensive, the taxes necessary to build them were spread among a large number of consumers, and the need to maintain the navy to protect trade seemed obvious. Territorial states had always needed large bureaucracies to raise the land taxes to pay for their standing armies. Taxes were levied directly on landowners to maintain an army that was used as much for internal repression as for defense against outsiders. The French system of tax farming before the Revolution meant that tax burdens could also be highly variable. Thus, taxes were much more likely to be resisted in territorial states. As territorial states followed trading states into capitalist production after the success of the industrial revolution in Britain, they had to regularize tax burdens. They thus began to enlarge, refocus, and professionalize their bureaucracies, as Weber made clear at the time: "Modern rational capitalism needs, not only the technical means of production, but also a calculable legal system and an administration based on formal rules" (Weber 1983, 28).

Such expanded bureaucracies initially drew on military models of management and then demanded more civilian control of the military as the national debt expanded (Tilly 1990, 206–7). The emergence of powerful state bureaucracies was, however, also a function of technical change in favor of fast land transport (the railroad) and instant telecommunication (the electric telegraph). In the late nineteenth century large, bureaucratically centralized territorial nation-states with only a small maritime component thus became technically possible. Railroads and electric telegraphs maintained such centralized power far better than had the canals, *routes nationales,* postal services, and mechanical telegraphs of the past. Most recently, Mann has noted four areas of growth in the period of state modernization at the end of the nineteenth century: (1) geographical size, (2) the scope of the state's functions, (3) administrative bureaucratization, and (4) political representation (Mann 1993, 358). Even the greatest trading state of the period could not ignore these shifts. Mann notes that in Britain in 1881 state expenditures for civil services were greater than those for military services, probably for the first time (376). By 1910 civilian personnel employed by the state made up 2.6 percent of all employed persons in Britain, whereas military personnel made up only 1.04 percent, a ratio of 2.5:1 (393). The civilian-to-military ratio at the time varied somewhat: for Austria-Hungary it was 3.7:1, for France 1.3:1, for Germany 1.5:1, and for America 2:8:1. By 1910, before the nationalization of British railroads, transport accounted for 12 percent of the state budget, and postal and telegraph services for 8 percent, both representing substantial increases over the past thirty years (379). Mann does not, however, effectively link rising civilian expenditure with high state spending on the transport of people and information.

There was, however, a more political component of late-nineteenth-century state bureaucratization and centralization, a result of the general adoption after the Congress of Vienna in 1815 of the French idea of national sovereignty based on territorial control. This idea of national sovereignty was in contradistinction to the older forms of sovereignty, which depended on such notions as the divine right of kings (pre-Revolutionary France); "enlightened despotism" (Napoleonic France); the king in Parliament (Britain after 1688); and the sovereignty of the people (America after 1776). The radically new French idea of the nation-state emphasized the nation's need to politically control territory as well as trade. As this notion became increasingly accepted in Europe, the linkage of control over territory with control over needed raw materials and foodstuffs for increasingly urban, industrial populations became clear. Trading states used economic linkages to tap into the surplus agricultural production of distant regions that had no need of political control, although they were sometimes described as empires. Territorial states were not set up in this fashion, and their radical expansion as nation-states produced the new imperialism of the late 1800s. The European powers and, increasingly, their offspring in America and Japan moved noticeably in this direction in the aftermath of the Berlin Conference of 1884–85. This led, of course, to the Grab for Africa, America's takeover of the declining Spanish Empire, and the first glimmerings of Japan's Greater East Asia Co-Prosperity Sphere. Even Britain, the most successful of the trading states, was forced to embrace the new imperialism to hold its own against such aggressive challengers as Germany and America. Such extensions of direct imperial control depended upon both a vastly increased bureaucracy and the rapid movement of people and, increasingly, information.

Better communications and bureaucratization also led to a radical alteration of the geopolitical balance in favor of territorial nation-states, an alteration first recognized by the British geopolitician Halford Mackinder early in this century (see Mackinder 1904). Mackinder, like many of his class and period, recognized the rise of Germany and America and the decline of Britain as a world power. He formulated a new geopolitics to persuade the trading states of the threat posed by any single territorial nation's rising to dominate the Eurasian heartland in terms of its railroads, aircraft, telegraphs, and national bureaucracy and thus acquiring the resources to build a fleet superior to those of the insular powers (fig. 1.2). The rise of Germany was thus clearly dangerous. The rise of America was far less worrisome, since America's territorial ambitions were scarcely Eurasian, and Britain's main relations were with the trading version of the American state. Through the Co-efficients dining club in London Mackinder was able to influence some of the most important members of Edwardian British society, including the Webbs, H. G. Wells, Balfour, Bertrand Russell, and the like (Blouet 1987, 134–37). He, in turn, was clearly influenced by such visionaries of air war as H. G. Wells. World War I demonstrates the influence of Mackinderian geopolitics on British geostrategic practice:

1. Ally with the weaker Continental power(s) against the stronger.
2. Become directly involved in a land war to resist the expansion of the stronger power(s).

3. Seek technical solutions to problems posed by such new technologies as aviation and submarine warfare, which challenged the "moat defensive."

Baugh (1987) characterizes this direct Continental involvement as the British adoption of the "French way of war," an essentially correct observation but one in which we must not lose sight of Mackinder. The British "moat defensive" was assaulted, not once, but four times in World War I: once by the German High Seas Fleet, once by submarine warfare, and twice by German aerial onslaughts, one by Zeppelins and one by heavier-than-air bombers. The last two produced a severe and blatantly sexual psychological strain. For the British, 850 years of freedom from penetration ended. As early as 1908 German cartoons showed a German Zeppelin penetrating Britain via the Thames estuary (fig. 1.3). As Schama (1996) points out, rivers have strongly feminine implications, especially for the British (373), and the Thames combines "the pastoral and the mercantile landscape. . . . Upstream, the union of Tame and Isis (who, in keeping with her Egyptian namesake, is now feminine) takes place in a fleecy arcadian world where zephyrs puff over the smiling water. . . . The climax of the journey is a second union: that of Thames and Medway, from which another, still mightier pregnancy is conceived. For within the womb of the swollen waters, salt and sweet, pastoral and commercial, floats the awesome embryo of the British Empire" (330). Scarce wonder

Fig. 1.2. Afro-Eurasia (Mackinder's World Island) United by Railways and Airplane Routes as Mackinder Saw It in 1919 (redrawn from Mackinder [1919] 1942, 112)

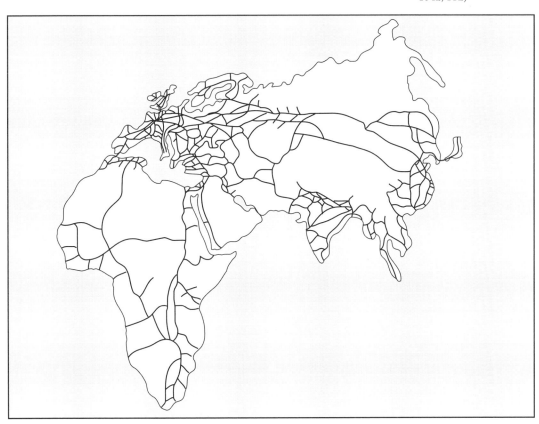

Fig. 1.3. German Cartoon, 1908, Showing a Zeppelin Penetrating Britain via the Thames Estuary (from *Simplicissimus* 13:546, by permission of the British Library)

Zeppelins Schatten

Auf dem Pulverfaß

that British propaganda in World War I portrayed Germans as rapists and the British countryside as virginal (Morris 1996). A tactless German national anthem stressing German masculinity and the missionary position ("Deutschland, Deutschland über Alles") probably did not help.

Had France and Russia both fallen, Mackinder's fears of German domination of the Continent would have been realized. Britain, despite all its geostrategic experience, thus committed a large land army to the defense of France to help bog the Germans down in a two-front war. The cost in lives of the resulting war of stalemate was an impressive testimony to the wisdom of the "moat defensive" geostrategy.

World War I raised several difficult questions, the most trenchant of which was how to avoid a stalemate in any future war. It was clear to those in power in Britain that the public would not again pay such a price for victory and that a return to some version of Mahanian geopolitics was necessary. It also sustained the vision of air war as an alternative to the bloodbath of the trenches. Wells' gloomy prophecy in *The War in the Air*

([1908] 1967) of a world sunk in a new Dark Ages had not come about, and the phallic Zeppelins of 1908 had been castrated by late 1916 (Penrose 1969, 173–74). Yet air war seemed, at least to the British, to have been very successful. The relatively minor heavier-than-air raids on London in 1917 by Gotha bombers caused near panic and pushed the British Cabinet into the arms of the strategic bombing enthusiasts who wished to retaliate in kind (Cross 1987). The very name Gotha, redolent of Gothic atrocities, probably helped perpetuate the "rape victim" hysteria that had earlier swept a Britain that felt dangerously exposed to aerial assault.

In the aftermath of World War I a raft of new geostrategies were thus proposed by those who Carver (1979) aptly calls the "apostles of mobility." All sought to break free of the appalling attrition of trench war by a combined air and ground advance coordinated by the new radio technology. Baugh captures precisely this point when he notes "that Britain fought the second World War essentially on blue water lines" (1987, 100). However, when he compares the Royal Air Force (RAF) to the British Expeditionary Force of 1914, he makes an elementary error. Bombing was more akin to the Mahanian geostrategy of blockade than it was to anything else.

The extreme strategic bombing enthusiasts were Giulio Douhet in Italy and Billy Mitchell in America, although neither propounded a geopolitical theory to replace those of Mahan and Mackinder. Both proposed that future wars would be won by the new geostrategy of bombing the enemy homeland, either to terrorize its citizenry (Douhet) or to destroy its infrastructure (Mitchell). Douhet's doctrines, which called for area bombing of cities to terrorize their inhabitants, eventually found their strongest adherents in Britain's RAF. This was not because of their theoretical purity but because the British realized after the Battle of the Heligoland Bight in December 1939 that flying unescorted bombers into German airspace was suicide. They had to retreat to bombing by night, and hitting a target smaller than a city at night was pretty much impossible in 1939.

Promulgating these doctrines required a massive investment in more than just airplanes. Radio was needed for coordination both between the ground and the air and between airplanes. Defending cities against an aerial attack that could come from almost any direction and height and at considerable speed initially seemed impossible. Britain thus spent much less on fighter airplanes in the 1930s than it did on bombers, large numbers of which would be needed to overwhelm the enemy. However, ongoing commercial research into the uses of long-range, high-frequency radio made it clear by the mid-1930s that a massive increase in research and development ought to make radio location of such imminent bombing attacks possible. What would emerge as radar offered, in conjunction with a new generation of fast-climbing and heavily armed fighter planes, new hope for the defensive geostrategy of Mahan, especially for a Britain facing the challenge of a rapidly rearming Nazi Germany.

The implications of radar for British geostrategy were quickly worked out in practice:

1. Retreat behind the renewed "moat defensive" made possible by radar.
2. Pursue a war of attrition based on strategic bombing of the enemy rather than a naval blockade, but with the same basic aim of forcing the civilian population to suffer.
3. Ally with the weaker Continental power against the stronger one to ensure the continued division of the heartland.

THE THEORETICAL STRUCTURE OF THIS WORK

As mentioned previously, I address here the capitalist world-system only as it has developed over the past 150 years. I do not deal with world empires, of which the Soviet Union was simply the most recent. I deal almost entirely with rather ephemeral electronic communications media and focus on Innis' modifiers *b* and *c*. Except for brief departures necessary to my argument, mainly into the geostrategic implications of electronic television technology in the mid-1930s, I am concerned only with two-way communications. I show that at least part of the struggle for hegemony between Britain, Imperial Germany, and America at the end of the nineteenth century and the beginning of the twentieth is best analyzed by looking at the geopolitics and technologies of their respective communications systems.

The Imperial German challenge was doomed to early failure by British dominance of global telecommunications through the fixed-route global submarine telegraph network laid with British capital and focused on London. Germany was simply unable to replicate the British global network. Further, Germany was unable to find a viable technical alternative in radio since German companies merely copied the British Marconi system of spark telegraphy. America was able to establish a continental telegraph network, but it had great difficulty projecting that network beyond its own boundaries in the face of Britain's effective monopoly of submarine telegraphs. America was, however, more successful in developing radio because American entrepreneurs and companies sought in continuous wave radio a more sophisticated technology than spark transmission. Because America's economy was similar to Britain's, and because British capital had free access to the American market, American companies frequently entered into agreements with British long-distance cable companies. At the same time, the British, in particular Marconi, recognized American improvements in long-distance, continuous wave radio and were prepared to buy American technology as soon as it became available. From a geopolitical perspective, this open-market system was obviously problematic in that it made the secure communications needed by the military much more difficult to achieve.

The U.S. Navy became obsessed with the need for communications secure from British oversight early in the twentieth century. It therefore searched long and hard for a technological end run around British dominance, beginning to achieve an alternative in the teens with continuous wave wireless telegraphy. The navy was able to convince American commercial interests that the British were reading all the cable traffic that flowed along British lines, whether encrypted in commercial ciphers or not, and that this was the reason for the poor performance of American commercial interests against their British competition. The navy was also able to convince the American government that Britain used its press services as a propaganda arm, distributing news along its cable network that was favorable only to British commercial and strategic interests. The result was that the navy created its own Poulsen Arc global network before and during World War I and afterwards attempted to turn it into a system that it could continue to control through the Radio Corporation of America (RCA).

Wireless telegraphy and, later, telephony lacked the nodes and fixed routes of the wired systems and thus could not be as easily controlled geographically. Messages in such wireless systems could, however, be easily intercepted, and unless they were securely encrypted, they could be as easily read. In two world wars no codes remained secure against determined cryptanalysis. The more successful codes merely survived longer, perhaps against less determined decryption efforts. Since mathematics was the basis of code breaking, it may be surprising to learn that the Germans did less well against British and American encryption than vice versa, especially in World War II. The reason is that Britain and the United States were simply more successful than Germany at engaging their mathematicians in the national effort, gave them more freedom of inquiry, and used them more efficiently. Nazi policies against Jews and intellectuals cannot have helped the German cause. Recent German work has certainly raised the issue of *Resistenz* to National Socialism. The great German aerodynamicist Ludwig Prandtl complained direct to Himmler about National Socialist activists' attempt to exclude Jews from the Society for Applied Mathematics and Mechanics, primarily because he felt the discipline would suffer (Trischler 1994, 81). German attempts at code breaking simply pale into insignificance beside the British and American successes.

Because of the American navy's successful campaign for national security, American wireless systems came to be controlled by a commercial monopoly granted on national-security grounds, RCA. The free market that should have come into being with such a geographically ubiquitous system was avoided. What should have emerged as an Innis Type 2 system came to resemble a Type 1 system. After 1945 this Type 1 monopoly was bolstered in two ways: in the 1950s American Telephone and Telegraph (AT&T) began to install a global network of submarine telephone cables, and in the late 1960s America began to launch communications satellites. The American submarine telephone cable network begun in the 1950s was a network of fixed nodes and routes analogous to the earlier British submarine telegraph network, although it carried higher-quality, verbal information. But AT&T was not allowed to extend its monopoly into satellite communication, although it intended to do so with a network of low-orbit satellites like its pioneering *Telstar* in the early 1960s. Instead, America set up an allegedly neutral carrier, Intelsat, headquartered in Washington, D.C., to manage the information flow through a system of satellites launched into much higher, geosynchronous orbits.

Given the technology of the 1960s and America's economic dominance, only the American taxpayer could afford such a communications technology. Communications satellites allowed America to develop a system that seemed to have some of the same advantages of geographic focus that the submarine cable network had for Britain. Intelsat, on the face of it, seemed to represent a system consisting only of nodes where information could enter. Any relatively cheap satellite dish could access the information, so almost all information had to be coded if it was to be secure. Early, low-orbit satellites like *Telstar* moved relative to the planetary surface and did not have a long life before their orbits decayed. Thus, they could only be used at certain times of day and over the limited area represented by their "footprint." Information from modern, high-orbit, geostationary satellites is available anywhere in their footprint 24 hr a day.

I suggest that the success of any one regional power grouping in the period of multipolarity we are now entering (NAFTA, the EU, Japanese-led Asia) can be predicted in part on the basis of its chosen communications strategy and that a likely outcome will be the failure of any one of these regional groups to achieve hegemony. Just as American radio communications challenged the British cable "monopoly" in the 1920s, European satellites and fiberoptic lines and Asian satellites are now challenging American satellite and submarine telephone cable dominance. America is no longer the only economy wealthy enough to afford satellites. Even China can do so, often for Western clients, although with a high failure rate. Satellites may be capital-intensive technology, but once in place they provide a ubiquitous, ephemeral communications medium. In the terms developed by Innis and Fox, they suit trading states far better than territorial ones, however much they have been bent to the geopolitical goals of the latter.

COMMUNICATIONS AND MACROTHEORIES

This work is also rooted in the two macrotheoretical perspectives of world-systems and long waves, both as modified by transportation technology, as I made clear in *World Trade*. Improvements in land transport in the early nineteenth century allowed the first commercial use of the continental interiors away from the narrow band of tidewater settlements that were connected across the seas and oceans by sail-powered ships. First steamboats, then railroads allowed this penetration. Little attention has been paid to the first technology, considerable to the second. Railroads were, however, accompanied by the first electronic communications system, the telegraph. The improved flow of information probably contributed as much as the improved transport of goods to the success of American commercial agriculture in the nineteenth century. Until recently the academic literature has almost completely ignored this contribution, regarding the telegraph as merely an appendage of the railroad. The telegraph played a key role when America was striving for efficiencies in the agro-industrial sphere. In the late nineteenth century the telephone was added to the telegraph, and a crucial dimension of expressive meaning was added to telecommunications. The telephone played three key roles in America's rise to hegemony. First, long-distance lines made possible by the carbon transmitter cemented together the commercial system of cities in the republic of the North (the term is from Hugill 1993, 76) between 1892 and 1915, a process begun using railroads and the telegraph. Second, De Forest's amplifier allowed a nationwide urban commercial network to develop between 1915 and 1956, a process reinforced by automobile transport and early commercial aviation. Third, the first transatlantic telephone cable, in 1956, extended the reach of American business overseas with a forcefulness not possible using the mail or the telegraph, a process reinforced by transatlantic-range airliners. All these changes in electronic communications reinforced the rise to dominance of the republic of the North as a capitalist core within the American system and its eventual successful hegemonic challenge to Britain and Germany.

Continental telegraphy and telephony were amazingly simple technologies, initially developed by trial and error and owing virtually nothing to the electronic theory of the late 1800s. This very simplicity may account for the lack of academic attention, but I attribute it to the perception that there is nothing exciting about what seems to be a static technology. Vehicles run by steam and, later, the internal combustion engine were exciting. In this regard, those of us interested in the historical geography of technology have a lot to blame Freud for. Railroads were also easy to research because they were operated by capital-intensive companies that left a considerable paper trail, and they were licensed by state and federal governments, which did the same. Steamboats were equally "sexy," but they ran on a public right of way and were usually owned by individuals rather than by corporations, so they have left little in the way of a paper trail. Telecommunications were run by large, government-regulated corporations that have published voluminous histories, but there has been relative silence from scholars.

The second macrotheoretical perspective is that of long-wave theory. In 1991 Brian Berry greatly advanced this theory with the publication of *Long-Wave Rhythms in Economic Development and Political Behavior*. Notable in his work is the insistence that the expansion of transportation infrastructures has been central to the expansion of industrial economies. He linked the work of Isard and Borchert to work on building, immigration, investment, capital life span, and other cycles (43, 81). In his conclusion Berry quotes approvingly from Hall and Preston's *The Carrier Wave* and argues that the information infrastructure may be just as important as the infrastructure of physical transport or even more important (187). Hall and Preston conclude that the three Kondratieff cycles before the one we are now entering were marked by the creation in a very short period of comprehensive and capital-intensive information and transport infrastructures: the telegraph in the late 1830s, the telephone in the 1880s, and cathode-ray tube (CRT) technologies such as television and radar in the 1930s. The 1980s are marked by the beginnings of the construction of a fiberoptic "highway" system.

The problematic technologies in this model are two that do not fit the Kondratieff model of innovation in the downswing of the world economy driven by the search to switch investment from less profitable older technologies, characterized by a great deal of competition, to potentially more profitable new technologies. Both problematic technologies developed in the midst of upswings. Long-range radio was first developed by Marconi in the first decade of this century, at which point it was an Innis Type 1 system. Attempts to use radio as a personal communication device, a precursor to today's cellular phones, began in the San Francisco Bay area only a few years later. Long-range radio matured in the early 1920s in Britain when the switch was made to higher-frequency, shortwave radio. This was a classic Innis Type 2 system, cheap and nonhierarchical, although the technology was bent to the purposes of the bureaucratized state as if it were a Type 1 system. The second technology that fails to fit the Kondratieff wave model is the programmable computer, another Type 1 technology in its early form as developed in the middle to late 1940s. In turn this technology evolved through the foresight of the Homebrew Club into the personal computer, a child of the San Francisco Bay region of the late 1960s that matured in the early 1990s. Although the members of the Homebrew Club, which included such pioneers of the personal computer as Steve Jobs, may not have foreseen the Internet, a classic Innis Type 2 communications system, their technology has obviously supported it.

These two macrotheoretical perspectives can be combined in the world leadership cycles proposed by Modelski and Thompson (1988) and Goldstein (1988). Two roughly fifty-year Kondratieff waves make up one century-long world leadership cycle. In each cycle one nation, the dominant geopolitical, technological, and economic power, is hegemonic.

GEOSTRATEGY, GEOPOLITICS, AND TECHNOLOGY

This work focuses on the switch from British to American leadership and contributes to discussions about two key geostrategic and geopolitical questions: Why did Germany lose the two-phase war of transition from 1914 to 1945, and why was the transition from British to American hegemony so slow? Since Kondratieff waves seem driven by technological change, it is reasonable to attribute technology with its fair share of the responsibility for such changes, but not all innovation clearly fits the Kondratieff model. Put succinctly, the nation that chooses or develops the best set of technologies and geostrategies to manipulate its environment, produce wealth, and project and defend its state power gets to be hegemon. Clearly, some nations have done a better job than others of innovating, especially innovating off the normal Kondratieff cycle.

Four telecommunications technologies are clearly important: telegraphy, telephony, radio, and radar. I focus on each in its turn. The first two may be wired or wireless systems and thus are linked to the third. The last two must be wireless. From a commercial point of view, the first three are obviously valuable for messages. Because of the advantage to competitors of reading such messages, commercial coding of such messages became normal. Commercial coding does not have to be very complex since most commercial information loses value quickly. The fourth technology improves safety in bad weather and at night for transportation not having a fixed route. Each technology has geopolitical implications because it changes the capacity of the state to control territory, to project its military power, and to defend its citizens. Telegraphic, telephonic, radio, and radar communications must be encrypted to have any chance of security, and encryption is never fully secure against determined decryption efforts. Military information needs to be protected longer than commercial information. Telegraphic messages are devoid of the expressive meaning inherent in human speech, thus subject to misinterpretation. Wired systems are spatially confining in war and are clearly implicated, along with railroads, barbed wire, and the machine gun, in the emergence of trench warfare in World War I. Wireless telecommunications allowed a return to wars of mobility, which, coupled with the technical advances in aviation, raised serious problems for defense until the development of radar. Radar encryption became necessary as the first electronics war evolved in the night skies over Nazi Germany. The Battle of Berlin in the winter of 1943–44 saw dynamic shifts on the basis of rapidly changing software technologies. In more recent air wars, especially those over Vietnam and the Persian Gulf, encryption and decryption of radar signals has become even more critical. The failure of the British radar to recognize as hostile the French radar used by the Argentineans to home their Exocet missile caused the loss of the destroyer *Sheffield* in the Falklands War.

Germany was a slow starter in all of these technologies but radar. The lack of political cohesion in the German-speaking world until after 1871 is probably partly to blame. In the mid- to late nineteenth century Germany imported and licensed telegraphic and telephonic technology. At the turn of this century Germany merely copied British wireless technology without radically improving it. Britain led in global submarine telegraphy by the late 1860s and in wireless telegraphy after 1900, although the lead belonged alternately to Britain and America until the early 1920s, when Britain moved ahead through the switch to higher-frequency, short-wave radio. America led in continental telegraphy in the 1840s but was unable to parlay that lead into global telegraphy in the late 1860s. America led absolutely in telephony from its inception and challenged Britain seriously in global wireless telegraphy from 1900 to the early 1920s. Given that Germany moved into the lead in all electronics after World War I, it is surprising that Germany did not do better with radar, but German radar hardware was perhaps too advanced in 1939, and its software not advanced enough, although well integrated into a geostrategy committed to the tactical offensive. British radar was technologically much less sophisticated but far better integrated into the "moat defensive" geostrategy that saved Britain in 1940. Britain recovered from this slow start to take technical leadership in the new microwave radar hardware after 1940, although practical leadership in the strategic (software) application of radar technologically inferior to German radar was achieved once the Chain Home defensive system was up and working, by late 1938. Despite some pioneering interest on the part of the U.S. Navy, America's proficiency in radar technology was so lacking that it had to get most of its technology from Britain in World War II. This relative leadership allowed Britain to hold off the American challenge rather longer than the macromodels would suggest. From the perspective of world leadership cycles, the transition to American hegemony should have been complete by about 1919. In actuality it took until 1945, after one of the most expensive and long-drawn-out wars of transition in history.

Of the four technologies named above, radar was the most significant during the second part of the war of transition. Ultimately, it is the German failure in radar that is most interesting, especially given the technical edge developed by the German electronics industry after 1918. Germany's substantial lead had disappeared by 1940. Ideology was allowed to drive electronics design in a particularly pernicious way, polarizing the research community into those who supported high frequencies on ideological grounds (and had a Party commissioner to back them up) (Kern 1994, 180) and practitioners who argued that existing lower-frequency systems worked well enough. Another essentially ideological failure was excessive centralization of decision making (at the level of the Führer) combined with an immensely wasteful decentralization and duplication of research and production among competing Nazi fiefdoms and an obsession with military secrecy that rivaled that of the American navy. Scientists, engineers, and users simply did not communicate with one another in a creative way. The High Command specified what front-line users needed, and scientists were expected to deliver—an example of the classic failure of command technology (McNeill 1982).

Whatever other faults it may have had, the British scientific establishment paid close attention to the needs of the military. For example, after meeting several times in 1939 with Sir Charles Wright, director of scientific research for the Admiralty, E. G. Bowen argued for the development of high-frequency radar systems on the entirely pragmatic grounds that it was the only way the size and weight of the system could be reduced to fit inside the nose of a twin-engine night fighter without drag-inducing external antennas (Bowen 1987, 143). It is thus to the organization of research and development, to the software rather than the hardware, that we must ultimately look for a complete explanation of British superiority in the high-frequency war. This superiority can be traced back to the Committee on Imperial Defence's decision to establish the Tizard Committee and that committee's subsequent willingness to consult top-level physicists such as Watson Watt about appropriate technology. The speed with which the British government acted on that advice to construct Chain Home is also impressive. The British government under Prime Minister Neville Chamberlain may have been appeasing Hitler with one hand, but with the other it was building the "moat defensive" that would ensure his downfall.

German failure was a result, however, of not turning fully to the defensive in late 1943, after the first really massive British raids on the Ruhr and Hamburg. What is even stranger is that the potential architect of defensive success, General Kammhuber, was ousted after the Battle of Hamburg, as if the remarkable success of British radar and electronic countermeasures (ECM) had been his fault. Here one must return to geopolitics. Nazi Germany subscribed to the same Mackinderian geopolitics as Wilhelmine Germany, merely updating them to cope with Neotechnic innovations in transport and communication. Hitler's Germany had to extend to the limits of the Axis and take complete control of the Eurasian landmass or it had failed. It was not only committed to offensive doctrines of war; it was utterly obsessed by them. Nazi Germany broke on the twin reefs of Britain's renewed "moat defensive" and the vastness and implacable winters of the Soviet Union, a "moat defensive" of a different sort and one that cost millions of human lives.

None of the major charges that can be leveled at Kammhuber and the Germans can be leveled at the British. Despite an almost pathological aversion to defensive doctrine in the 1920s and early 1930s, when the strategic offense dominated British air force thinking, reality intruded in the mid-1930s and the British developed an effective defensive system with the right mix of radar, ground control, airplanes, and weaponry. When the night blitz began after the Battle of Britain in late 1940, the British were not caught short. They had reasoned by 1936 that the failure of a daylight strategic offensive must, of necessity, lead to a night offensive and had then taken the necessary steps to develop airborne intercept (AI) radar. Although Chain Home was crude compared with German systems, reasonably good relations between such officers as Air Chief Marshal Sir Hugh Dowding, in charge of Fighter Command, and the scientists allowed a clear-sighted scientific program to prosper and, in the form of centimetric radar, to produce technological leadership over Germany by 1941. This free exchange of knowledge coupled with a sense of urgency to produce remarkable progress. If the RAF had waited until Chain Home was complete to work out tactics, "there would not have been time to elaborate the new tactics and train the crews" in time for the day Battle of Britain (Crowther and Whiddington 1948, 10). In Britain a remarkable intimacy developed between the devisers, producers, and consumers of the new system of electronic warfare. The Telecommunications Research Establishment, under A. P. Rowe's direction, instituted a system of "Sunday Soviets," meetings so called because of the complete equality and outspokenness they encouraged in addressing vital issues of science at war (84–89). These frank discussions of problems and possible solutions among all concerned, from the most theoretical physicists to operational aircrew, were in direct contrast to the German use of command technology, in which the big electronic firms were told what the military staff officers thought their aircrew wanted. Whatever the British lacked in electronic hardware, at least at the start of the war, they more than made up for in software. They thus came to develop better doctrine, systems of application, and hardware in the most crucial area of research and development in the whole of World War II.

CHAPTER TWO

Telegraphy and the First Global Telecommunications Hegemony

Two mighty lands have shaken hands
Across the deep wide sea;
The world looks forward with new hope
of better times to be;
No more, as in the days of yore,
Shall mountains keep apart,
No longer oceans sunder wide
The human heart from heart,
For man hath grasped the thunderbolt,
And made of it a slave
To do its errands o'er the land,
And underneath the wave.

—*British Workman*, 1858

MECHANICAL FOREBEARS

The telegraph was the first effective mechanism for telecommunication. In the classical era, chains of beacon fires or smoke columns were common, although the messages they transmitted were minimal. A crude and ineffective mechanical telegraph "raising and lowering wooden beams from a tower" can be traced back to the Romans, although they generally used fire and smoke (Wilson 1976, 3). The Frenchman Claud Chappe experimented with electric signaling, but he realized that it was beyond the technology of the late 1700s. He then built the world's first really effective telegraph in Revolutionary France by adding a telescope in order to see mechanical arms at a distance, coining the name *télégraphe* to describe his invention. Chappe's "T-type apparatus for long-distance visual communication" had its first really successful trial in 1793 (5, 122; quote from 5), and the French went on to build a nationwide system. Chappe's invention was seen in explicitly geopolitical rather than commercial terms. Joseph Lakanal, a scientist and member of the Committee of Public Instruction, was detailed as one of three members of the committee to observe Chappe's invention. "Lakanal wisely observed that the establishment of the telegraph was the best answer to publicists who thought France too big for a republic. Perhaps he foresaw that the telegraph would forge close links between the central authority and its representatives in the departments that had replaced the provinces" (122). The geographical results of such insight were evident in the comprehensive national network of mechanical telegraph lines completed in France by 1846 (fig. 2.1).

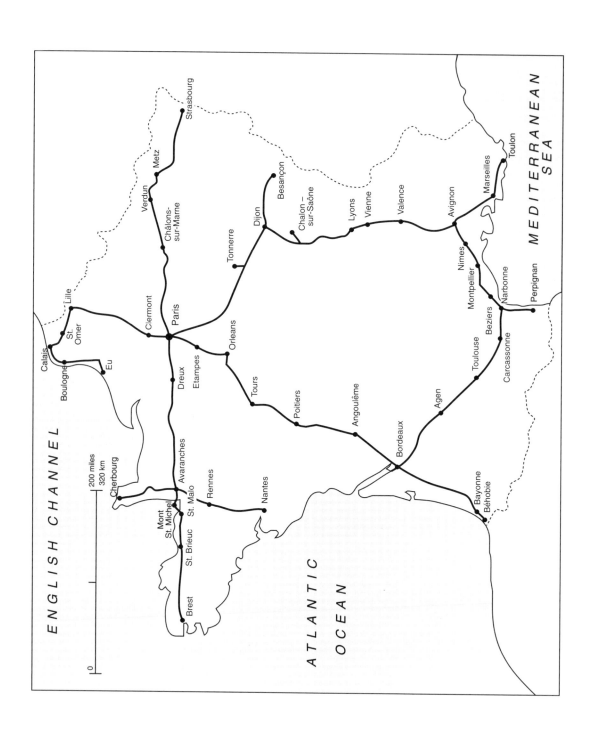

ENGLISH CHANNEL

MEDITERRANEAN SEA

ATLANTIC OCEAN

Strasbourg
Metz
Verdun
Besançon
Châlons-sur-Marne
Dijon
Chalon-sur-Saône
Tonnerre
Lyons
Vienne
Valence
Marseilles
Toulon
Avignon
Lille
Clermont
Paris
Nîmes
St. Omer
Orleans
Montpellier
Narbonne
Perpignan
Calais
Béziers
Boulogne
Eu
Dreux
Etampes
Toulouse
Carcassonne
Tours
Poitiers
Angoulême
Agen
Bordeaux
200 miles
320 km
Cherbourg
Avaranches
Rennes
Nantes
Bayonne
Béhobie
Mont St. Michel
St. Malo
St. Brieuc
Brest

0

Other polities copied the mechanical telegraph, although they did not pick up on Lakanal's geopolitical insight. In Britain in the mid-1790s, under the threat of war and possible invasion from Revolutionary France, and with knowledge of the success of Chappe's telegraph, the Admiralty established a chain of signal stations to better connect London to the fleet in its major eastern and southern ports (Wilson 1976, 11). The initial system used shutters in short towers. The system was cheaply and quickly installed but hard to read at a distance. It fell out of use following the success of the British navy at the Battle of Trafalgar. Fears of invasion simply faded away. The Hundred Days was too short a time to restore the shutter system, and after Napoleon was soundly defeated a second time at Waterloo, the old shutter system was replaced with a semaphore system using mechanical arms. This system was modified and improved through the 1820s and 1830s, although it remained a primarily military system (46). A small number of purely commercial systems were built in Britain in the aftermath of the Napoleonic Wars, the most significant of which was the line to notify shipowners and merchants in Liverpool of shipping arrivals in the Mersey estuary (69).

All such systems were restricted by line-of-sight operation and the range of telescopes in good weather and daylight. A large number of repeater stations were needed, and human operators introduced errors in repeating messages. Transmission rates were abysmal: the French system, the best in Europe, could manage two characters per minute (min) when the light was optimal, one per min in poorer light. The Prussian army officer Major August O'Etzel, observing the operations of the French system to help improve the performance of the Prussian telegraph between Berlin and Coblenz, noted that "on some days up to six messages of 20–30 words [were] transmitted" (Wilson 1976, 144).

THE ELECTRIC TELEGRAPH

Almost from its inception the electric telegraph broke free of the climatic, temporal, repeater, range, and rate limits that afflicted its mechanical forebear. True, lines would sometimes be destroyed by ice, but human agency was far more destructive. In many parts of the world telegraph lines would become fair game for the metal they contained.

From the first, crude installations of the 1830s and 1840s it was obvious that electric telegraphs had substantial range. The progress of land-based systems became rapid after the successes of the 1840s. In America the telegraph expanded alongside the railroad in a program of joint conquest, its principal utility being that it allowed railroads to avoid both the high accident rate on single-track systems that had to accommodate two-way traffic and the high capital cost of double-tracking. With telegraphy trains could proceed along single-track systems to await passing traffic at the appropriate passing loops instead of having to stick rigidly to an unreliable published schedule or risk head-on collisions (Hugill 1993, 320). Western Union's first great success was alongside the multiple trans-Appalachian railroads completed in the early 1850s (Hugill 1993; Thompson 1947; Vance 1992, 1995). On a more global basis, the first transatlantic submarine cable was completed in 1858, the first successful one in 1866. By the end of the nineteenth century the globe had been crisscrossed using a technology only a little more than a half-century old. Moreover, most of this new information infrastructure was installed with private capital, necessitating little of the massive government spending and subsidies that had accompanied the equally spectacular growth of the railroads.

The technology of electric telegraphy was relatively simple, despite fairly consistent attempts to complicate matters. Software design proved far more critical and time-consuming than designing and developing the hardware itself. Developing submarine cables and the ships to lay them was more difficult, requiring a theoretical understanding of the way electrons flowed along such long lines. Even so, the theory was formulated and the cable ships were in place by the mid-1860s.

The most complicated problem with the electric telegraph was not technical but geopolitical. Britain built the first global information infrastructure, establishing itself securely as the information hegemon by the late 1800s at the geographic center of a global network of submarine cables. The other polities never caught up, no matter how hard they tried. Only with the switch to other forms of telecommunications did Britain's dominance begin to slip, but its information hegemony did not end until the late 1950s.

TECHNOLOGY

In hardware terms the electric telegraph is a simple device. Installed overland, it was a single, uninsulated iron or copper wire strung along a line of glass or ceramic insulators mounted well off the ground, usually on the crossarms of wooden poles. The return circuit was, quite literally, the ground. Even feeble amounts of electricity, generated chemically by lead acid batteries, could drive very low frequency signals, generally below 30 hertz (Hz), very long distances (Fagen 1975, 195). Since the signal was a simple on-off signal, it could tolerate a high level of noise. In modern terms it was a digital signal—there or not there—unlike the later telephone, which had to send much more complex wave forms, analog signals from which human speech could be reproduced. Analog signals could not tolerate much extra noise, a problem first studied seriously by Shannon (1984).

Submarine cables presented more of a problem. The wire had to be both insulated from the sea water, itself a conductor because of its salt content, and armored against sea creatures, who found electrical signals attractive and ate their way through the first generation of merely insulated cables. A primitive submarine cable was installed across the English Channel as early as 1850, but the cable broke almost immediately. A stronger cable armored with galvanized iron wire succeeded in 1851, convincing companies and investors alike that the Atlantic could be bridged.

THE FIVE TECHNICAL PERIODS
OF SUBMARINE CABLE CONSTRUCTION

*Telegraphy and
the First Global
Telecommunications
Hegemony*

29

Five distinct technical periods of submarine cables can be identified: experimental, 1857–58; successful but short-lived, 1865–69; low-speed lines, 1873–82; high-speed lines, 1894–1910; and very high speed lines, 1923–28. The first two are certainly the most interesting from a technical point of view. Thereafter the process was one of development rather than innovation.

The Experimental Period, 1857–1858

The failures of the experimental period resulted more from an inadequate theoretical understanding of transmission than from inadequate insulation or armoring, although these had their place. The 1857 Atlantic cable was a complete failure. The 1858 cable transmitted for about three months, then failed in deep water and could not be repaired (Albert 1934, 6). (For the successful but short-lived period, the use of much heavier duty cable was planned based on a better theoretical understanding of transmission as well as a more accurate appreciation of the hostile environment of the seafloor.) Rubber and tar were used to insulate the first, primitive cables of the 1840s. These were quickly replaced by gutta-percha, which was found to hold up much better. Rubber and gutta-percha are isomers. They have the same molecular formula but the atoms are arranged differently in the molecule. Both are natural latexes from tropical trees, but because the molecules of gutta-percha have a *trans* (opposite-side) structure, it is far less elastic than rubber, whose molecules have a *cis* (same-side) structure. Gutta-percha retains its plasticity over time as well as under the extreme pressure and low temperatures that characterize the seafloor. The best gutta-percha came from Malaysia and the Indonesian archipelago, Borneo in particular (Bright [1898] 1974, 258). British control of this trade, through the port of Singapore, was marked, and it was crucial to British dominance of submarine cable manufacturing. Copper wire weighing up to 160 kg/km was first covered with gutta-percha insulation and then wrapped in "brass tape against boring insects, tarred hemp to cushion against blows, and steel wire around the outside" (Headrick 1988, 102). Such technology was extremely expensive—up to £200, or $1,000, per km—and extremely heavy. A cable designed to lay on the ocean floor weighed 1–2 tn/km, but one designed for the continental shelves or shallow seafloors with heavy tides and fishing activity could weigh up to 12 tn/km (102). The 1865 cable broke after two-thirds had been laid, but it was strong enough to be picked up and completed after the 1866 cable was successfully put into service (Albert 1934, 6).

In 1857 and 1858 part of the problem was that two ships carrying the needed cable, one British and one American, had to meet in the middle of the Atlantic. By 1865 Brunel's huge *Great Eastern* was available, the only ship in the world able to carry the nearly 4,200 km and 9,000 tn of cable needed to bridge the Atlantic (Oslin 1992, 173). The *Great Eastern* dominated cable laying for 10 years, after which more specialized ships begin to appear (Haigh 1968, 59). British ships dominated at least through the end of the century: as late as 1904, 28 out of 41 cable ships were British, 5 were French, and the other 8 belonged to 6 other countries (Headrick 1988, 103).

After the 1858 cable had been installed, Charles Bright, the chief engineer, and Professor William Thomson, the chief electrician on board ship, turned it over to Edward O. W. Whitehouse (one of whose middle initials stood, appropriately, for Wildman), a former surgeon who served as the Atlantic Telegraph Company's chief electrician on land (Bright [1898] 1974, 25, 52). Although the cable initially worked satisfactorily, it was slow, the maximum rate of transmission being three five-letter words per min (51). It failed far more quickly than it ought to have because Whitehouse believed that currents of very high intensity (i.e., a high potential) were the best for signaling. He ordered the construction of enormous induction coils, 5 ft long, excited by a series of very large, potent electric cells. These cells yielded currents with an estimated potential of 2,000 V (52). The loss of this cable resulted in a loss to the shareholders of some £100,000 ($500,000) (Black 1983, 27).

As experiments by Bright later showed, such high currents quickly broke down the gutta-percha insulation. To put it simply, the 1858 cable failed because of Whitehouse's inadequate knowledge of electrical theory. The *New York Leader* asked, appropriately enough in the circumstances, "Have we a pack of asses among us and are they specially engaged in experiments over the Atlantic Cable?" (quoted in Oslin 1992, 172). William Thomson, later Lord Kelvin, believed that the cable would have worked quite well had it not been so abused (Bright [1898] 1974, 53). Thomson went on to construct a viable mathematical formula based on the capacitance and resistance of the cable to predict how far a signal could be propagated. Thomson saw the flow of an electric wave in a conductor as analogous to the flow of heat in a continuous medium (Wasserman 1985, 25). According to Sloan, "Thomson's understanding of a cable's transmission characteristics, arrived at through the use of Fourier's mathematics, suggested a way to create the signal that would make its information content more distinguishable more quickly at the receiving end" (1996, 179).

Thomson's theory was adequate for transmitting low-frequency digital signals within the confines of submarine cables, but later attempts to apply Thomson's theory to the propagation of high-frequency analog telephone signals proved disastrous. Oliver Heaviside's much more thorough understanding of Maxwell's 1864 theory of electrodynamics was needed first (see chapter 3).

Another technical problem was caused by an inadequate understanding of the topography of the ocean floor. In the absence of any maps of the ocean floor, the general assumption was that it was smooth. The first cables were laid with little or no slack, so that they strung out between peaks in the ocean floor and soon broke. Lieutenant Matthew Fontaine

Maury, head of the U.S. National Observatory, was the first to properly map the floor of the Atlantic, based on data from the U.S. brig *Dolphin* in a voyage of 1853. The development of the Brooke sounder by Lieutenant J. M. Brooke, of the U.S. Navy, allowed Maury to map samples brought up from the ocean floor. Maury determined that there was a substantial plateau with no strong currents or disturbances between much of New-foundland and Ireland. In a letter to the secretary of the navy dated 22 February 1854 Maury named it Telegraph Plateau (McDonald 1937, 31). He described it as placed there especially "for the purpose of holding a submarine telegraph, and of keeping it out of the way" (quoted in Bright [1898] 1974, 29–30).

The 1858 cable failed after a very small number of messages had passed through it. Sources differ on the exact number. George Saward, secretary of the Atlantic Telegraphy Company, claimed that the cable had carried 271 communications (Coates and Finn 1979, 8, 16). Charles Bright gives the number as 732 (Bright [1898] 1974, 50). Even at the high cost of $10 per five-letter word, news agencies as well as governments were quick to utilize it (Coates and Finn 1979, 79). Contemporary atti-tudes to the initial success of the 1858 cable were positive; a placard in the New York parade celebrating the achievement read:

REJOICE ALL NATIONS
Married,
On Thursday, August 5, 1858
In the Church of Progress,
At the alter of Commerce,
The Old to the New World.
May they never be divorced!

(Coates and Finn 1979, 130)

Such sentiments were, however, at odds with the refusal of American investors to invest in the 1857 and 1858 cables, however much these ear-ly efforts were a joint product of both British and American enterprise and governments. Cyrus Field sold 250 out of a total of 350 shares at £1,000 a share ($5,000) in two weeks in London, Liverpool, Glasgow, Manchester, and elsewhere in Britain. Yet, of the 88 shares he reserved for sale in his native America he disposed of only 21, leaving him with a personal commitment of $335,000 (Coates and Finn 1979, 9). After the failure of the 1857 attempt another £100,000 was raised from the origi-nal backers (14). The collapse of the 1858 venture, taking with it £100,000 ($500,000) of the shareholders' money (Black 1983, 27), caused numer-ous inquiries, none of which discovered any attempt to defraud. Never-theless, when the time came to raise money for the 1865 attempt, in-vestors were much harder to find, even in Britain. Of the £600,000 ($3 million) capitalization approved by the company directors only $306,000 had been raised by May 1863 on both sides of the Atlantic. Of this sum, Field managed to raise $66,000 in the major cities of the northern Unit-ed States, a remarkable achievement in the middle of the Civil War (Coates and Finn 1979, 18, 21–22).

The Successful but Short-Lived Period, 1865–1869

The problem of financing the 1865 cable was not solved until John Pender, a textile merchant in Manchester and Glasgow, formed the Telegraph Construction and Maintenance Company (TCM) in 1864 by merging the major constructor of the cable, Glass, Elliott and Company, with the company that manufactured the insulation, the Gutta-percha Company (Coates and Finn 1979, 22–23). When laid, the 1865 and 1866 cables were operated by Pender's first great operating company, the Anglo-American Telegraph Company.

Most of the major investors in Pender's companies would come, like Pender himself, from the textile industry (Headrick 1991, 36). This should not be surprising since the textile industry was the first truly international industry. By the mid-1800s most of the world's cotton was being grown in the slave states of the American South, warehoused in the great British cotton port of Manchester, spun and woven in the textile towns of Lancashire and around Glasgow, and marketed worldwide. The creation of TCM and Anglo-American marked the beginning of Pender's inexorable rise to dominance over the global submarine cable business. By 1892 his Eastern and Associated Telegraph Companies (more simply, Eastern) would own 45.7 percent of the world's total cable mileage and nearly 70 percent of Britain's. Eastern became the largest multinational corporation of the nineteenth century (39).

Yet raising money for such cables initially was not easy, even in wealthy, liberal Britain. There was simply a limited amount of capital available for all the profitable global infrastructural investments of the Victorian world, not to mention booming industry and profitable overseas land development. Atlantic cables had to compete with Argentinean railroads, American steel mills, Appalachian coal mines, and Australian sheep stations, to name but a few of the options. Perhaps the biggest benefit to cable investment was the sudden release of British capital made possible by the British government's nationalization of the domestic telegraph companies in 1868. The Telegraph Purchase Bill liberated "something like £8,000,000 sterling [$40 million] for reinvestment by those who looked favorably on electric telegraphs as a subject of safe and sure remuneration" (Bright [1898] 1974, 110).

Financing for most of the world's cables continued to come from private companies such as Eastern throughout the nineteenth century. By 1892 just under 90 percent (by length) of the world's cables and all the long-distance ones belonged to private enterprise. The average length of all privately owned lines was 662 km, whereas that of government lines was 29 km (calculated from Headrick 1991, 38–39 [tables 3.2 and 3.3]).

Britain's monopoly in manufacturing and laying of submarine cables in the nineteenth century was even greater than its monopoly in ownership. TCM made two-thirds of the world's cables before 1900. Most of the other third was made by three other British companies: Siemens Brothers; India Rubber, Gutta-percha and Telegraph Works; and W. T.

Henley Telegraph Works (Headrick 1991, 31). Despite its German name, Siemens was a British company. Wilhelm Siemens settled in London in the early 1840s, changed his name to William, and was eventually knighted. He remained in close contact, however, with his German relatives and their principal engineering company, Siemens und Halske (Siemens 1957a, 11, 17). The French made a few short cables after 1881, and the Germans began to manufacture them in the 1890s. Americans depended on Britain until the 1920s and beyond.

The control of cable-laying ships was also almost entirely in British hands. Until 1873 only the 18,912 tn *Great Eastern* could hold enough cable, submerged in water to prevent the gutta-percha from drying out, to cross an ocean. But the *Great Eastern* had not been built to carry cable; it was a converted passenger ship, with some of its boilers removed to make room for the cable tanks. When Hooper's Telegraph Works of London had the *Hooper* built, it was only the second ship designed specifically for cable laying and repairing and the first really large one (Haigh 1968, 78). It was also the second largest ship in the world at the time, though at 4,935 tn it weighed far less than the *Great Eastern*.

Once Thomson had solved the basic theoretical problem of transmission, the single biggest problem was how to raise enough capital to pay for the cable and the ships to lay it. All the other problems could be solved by a fairly straightforward process of technology development, most of which simply focused on improving the speed and accuracy of transmission. Because of the rapid advances in submarine cable construction, cables were rendered obsolete soon after they were installed. Yet the later cables had such a long life and represented such a huge capital investment that it was important to use them as long as possible, hence the focus on improving transmission.

Because of the feebleness of signals on very long submarine cables, Thomson had developed the mirror galvanometer by 1858. This device required one clerk to read the message and one to write it down, left no direct record, and was subject to considerable human error (Headrick 1991, 32). Contemporary land lines used a printing device, which avoided all these problems. In the early 1870s Thomson introduced his siphon recorder, which allowed a permanent record of submarine cable signals. Automatic transmitters were developed that speeded transmission using prerecorded, punched paper tapes. In the late 1870s duplex telegraphy was introduced, allowing a single cable to simultaneously transmit in both directions, although this technique could be used only on newer cables in perfect condition (see Bright [1898] 1974, 526–675, for a summary of these improvements). The consequence of all these improvements and many more was that transmission speeds rose steadily and costs per word steadily dropped.

Whereas the 1858 cable struggled to reach a transmission speed of 3 five-letter words per min, the 1865 and 1866 cables initially transmitted about 8 words per min, increasing to 15–17 words per min "as the staff became more accustomed to the apparatus" (Bright [1898] 1974, 105). A major reason for the increase was Thomson's demonstration that signaling speed depended on conductor resistance as well as electrostatic inductive capacity. Increasing the diameter of the copper core of the cable substantially lowered the resistance without excessively increasing the capacitance (81).

The 1865 and 1866 transatlantic cables were joined in 1869 by an identical cable installed for a nominally French company between Brest, the island of St. Pierre, and Sydney, Nova Scotia. The cable was manufactured in Britain, part by TCM and part by W. T. Henley, laid by the *Great Eastern,* financed largely by British investors, and operated chiefly under British direction and management (Bright [1898] 1974, 107). Like the earlier British-owned cables, the 1869 cable had a short working life. The hemp wrapped around the ten steel wires that made up the outer armor was subject to decay. Once the hemp decayed, the steel wires followed suit, and the cable could no longer be lifted for repairs. In addition, the specific nature of the armor was such that teredo worms could easily burrow between the steel wires and damage the copper core (84). The 1869 cable marked the end of the first period of cable laying.

The Period of Low-Speed Lines, 1873–1882

The 1866 cable finally failed irretrievably in 1872, the 1865 cable in 1877, and the 1869 French cable in 1894 (Bright [1898] 1974, 108, 129). The failure of the 1866 cable ushered in the third technical period, the period of successful but low-speed lines. Anglo-American Telegraph installed the first of these in 1873 and 1874. The failure of the 1865 cable resulted in a third Anglo-American line in 1880 (129). The first genuine competition for Anglo-American was the Direct United States line laid in 1875 by Siemens. Like the other cables of the 1870s, the Direct transmitted 16 words per min, but in 1878 the Direct was converted to duplex working, with 32 words per min now possible overall (123, 129). Siemens installed similar cables for two other operating companies, one French, one American. In 1879 a second French line was installed, promoted and financed by "several large Parisian banking houses" (132). In 1881 and 1882 Siemens installed lines for Jay Gould's American Telegraph and Cable Company to complete the third technical period. These last two lines were then leased by and worked in conjunction with Gould's Western Union, which dominated the North American continent. However, both the French and the American lines entered the Joint Purse agreement, set up by Anglo-American in 1871. The Joint Purse was a transatlantic telecommunications cartel, although the Gould lines participated on their own terms (Headrick 1991, 35) and the French withdrew in 1894 (Bright [1898] 1974, 132). Under the terms of the Joint Purse agreement all companies were to charge the same rates and divide the income in proportion to their traffic (Headrick 1991, 35). The Joint Purse had the additional advantage of providing continuous service when a specific line was being repaired, most breaks being caused by fishing trawlers operating on the continental shelves.

The Period of High-Speed Lines, 1894–1910

The fourth technical period was characterized by much heavier duty lines capable of very much higher transmission speeds than those achieved in the period 1873–82. Once again the diameter of the metal core of the cable was increased to lower resistance. Duplex working combined with automatic transmission on these improved cables to nearly quadruple working speeds. Bright noted that simplex working of the third-generation cables produced maximum speeds of 25–28 words per min, "the former, say, by average manual transmission and recording, and the latter by automatic transmission. The actual speed obtained by automatic transmission with the [1894 Anglo-American] cable is as high as forty-seven (or even up to fifty) five-letter words per minute" ([1898] 1974, 142). Duplex working, of course, doubled these figures for overall working.

The Period of Very High Speed Lines, 1923–1928

The fifth phase of cable laying, in the 1920s, saw further increases in transmission speeds, from 2×50 words per min to 5×60 in 1924, 4×80 in 1926, and 5×50 in 1928 (Coates and Finn 1979, 160–61). The last cables to carry telegraphic signals across the Atlantic floor were not designed as telegraph cables per se. The first telephone cables of the 1950s used the frequency-division multiplex carrier technology devised by American Telephone and Telegraph to carry low-frequency telegraph signals and high-frequency voice signals simultaneously.

As capacity increased, costs fell substantially. Coates and Finn calculated an average decrease of 15.4 percent per year in the costs of transatlantic telegraphy from 1866 to 1888 (1979, 87–89). The 1858 rate of $10 per word continued on the 1865, 1866, and 1869 cables until 1872. The rate dropped to $5 per word in 1872 and $2.50 shortly thereafter, then to 50¢ by 1884 and 25¢ by 1888. There is, regrettably, little consistency in the published data on rates. Coates and Finn (1979) do not agree with Bright ([1898] 1974, 143–44) or with Anderson (1872). Anderson recounts a tariff of $3 per word in the first six months of 1871 and $2 per word in the last six months (35). Since Anderson was primarily concerned with measuring the statistical effect of tariffs on usage, his figures cannot be ignored. Bright's central involvement with the cable gives him some claim to primacy, however, so the situation rests there, unsatisfactorily as to absolute rates but consistent as to absolute decline. Coates and Finn cite Bright, although not on rates, but not Anderson.

GEOPOLITICS

Once the relatively simple technical problems were solved, by the early 1870s, and telegraphy was shown to be an attractive proposition for investors, other considerations came to the fore. The most significant of these was the competition between polities for world power, one aspect of which is known as geopolitics. Telegraphy's ability to control space for economic and military purposes in the interests of the state made it as pure a form of geopolitics as any, as Lakanal had understood as early as the French Revolution. Once electric transmission had solved the many operational problems inherent in the mechanical systems, polities were quick to recognize its utility, despite all the ebullient forecasts of improved commerce, brotherhood, and peace. Although the British took and held a virtual monopoly from 1858 on, such a monopoly was not fore-ordained. Nor, once achieved, did it go unchallenged.

American Attempts

In 1854, before a transatlantic submarine cable seemed feasible, a land link was proposed between the Old World and the New, crossing only about 100 mi of shallow water at the Bering Straits (Bright [1898] 1974, 113). Returning in 1859 from his second trip to Russia and with considerable experience of the Amur River basin as a likely route, Perry Collins sent American Secretary of State Lewis Cass a detailed proposal for just such a telegraph (Collins 1962, 33). In part because of the support of the California members of Congress, who naturally believed that their state's future lay as much with the Pacific Rim as with North America, Collins had been appointed American commercial agent of the United States to the Amur River region in 1856 (21). It is not impossible, as Coates and Finn claim (1979, 51), that this was "most likely as a cover for his investigations into the possibility of the Siberian telegraph." The failure of the 1858 submarine cable and the completion of the transcontinental line across the United States in 1861 encouraged Collins to agitate seriously for his Siberian line. Since Congress refused to act, Collins tuned to Hiram Sibley, who, with Thomas Ezra Cornell, had founded Western Union. Sibley saw the Bering Straits route as a viable competitor to the transatlantic submarine cable, especially in light of the expensive failure of the 1858 attempt and the consequent difficulties the British were having in raising money to finance another attempt. Sibley set up the Western Union Extension Company in 1864 with a nominal capitalization of $10 million—100,000 shares of stock at $100 a share (Coates and Finn 1979, 51). Unlike the British submarine stock, that of the Western Union Extension Company was taken up immediately, and at the height of the Civil War, most of it by the original stockholders of Western Union, albeit only at an initial assessment of 5 percent. So confident were investors that "in less than two months its stock sold for seventy-five dollars per share, with only one assessment of five dollars paid in" (Kennan 1910, 2–3). The Western Union Extension, usually referred to as the Collins Line, was thus capitalized at more than three times the $3 million capitalization approved for the 1865 transatlantic cable, only $1,530,000 of which had been raised by the summer of 1863.

The intense competition over connecting America to the rest of the world resulted in part from the ending of internal competition within America that came with the signing of the Treaty of Six Nations in 1857

(Thompson 1947, 310–30). In 1855, in his first venture into the telegraphy business, Cyrus Field had achieved three ends. First, he had forged the powerful, new, American Telegraph Company out of a large number of small, inefficient lines operating in New England and down the Atlantic seaboard. Second, he had developed a special relationship with the New York Associated Press and agreed to give them priority in news service (Coates and Finn 1979, 48–49). Third, he had gained control over the lines into the Maritime Provinces of Canada through his purchase of the New York, Newfoundland, and London Telegraph Company. He saw the success of the Atlantic cable as crucial to his great plan of controlling what he believed would be an immensely profitable monopoly of traffic between New York and London. Western Union, shut out from access to the proposed transatlantic cable by the terms of the Treaty of Six Nations, had little choice but to pursue an overland strategy across the continental United States, up through Canada into Russian Alaska, thence across the Bering Straits and Siberia into Europe. The failure of the 1857 and 1858 cables and the linking of New York to San Francisco in 1861 seemed to confirm the wisdom of this strategy, although it was abruptly abandoned with the success of the 1866 cable.

This early competition for traffic between the Old World and the New at first seems to have been primarily commercial. Most interpretations of American history point to America's coming of age as a global power as occurring at the end of the nineteenth century. Western Union projected a demand of 1,000 messages per day for its overland route. At $25 per message this implied an annual income of over $9 million, which more than justified the $10 million investment. Yet a powerful element of geopolitics was present even this early. The Civil War saw the first rise of the United States, or at least its northern component, the republic of the North (Hugill 1993, 165–66), to the status of a world rather than a regional power. In the aftermath of the Civil War the republic of the North possessed by far the largest and most powerful navy in the world and demonstrated a willingness to use it toward geopolitical ends (Hugill 1990).

The dispatch to Europe of Assistant Secretary of the Navy Gustavus Vasa Fox aboard the U.S.S. *Miantonomoh* to thank Russia for its support in the Civil War and, presumably, to negotiate the Alaska Purchase was not the act of a regional power. More significant, however, was the obvious threat the *Miantonomoh* posed to British naval dominance. In the aftermath of the *Miantonomoh*'s visit the chief constructor of the British navy, E. J. Reed, totally rethought British battleship design to come up with the three ships of the Devastation class, all intended to counter the *Miantonomoh* and her presumed sister ships and successors, the larger and much more seaworthy Kalamazoo class. In any case, the *Miantonomoh* was a clearly visible, demonstrated threat in 1866, and the *Devastation* only an implied answer until it entered service in 1873; and the *Thunderer* and the *Dreadnought* took even longer. In the meantime Britain reopened the claim by the republic of the North to the damages caused by the Confederate cruiser C.S.S. *Alabama,* built in a British shipyard to prey on federal shipping with great effect until it was sunk by the U.S.S. *Kearsage* off the coast of France in 1864. Some six months before the *Miantonomoh*'s arrival in Britain, Parliament had openly refused to consider the *Alabama* claim, citing the Neutrality Laws. When Britain's naval hierarchy expressed, in the pages of the *Times* (London), its belief that no existing British warship could match the *Miantonomoh,* the British foreign secretary advised Parliament to appoint a royal commission to look into the working of the Neutrality Laws. When the commission reported that revision was justified, the *Alabama* claim was reopened and eventually paid (Hugill 1990).

As figure 2.2 indicates, Western Union had designs on more than the European telecommunications business. A line around the southern margins of the Eurasian landmass would have added China, Japan, India, and the states of the Persian Gulf to Western Union's telegraphic empire. Short submarine cables could then have been extended to the Philippines and even as far as Australia. Francis W. Pickens, the American ambassador to Russia, noted that such a telecommunications system would provide America with "control of the world" (quoted in Collins 1962, 32). Had Western Union's plans been carried out, New York would have been connected to Hong Kong and Australia before London was.

The proposed connections to China and Japan were very much in line with America's imperial ambitions in that part of the world following the second Opium War, in 1856–58 and 1859–60. Although Britain was the main European power to force China to its knees in the Opium Wars, America was a signatory to and beneficiary of the Treaty of Tientsin. The treaty settled the first half of the war, in 1858, and opened to Britain, France, Russia, and the United States eleven treaty ports in addition to those opened to Britain after the first Opium War, in 1839–42. In 1854 Commodore Matthew Perry, who had commanded the Gulf Fleet in the Mexican War and helped take Veracruz, forced Japan to open its doors to American trade. Although the proposed connections to Hong Kong, Australia, and India might appear out of place in the context of 1862, when British global power in retrospect seems to have been indisputable, they were certainly in line with American imperial ambitions of the period.

Despite the best efforts of a veritable army of men, most of them ex-soldiers mobilized under near military discipline in the immediate aftermath of the Civil War, progress along the rugged Pacific Northwest coast and in the harsh winters of the Amur River region of Siberia was appallingly slow (Collins 1962). The British government clearly did not perceive the Collins Line as much of a geopolitical threat, since it had no qualms about approving the route from the Oregon Territory through British Columbia into Russian Alaska. In the event, of course, none of this massive activity came to much after the submarine cable of 1866 was shown to work. During the next 30 years American's imperial ambitions would, in any case, focus on the completion of the process of expansion across central North America. Only in the mid-1890s against former Spanish colonies would America embrace overtly external, as opposed to its traditional continental, imperialism, and only then would Americans' obsession with long-distance telegraphy return.

The British Achieve Dominance

The considerable success of the 1865 and 1866 cables, financed, manufactured, and laid by the British, completely changed the investment climate for submarine cables, and the British government's 1868 buyout of
domestic lines created a huge reserve of capital to be invested in new en-
deavors. The main British focus thus moved from the Atlantic to the rest
of the world, linking first India, then the rest of the empire, and then the
world to London.

Like the Western Union scheme for a telegraph linking Russia and
America, the first line linking India to Britain followed a largely overland
route. It connected the already well elaborated and largely state-owned
European network (fig. 2.3) to India through Turkey, the Balkans, and
the Middle East. Like the Western Union scheme, it followed a failed submarine attempt, in this case that of the Red Sea and India Telegraph
Company line from Suez to Bombay, completed in February 1860 and
dead by March of the same year (Headrick 1991, 20). The failure of the
Atlantic cable of 1858 had cost £100,000, that of the Red Sea cable cost
investors in submarine cables a further £160,000 ($1.3 million in all)
(Black 1983, 27). The Indo-European Telegraph linked Britain to India
by January 1865, but it was expensive and slow, and the frequent retrans-
mission of messages (12 to 14 times), often by people whose command
of any written language, let alone English, was minimal, proved disastrous. A message containing 20 words cost $25 and took an average of five
to six days to reach its destination, arriving full of errors, when it was not
simply lost. Nevertheless, it was a great deal faster than a ship. The news
agency Reuters began to use the Indo-European Telegraph extensively,
as did commercial interests, but the British government dared not. By the
terms of its agreements with the countries through which it ran, the Indo-
European had to give local governments priority. Foreign telegrams were
thus simply put aside for quieter times, and there was a constant fear of
espionage. The British were particularly concerned about the section of
the line passing through Russia, an enemy in the recent Crimean War
and one with well-known designs on India (Headrick 1991, 21).

The solution was an all-British submarine cable, laid in three sections
by three of Pender's operating companies, manufactured by Pender's
TCM, and laid by the *Great Eastern*. In 1868 Pender's Anglo-Mediter-
ranean Telegraph Company laid the Malta-to-Alexandria leg. In 1869 his
Falmouth, Gibraltar, and Malta Telegraph Company connected Britain
and Egypt via Malta. In 1870 the Suez-Bombay cable finally gave Britain
a politically secure connection with India. Unlike the earlier Red Sea cable, the Suez-Bombay link was properly laid. It was paid for entirely by private investors, from whom Pender easily raised $6 million. Given the bad
taste the collapse of the Red Sea Company in 1859 had left in the mouth
of the British government, which had guaranteed stockholders £36,000
($180,000) a year for 50 years, the success of Pender's Eastern, formed
in 1872, was significant. It seemed, at least by the early 1870s, that imperial telecommunications could now be safely left to private investors, submarine cables, and Pender's discretion (Headrick 1991, 20, 24, 36). The

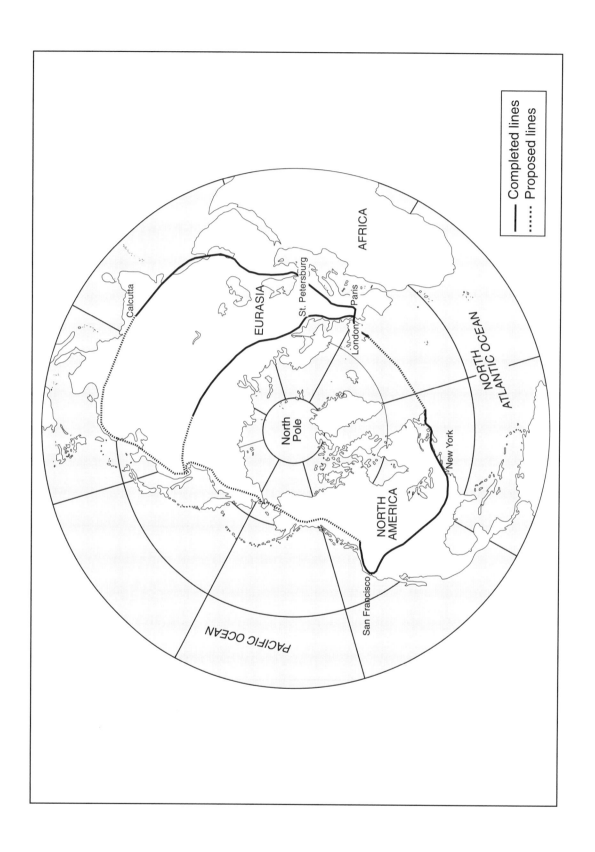

| Completed lines |
| .... Proposed lines |

final link in Pender's imperial chain, the Eastern Extension Australasia and China Telegraph Company, familiarly known simply as the Eastern Extension or Eastern, was formed in 1873. Pender went on to found numerous other companies that served Latin America, the west coast of the Americas, and Africa, but Eastern and the Eastern Extension were the jewels in the crown (36).

In fact the need for subsidies returned very quickly. Sir Charles Bright pointed out in a letter to the *Times* printed on 18 April 1928 (reprinted in *Zodiac* 20 [June 1928]: 338) that such subsidies were considerable. As early as 1879 the British government agreed to pay the Eastern Extension £32,000 a year for 20 years to subsidize the line to Australia and £55,000 a year to subsidize the lines down the east coast of Africa. In 1885 the government began paying £19,000 a year for the lines to the African west coast, and in 1893 it began paying £28,000 a year for the line from Zanzibar to Mauritius, in both cases for 20 years. At their peak, therefore, from 1893 to 1899, the subsidies for these lines alone reached £134,000 ($672,000) a year, on top of the continuing subsidy of £36,000 a year ($180,000) for 50 years to the shareholders in the failed 1858 Red Sea cable.

Only the French attempted to combat Britain's tightening control on the first global telecommunications network in the period from the 1870s through the 1890s. These schemes invariably failed either to attract crucial government subsidies and guarantees or to make a profit. France had a global empire but not a global trading system, and cable profits were clearly tied to trade. After 1870 France's principal geopolitical concern was its neighbor Germany, not a far-flung empire, and any public expenditure that did not strengthen France against Germany was politically worthless (Headrick 1991, 42–43).

Anglo-American Joint Dominance

Through the 1890s the only country able to compete with Britain in financing submarine cables was America, whose tremendous profits from its internal telegraph system were available to help fuel external expansion. Yet American finance was open to British investors, and British investment was heavy in American telegraph stock just as it was heavy in American railroads. For much of the later nineteenth century British industrial growth and the British industrial population needed to import raw materials, fiber for the most part, and foodstuffs. Commodities such as American cotton and wheat, Australian wool, and, after refrigerated ships were developed in the 1870s, Argentinean beef flowed into Britain in ever increasing amounts. Submarine cables, better steamships, railroads, and land-based telegraph systems coordinated the financing, production, shipping, and warehousing of all these necessities and more, yet the very coordination itself was also a profitable activity.

And this was merely open investment in widely traded stock. There was a second, secret level of British investment, especially after the upsurge of nationalism at the end of the nineteenth century. One aspect of this

Fig. 2.2. Proposed Routes of Western Union's Russian-American Telegraph and the British Atlantic Cable, 1865 (redrawn from Thompson 1947, 425) *(opposite)*

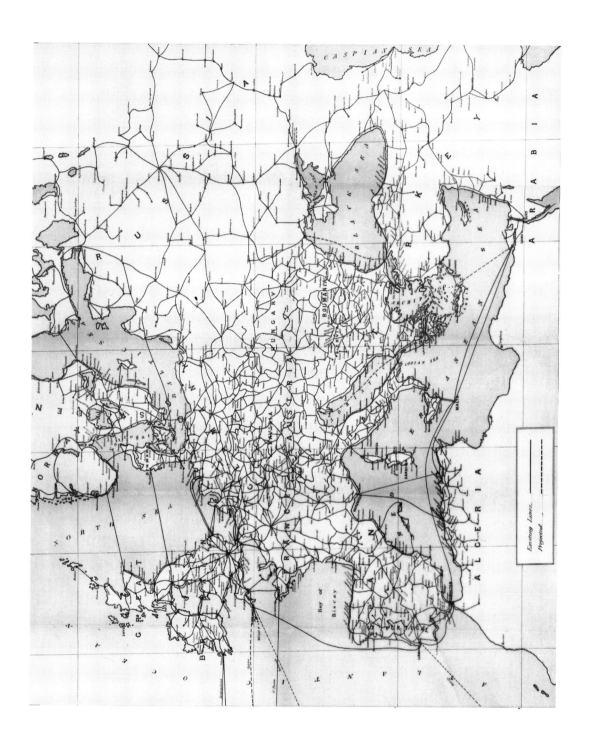

secret investment became public in the 1920s when Clarence Mackay admitted in the Cable Landing License Hearings before the U.S. Congress that 50 percent of the shares in the nominally American Commercial Pacific Cable Company, the dominant line between the American west coast and Asia, had been secretly owned by Eastern since the founding of Commercial Pacific in 1901 (Headrick 1991, 100–101). The increase in American financing of the Atlantic cables in the late nineteenth century, which resulted in America's total financial control of the Atlantic cables by 1911, never concerned the British government. Many cables were only leased to American companies in any case. More to the point, the leased cables all passed through Britain, and the British government could still "censor and scrutinize the traffic as easily as under British management" (102).

In the Committee for Imperial Defense's 1911 map of the world cable system (fig. 2.4) Britain's geographic centrality is crystal clear. The world cable networks were carefully classified by ownership; and the discussion of which cables should be cut in the event of war, and in which order, is extensive. The opening section of the nearly 250-page report of the standing subcommittee dwelt on recent rapid changes in the international situation. The report thus considered only the steps to be taken in the event of war with Germany or with the Triple Alliance, with or without Turkey. That France and Russia would join with the British Empire was considered likely: "Generally speaking, if France and Russia were in alliance with this country, it would be possible to isolate Germany from practically the whole world outside Europe, by cutting the cables to the Azores, Tenerife, and Vigo and the three cables landing on Yap Island" (PRO, CAB 16/14, 1, 14). The subcommittee recognized, however, that cutting the German cable between the Azores and New York would be strongly resented by the United States. Among other things, "that country [the United States] has definitely adopted the principle of the inviolability of cables connecting neutral ports" (PRO, CAB 16/14, 12).

In a revealing set of notes in appendix K of the report is a translation extracted from Dr. A. Roper's *Die Unterseekabel,* published in Leipzig in 1910, to the effect that "the course and result of a naval war may depend on the opportune arrival of commands and information by submarine cable" (PRO, CAB 16/14, 223). The same appendix also quotes from a 1900 essay by Captain George Owen Squier, of the United States Signal Corps, that in time of war submarine cables would be "more important than battleships or cruisers."

Building international tension from the 1890s on, British mistrust of Germany, German mistrust of Britain and the United States, and American, or at least the U.S. Navy's, mistrust of Britain combined to convince all three polities that total control of their own telecommunications systems was an absolute necessity. This tension came to a particular head with the competition to construct the last major submarine cable, that linking Asia and America across the Pacific. The problem was that the distances were so great that one or more repeater stations were needed, which made otherwise irrelevant islands suddenly important. Nevertheless, in view of the fact that all the lines running from Britain east to Asia

Fig. 2.3. European Telegraphs in the 1870s (photo courtesy of the Library of the Institution of Electrical Engineers, London) *(opposite)*

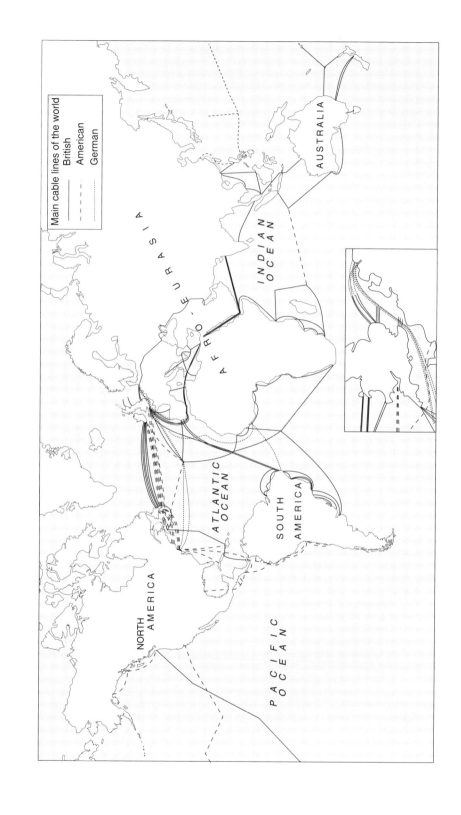

Main cable lines of the world
British ——————
American ——————
German ....

NORTH
AMERICA

SOUTH
AMERICA

AFRO - EURASIA

AUSTRALIA

ATLANTIC
OCEAN

PACIFIC
OCEAN

INDIAN
OCEAN

and Australia passed through possibly hostile countries or through shal-low seas where the cables could easily be dredged up and destroyed, the British Admiralty became convinced of the necessity of a deep-sea cable from the west coast of Canada to Australia. The completion of the Cana-dian Pacific Railroad (CPR) across Canada in 1879 confirmed the utility of a transpacific cable strategy (Barty-King 1979, 93–94). It was, in fact, one of the main architects of the CPR, Sandford Fleming, who pushed hardest for the "All Red Route" for imperial cables.

In the 1880s, soon after the CPR was finished, Fleming faced great op-position from Pender, whose Eastern and Eastern Extension monopoly of traffic to the east from Britain would have been broken by Fleming's scheme. Fleming was easily defeated. But by the 1890s the international climate had changed, and Eastern's cables were compromised. In 1892, shortly after the Royal Navy had completed its hydrographic survey of the proposed Pacific route, H.M.S. *Champion* seized Johnston Island, south-west of Honolulu, as a possible landing site for the Pacific cable. The Hawaiian government, however, claimed the island as its own, and the British government "felt obliged to restore it" (Barty-King 1979, 94). In 1894 Fleming went to London to press his case for a second route using Necker Island as the mid-Pacific repeater station. When the British gov-ernment demurred, Fleming took action, planning a private annexation by a retired naval officer from Toronto whom he enlisted to sail to Hon-olulu, "from where he was to book a steamer and proceed to Necker, un-furl the imperial flag, and leave behind evidence of his visit to legitimize the anticipated claim of British sovereignty." The British government got wind of the plan, and the Foreign Office "revealed the plot to the Provi-sional Government of the Hawaiian Islands, headed by U. S. national San-ford B. Dole [who] had recently overthrown the government of the in-digenous Hawaiian Queen" (Thompson 1990, 69). Dole's provisional government immediately annexed Necker (Barty-King 1979, 102).

In June 1894 the Dominion of Canada made the All Red Route a prin-cipal topic of the colonial conference held in Ottawa (Barty-King 1979, 102; Johnson 1903). The conference stressed the need for communica-tions among the countries of the outer empire—Canada, New Zealand, Australia, and South Africa bypassing London if necessary. In 1897 the Imperial Cable Committee recognized that the needs of the outer em-pire and British imperial security had essentially merged. On 29 June 1899 a deputation from the Eastern and Eastern Extension Companies visited the Chancellor of the Exchequer and the secretary of state for the colonies. Joseph Chamberlain, then secretary of state, accepted the ar-gument that the completion of the American line from Hawaii would give American traders an advantage in the Australian marketplace. Lord Tweeddale, the chairman of Eastern, noted that depending on an Amer-ican line in time of war would not be safe: "Although we have enjoyed freedom from war since we have been established—a state of war affect-ing our cables—there is no reason to think that we shall have that im-munity in the future" (Cable and Wireless Archives).

The recommendation to fund the Pacific submarine cable was finally published that same year, 1899, with considerable controversy about the

Fig. 2.4. American, British, and German Submarine Cables, 1911 (PRO/CAB, 16/14 map appendix) *(opposite)*

delay (Barty-King 1979, 116). Foot-dragging over construction continued. The congress of chambers of commerce that met in London in June 1900 resulted in several resolutions, notably those of the chambers of commerce of the cities of Ottawa and Vancouver that the Pacific cable be hurried to completion (Cable and Wireless Archives).

When the line was finally built, it had to be much longer than the earlier routes proposed via the Hawaiian Islands. The British were forced to annex the Fanning, Christmas, Penrhyn, and Suwarrow Islands. The new route cost some $2.25 million more than estimated for the Necker route (Thompson 1990, 69). Meanwhile, of course, American enterprise had laid a cable along the more direct route, starting at San Francisco instead of Vancouver and proceeding via Hawaii to Australia. This was laid by the Commercial Pacific Company, which, it later turned out, was half owned by Pender's Eastern group, TCM providing the cable itself. Once again, the security paranoia of the U.S. Navy seems justified inasmuch as one transpacific line was all British and the other was clandestinely half-British.

British Dominance Pays Off: World War I

World War I was a defining moment for the British submarine cable system. When war broke out in late 1914, Britain immediately cut all German cables, just as the Committee for Imperial Defense had specified in 1911. This began the crucial process of isolating Germany from American sympathy. Attacks on Germany's rapidly growing system of wireless telegraph stations, especially the crucial African relay stations at Kamina and Dar-es-Salaam and the Pacific relay station at Nauru, followed, reducing Germany to its master station at Nauen, just outside Berlin, and the American receiving station at Sayville, New York. As *Wireless World* *(WW)* noted at the time, shutting down the African stations silenced German naval commerce raiders in both the South Atlantic and the Indian Ocean and led directly to the sinking of the passenger liner turned cruiser *Kaiser Wilhelm der Grosse:* "No German ship afloat in the far-distant waters has a wireless installation of the necessary strength to enable them to communicate direct with Germany. . . . Britain's supremacy in wireless is proving a big factor in the situation, especially as the enemy's long-distance communications have been destroyed" (*WW* 2 [October 1914]: 411). Thus most of Germany's wired and wireless telecommunications were silenced within weeks of the outbreak of war. Only the station at Windhoek, in German Southwest Africa, took longer, falling to South African troops in May 1915 (Headrick 1991, 141).

German attacks on British telecommunications were far less successful, although German commerce raiders generated considerable excitement, if not many results, in the Pacific. Attacks on Fanning Island by the cruiser *Nürnberg* and on Cocos-Keeling Island by the cruiser *Emden* disrupted the Pacific cables briefly. At Cocos-Keeling the German raiders destroyed dummy cables laid as decoys but not the main cable to Singapore and lost the *Emden* to the Australian cruiser *Sydney,* who had been notified of the *Emden*'s raid by wireless telegraphy as a result of the cable operator's signaling the start of the attack. Destruction of the German radio stations also forced such cruisers to report to Germany via shore cable stations, which restricted their mobility.

American neutrality from 1914 to 1917 might seem to have favored Germany, but the neutrality was mainly on paper. All communication between Nauen and Sayville had to be in "clear," that is, uncoded. The German radio station at Tuckerton was simply commandeered by the U.S. Navy on 9 September 1914. That at Sayville was allowed to continue until July 1915, when it was taken over on the pretext that a new alternator and antenna made it a new station, necessitating a new license. The British-owned American Marconi station at Siasconsett was taken over in September 1914 for transmitting an "unneutral" message. Unlike the German stations, however, it was returned to private control in January 1915, although it could only transmit in "clear." "German protests that the British still had their cables and could send coded messages without censorship or interference fell on deaf ears. During the 'neutrality period' the United States was not so much neutral as opportunistic" (Headrick 1991, 144).

The Germans were not totally devoid of cables, but they lacked easy access to secure, direct communications between two points. They were thus forced to depend heavily on coded signals on "neutral" cables to maintain security, and their belief that their codes were inviolable became one of their all too numerous Achilles' heels. Immediately upon the outbreak of World War I the British Admiralty began its attack on German codes. Sir Alfred Ewing, a Cambridge professor interested in ciphers who had become director of naval education in 1908, was asked to take on the task (*Zodiac* 20 [January 1928]: 200). By late 1914, operating out of Room 40 of the British Admiralty and armed with a German naval signal book provided by the Russians, who had acquired it after sinking the German cruiser *Magdeburg*, Ewing and his cryptanalysts had begun to read German naval signals (James 1956, 29). In April 1915 a German diplomatic code book was acquired in Persia, and the British began to read German code number 13040 (James 1956, 69; Headrick 1991, 167). In 1915 the British also realized that German diplomatic traffic was being sent via Stockholm as if it were Swedish diplomatic traffic. Complaints to the Swedish government stopped this, but the Swedes merely sent German cables via Buenos Aires instead, on a 7,000 mi detour called the Swedish Roundabout. By early 1916 Room 40 knew of the detour but preferred to read the messages rather than force the Germans to use yet another route, a policy they continued for the rest of the war.

Perhaps the single most significant result of this policy was the acquisition of the Zimmerman telegram. This telegram, named for German Foreign Secretary Arthur Zimmerman, together with unrestricted submarine warfare, brought the United States into World War I. Because of the slowness of the Swedish Roundabout, the German ambassador in the United States, Count Johann Heidrich von Bernstorff, had received permission from President Wilson in late 1916 to communicate directly with Berlin on American channels in German code. Such communication was supposed to originate in plain language, but on this occasion the German Foreign Office was adamant that the American ambassador in Germany, James Watson Gerard, accept it in code. Given Wilson's orders, Gerard had little choice, however much the State Department disapproved of the procedure. James politely describes this lunacy as a "piece of effrontery [that] can never have been equaled in the history of political intrigue" (1956, 137). Room 40 intercepted Zimmerman's telegram to Bernstorff as it passed through London. Since it was coded in a new German code, 0075, which Room 40 had only partly decrypted, Ewing and his cryptanalysts could not read it all. Nor could they protest to the "neutral" United States, since that would have confirmed the U.S. Navy's suspicion that its traffic was being intercepted. The part of the message that Room 40 could decrypt instructed Bernstorff to send the message on to the German ambassador in Mexico City, von Eckhardt. Room 40 guessed that Eckhardt did not have code 0075 and that the message would have to be retransmitted within North America in code 13040, which it was. When the recoded version reached London, it could, of course, be fully decrypted. The telegram contained Zimmerman's offer to Mexican President Venustiano Carranza of "generous financial support and an undertaking on our part that Mexico is to reconquer the lost territory in Texas, New Mexico, and Arizona" if Mexico would join Germany in a war on America (141).

Accounts of how the telegram was presented to America vary. James notes that the British presented the telegram as the product of espionage in Mexico City (James 1956, 144), in part not to excite the paranoia of the U.S. Navy and in part to avoid compromising Room 40. German sympathizers tried hard to pretend that the telegram was a forgery until Zimmerman himself confessed to its authenticity. Ewing commented in 1928 that "it was . . . communicated very confidentially by Lord Balfour [the British foreign secretary] to Mr. Page [the American ambassador], and through Page to Wilson" (*Zodiac* 20 [January 1928]: 200). Equally to the point, the British were able to dupe the Germans into believing that the American secret service had obtained the telegram in Washington (James 1956, 153–54). As Headrick rightly says, "So tight was Britain's grip on world communications that it could not only block or read, at will, the most secret messages of its enemies, it could even use that information without revealing its sources. Never before or since in history has communications power been so concentrated and so effective" (1991, 169).

Britain owed much of its success to Pender and the Eastern group. In part this success was commercial; in part it was the result of substantial subsidy from the British government. French, German, and American attempts to produce parallel submarine cable networks in the nineteenth century were commercially unsuccessful, and Eastern simply bought any shares that became available in such cables. The nominally American transpacific cable thus turned out to be half-owned by Eastern. Eastern also weathered the technical competition from German and American low-frequency, long-wave radio in the first quarter of the twentieth century.

Ironically, the challenge that most nearly brought Eastern low was the internal one posed by the British Post Office's adoption of Marconi's high-frequency, short-wave beam antenna radio telegraphy in the late 1920s (see chapter 5). Beam stations were radically cheaper than submarine cables, allowing substantially reduced rates. "Within six months the Post Office beam stations captured 68% of all Eastern Telegraph and Eastern Extension traffic, and more than half of that of the Post Office's state-owned, state-controlled Pacific cable" (Barty-King 1979, 203).

As Eastern's business slipped away, it called in its political chips. Vyvyan notes that Marconi, which built and operated the beam stations for the Post Office, originally proposed the merger from what seemed a position of strength, but Eastern would not accept any deal that did not turn the Post Office beam stations over to Eastern (1974, 205–6). Imperial and International Communications, renamed Cable and Wireless in 1934, thus came into being on 30 September 1929. Twelve of the board members were from Eastern and 8 from Marconi. Although the distribution of stock was much more favorable to Marconi, at least in the opinion of the London *Sunday Times* (quoted in *Zodiac* 20 [April 1928]: 279), Marconi took this hard and withdrew from his own corporation, eventually returning to Italy and becoming an apologist for Mussolini and the Fascists. The merger produced a single company that controlled thirteen cable ships, 253 cable and wireless stations, more than half of the world's cable systems, and nearly all of the world's high-frequency radio stations. On the map that resulted from the merger London was once again securely at the center of global telecommunications (fig. 2.5). R. N. Vyvyan, one of the Marconi engineers responsible for the beam system, noted in 1933 that the reason for the merger and the impetus for placing the Post Office beam stations under private control was "the importance of the cable system to the British Empire from a strategic point of view" (1974, 93). As early as March 1928, Roland Belfort, the author of *Cables of the Empire* and other books, noted that "in America it is already being plainly stated that the Marconi-Eastern combination has been formed solely to fight the American radio-cable organisation" (1928, 324). Belfort clearly believed that only a complete merger of all British and British imperial radio and cable interests, public and private, would have any chance of fighting off "the American cable-radio magnates, and a number of foreign governments" (324). The *Electrician*'s editorial comment on the report of the Imperial Wireless and Cable Conference took precisely this view (101 [3 August 1928]: 116). In the event, however, British dominance was retained not just by the merger of all British telecommunications interests. The Great Depression hit America hard, and no concerted governmental effort developed to back American enterprise to the extent that the British government backed Cable and Wireless.

One measure of the retention of British power is that not until late 1943 did the American government have access to a transatlantic telegraph cable that did not pass through Britain, and that

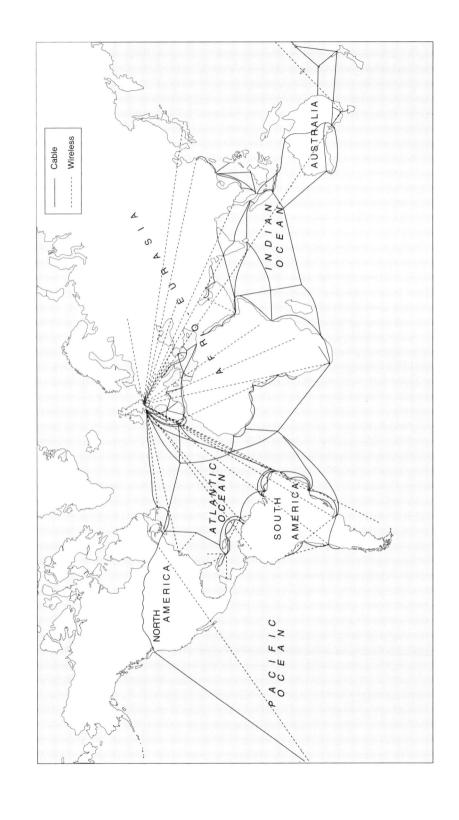

Cable

Wireless

NORTH
AMERICA

SOUTH
AMERICA

ATLANTIC
OCEAN

PACIFIC
OCEAN

EURASIA

AFRICA

INDIAN
OCEAN

AUSTRALIA

one passed through Gibraltar. Once Britain agreed in mid-1943 to share all decrypted materials up to and including the Ultra decrypts of German Enigma traffic, the amount of diplomatic intelligence received by the American government rose by 50 percent (Smith 1993, 161).

CHAPTER THREE

"The Whole World Kin"
Telephony and the Development of the Continental Polity to 1956

"Hello, Winnemucca!"
"Winnemucca talking. Three-twenty A.M."
"Hello, Cheyenne!"
"Cheyenne on the line. Four-twenty."
"Hello, Chicago!"
"Chicago here; five-twenty."
And so through Harrisburg and New York and on down the east coast:
"Hello, Havana!"
"Havana on. Six twenty-one, A.M."
Like reaching with a magic wand out across the earth; space annihilated,
the whole world kin.

—David Woodbury, *Communication*

In the history of telecommunications the development of the telephone occupies an oddly localized, national niche in an otherwise strongly international story. Put simply, the telephone as initially developed in the late 1870s was a very short range device, suitable for telecommunications within a city but not between cities. Its early history is almost entirely American. Inter-city communication grew slowly in the 1880s and 1890s but much faster after 1900 because of technical improvements and much faster again after 1915. Yet telephone connections between North America and the rest of the world were not possible until the 1920s, not thoroughly reliable until the mid-1950s, and not very affordable until the late 1960s. These technical improvements came at considerable cost in both research effort and capital investment. Most of the basic research was achieved by Bell Laboratories or its predecessors, company units set up by AT&T to use some of its monopoly profits to achieve its long-term goal of better, cheaper, longer-ranging service.

Just as the technical structure of the early telephone industry mitigated against internationalism, so did the early structure of patent, licensing, and financial agreements. These agreements are best described as Byzantine, and I leave them for the most part to business historians, for whom they seem to hold an endless fascination. They had, however, some positive consequences. The terms of the settlement of the lawsuit with Western Union in 1879 ensured that Western Union would stay out of the telephone business and Bell would stay out of telegraphy, which meant that in the late nineteenth century British investors paid little attention to American phone companies. Shares in American telegraph companies, of course, sold well on the London stock market, since British investors were interested in tapping the surplus production of the periphery of the world-economy. On the domestic front, the American government's mounting antitrust sentiment persuaded Bell to concentrate on its long-distance business after 1900. This business could be reasonably represented as a "natural monopoly." As long as they fed long distance traffic to Bell, local companies could be independent and the illusion of competition could be preserved. In 1899 American Bell Telephone thus conveyed all its assets to AT&T, which founded the Department of Long-Distance Lines in 1900, renamed the Long Lines Department in 1917 (Fagen 1975, 34–35). In 1909 AT&T came close to monopolizing American telecommunications when it acquired a substantial interest in Western Union. This was, however, a violation of the antitrust laws, and AT&T sold its Western Union stock in 1913 (36).

AT&T's concentration on its long lines after 1900 emphasized the need for a research unit under the company's control. This unit emerged from Western Electric and had a complex history until it was formalized as Bell Telephone Laboratories in 1925 (Fagen 1975, 42–57). Their strength in research gave Bell and AT&T a commanding position in the licensing of telephone technology outside North America. AT&T had little to do with telephone traffic in industrialized polities outside North America because such companies were usually part of nationalized telecommunications monopolies operated by the state.

Unlike the historical geography of the telegraph, which was technically capable of global coverage once the relatively simple problems of manufacturing reliable submarine cables and developing a viable theory of transmission were solved, that of the telephone is complex. Compared with the transmission of telegraph signals, the transmission of telephone signals required huge amounts of power. Although it was possible to construct a workable theory of transmission on submarine telegraph cables early on based simply on the work of Michael Faraday, understanding the propagation of telephone signals required vastly more theoretical understanding based on the much more mathematically demanding work of James Clerk Maxwell.

In the case of the telephone, therefore, increasing range and decreasing cost have been unusually dependent on continual technological improvements, which themselves required a high level of theoretical knowledge. The telegraph achieved a remarkable level of maturity within 22 years, the time separating the Baltimore-Washington line (1844) and the first successful Atlantic cable (1866). Although the first working Atlantic cable was, remarkably, completed in 1858, there was no adequate theory of how signals were propagated along it, and it failed almost immediately. In the case of the telephone, almost 80 years separated the primitive urban services (1877) and the first Atlantic cable (1956). The telegraph connected all corners of the globe by the first decade of the twentieth century, some 60 years after its inception and with little direct government spending. Not until the massive American satellite program of the late 1960s and early 1970s did the telephone achieve similar geographic ubiquity. Recent developments in cellular telephones suggest that almost every person on the planet will have an individual telephone without wired connections in the foreseeable future. In this sense the history of wireless telecommunications, at least in the use of extremely high frequency transmission, will merge with that of wired telephony. For the sake of clarity in my geopolitical arguments I keep them separated here.

"The Whole World Kin"

55

ALEXANDER GRAHAM BELL'S "GRAND SYSTEM"

As early as 1877, in a prospectus designed to promote business interest in the telephone, Bell stated, "In the future, wires will unite the head offices of the Telephone Company in different cities, and a man in one part of the country may communicate by word of mouth with another in a distant place" (quoted in Fagen 1975, 23). A central requirement of this "grand system" was that it be user-friendly, requiring no special skills on the part of the user (60). This goal was restated with an even greater emphasis on long-distance service following the restructuring of Bell when its initial patents expired in the early twentieth century.

In 1907 the president of AT&T, Theodore Vail, set a goal of "universal service at affordable rates" (O'Neill 1985, 4). Two obvious questions come to mind: Universal in what sense? and affordable by whom? If universal meant merely that lines were to be available throughout the United States, that goal was achieved by 1925 (3). By then rates were still substantial but affordable by at least most major businesses. But the average waiting time for toll calls was still 2.5 min, and operators still had to call back when a line became available. Direct distance dialing (DDD), the ability to pick up a phone, dial a number, and be connected anywhere in the United States, simply did not exist until 1951 and was not universal even within the Bell System until the late 1960s (723). The independent phone companies seem to have lagged slightly behind in implementing DDD, although the largest independent company, the General Telephone and Electronics Corporation (GTE), was converting rapidly to DDD by 1962 (McCarthy 1990, 104).

Direct international dialing by Bell System operators began in 1963, and by customers in 1970 (O'Neill 1985, 723). New York and the cities of Megalopolis (the region of coalescing urban areas along the Atlantic seabord between Washington, D.C., and Boston) were served first; and the cities of the Manufacturing Core (the region from the Atlantic to Chicago, where manufacturing predominates) were served by the mid-1970s. It took longer for international dialing to spread to GTE's network and especially to the rural South, which was not so served until the mid-1980s.

Affordability is harder to assess. Toll calls were not commonly made by households until well after World War II; business was by far the major customer.

In 1925, a three-minute station-to-station, coast-to-coast call cost $15.95 during the business day, $8.00 after 8:30 pm, and $4.00 after midnight (in 1925 dollars). Even at the lowest rate, it required 7 1/2 hours of the average factory worker's pay to place such a call. In 1983, a three-minute subscriber dialed coast-to-coast call on Bell facilities cost $1.72 during the business day, $1.04 after 5:00 pm, and $0.69 after 11:00 pm and for most of the weekend. At the lowest rate it would have cost just five minutes work at a factory worker's average pay to place such a call. (O'Neill 1985, 786).

The affordability of international calls is harder to assess. Before World War II, wireless telephony across the Atlantic was terrifyingly expensive. Even the first wired system, in the 1950s, was beyond the purse of household consumers. In recent years international rates have traditionally varied more than domestic rates, but about $1 per min has become a reasonable off-peak norm in the past decade. The breakup of AT&T has further confused both the domestic and the international pictures, with numerous companies claiming to offer the lowest rates. By the standards of all but the lowest-paid workers in the industrial world, however, costs for wired telephony have now become nearly universally "affordable."

Figure 3.1 shows that from a simply geographical viewpoint Vail's goal of national interconnection was achieved within the continental United States by 1925, by which time "it was possible to call almost everywhere . . . on a circuit of good quality" (O'Neill 1985, 3). Wireless telephone circuits and beam antennas allowed this goal to be more or less realized at the international level by the late 1930s (fig. 3.2). At the continental level, the map of Europe shows a similar interconnectedness to that of America by 1926, with submarine cables or land links between all the major polities (fig. 3.3).

Such maps of telephone circuits from this period speak, however, mainly to the geographical pattern of the system, only partly to its capacity, and not at all to its quality. Within the continental United States the quality was generally very good by 1925 because of AT&T's creation of a "natural monopoly." In Europe, despite AT&T's attempts to achieve a similar degree of quality by exporting its technology, national interests resulted in variations between systems and thus some problems across political boundaries. At the transoceanic level, the short-wave wireless telephony of the 1930s suffered from severe fading and noise. Sunspot activity could render circuits unusable at times. Figure 3.3 indicates where low-capacity open-wire circuits were about to be replaced with high-capacity cable links within Europe. Transoceanic wireless capacity was even more severely limited than capacity on open-wire systems. In figures 3.1, 3.2, and 3.3, high-capacity, high-quality systems were limited to the northeastern continental United States and Canada and to major European cities and industrial areas. Within Europe as a whole links between polities and even between major regions within a given polity were of minimal capacity.

European polities, even capitalist Britain, had nationalized telecommunications early on. There was no goal of ubiquity for the telephone: the telegraph served such a purpose. The telegraph was cheap, and message boys on bicycles and, later, motorcycles delivered telegrams to the door of the recipient. The telephone was seen in Europe as a tool of business and industry, necessary only in major cities and industrial regions, not a device for the household. In this regard the separation of the telegraph and telephone companies insisted upon by the antitrust movement in the United States served AT&T and private and business subscribers very well. It kept the highly capitalized telegraph companies from buying out their competition, and it helped create a large telephone market to drive down rates.

EQUIPMENT

A major consequence of the American antitrust movement for Bell's early history was the separation of the manufacturing arm from the carrying arm of the company. Similar splits came to characterize other American telephone companies, GTE in particular. Such a split allowed American companies to expand into the rest of the world through their manufacturing subsidiaries even when nationalized telephone companies prevented them from providing a carrying service. This split originated in the Western Union Telegraph Company's entry into the telephone business in 1877. When Western Union agreed to relinquish its telephone interests in late 1879 because of its infringement on Bell's patents, Bell bought out the company that had been manufacturing telephones for Western Union, the Western Electric Manufacturing Company of Chicago. The latter company was reorganized in 1882 as the Western Electric Company to produce high-quality equipment that was interchangeable throughout the Bell System (Fagen 1975, 32).

Western Electric established its first international factory in Antwerp in 1882 and by 1918 controlled much of the world's telephone manufacturing business. This business was reorganized and consolidated in 1918 as the International Western Electric Company, then sold to International Telephone and Telegraph (IT&T) in 1925. At that point all corporate connections to Bell and AT&T ceased, although technology licensing continued (33).

The only American company to rival AT&T in the sale of telephone technology overseas was the Strowger Automatic Switching Company of Indiana, which developed the first automatic exchange and made the dial telephone possible. Bell remained wedded to the concept of human operators, when Strowger's exchange was far more efficient, accurate, and reliable. By substituting capital for labor, Strowger's exchange conformed better to the developing norms of Neotechnic society and their stress on, and control of, labor costs. The dial telephone and the Strowger exchange also greatly advanced the achievement of a user-friendly system.

TECHNOLOGY AND THE TELEPHONE

From a historical geographic viewpoint, the key to the success of the telephone is switching and transmission technologies, not corporate structure or nationalization. Only the appropriate switching and transmission

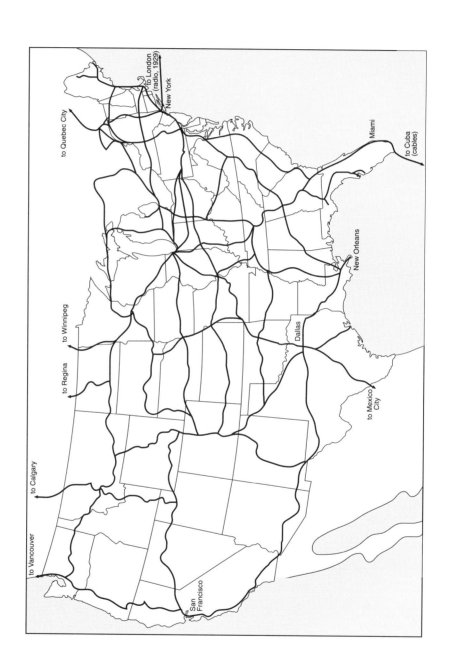

to Vancouver

to Calgary

to Regina

to Winnipeg

to Quebec City

to London
(radio, 1929)

New York

Miami

to Cuba
(cables)

New Orleans

Dallas

to Mexico
City

San
Francisco

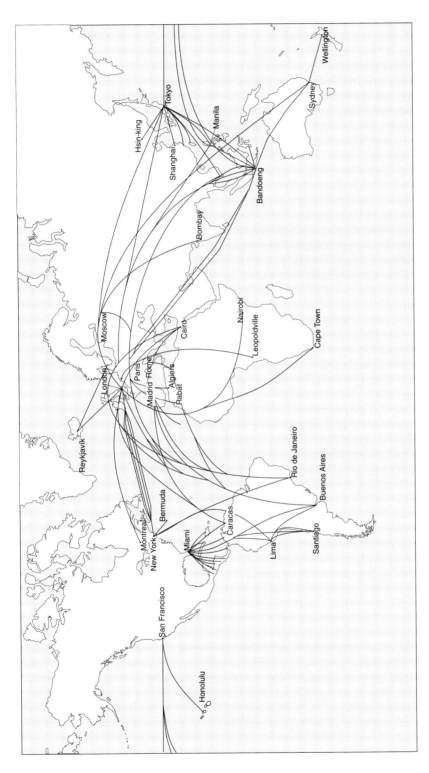

Fig. 3.1. Long Lines
Network of the Ameri-
can Telephone and
Telegraph Company,
1925 (redrawn from
Fagen 1975, 346)
(opposite)

Fig. 3.2. International
Wireless Telephony,
1937 (redrawn from
Bown 1937, fig. 1)

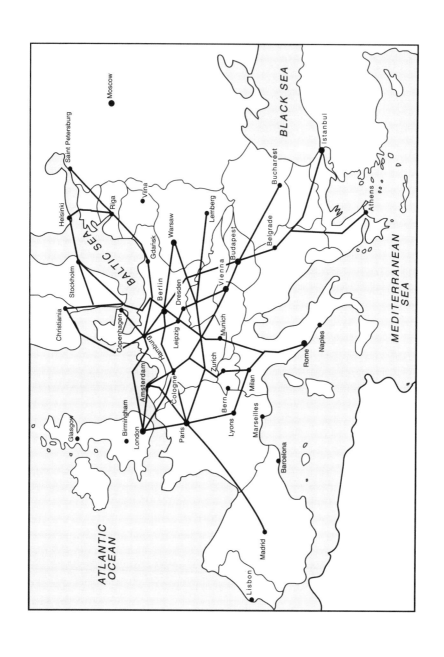

technologies could provide the geographic ubiquity sought by Bell's "grand system."

Switching Technology

Without switching technology, telephone systems could never have come into being. Before switching, each telephone line between two users had to be a discrete, thus private, line. Although switching through an exchange had become common in certain forms of telegraphy (Garnet 1985, 21; Lipartito 1989, 69), when applied to the telephone the exchange produced two problems. The first of these was error on the part of the human operators, usually young women, whose job was to switch one line to another. It was easy to mis-hear a number or put the plug into the wrong hole. This problem would be solved by the automatic exchange and the dial telephone. The second problem was much more severe. Connecting one line to another caused *attenuation,* a severe loss of signal strength. Since distance also caused attenuation, switching had serious implications for transmission technology.

Automatic switching was one of the few fundamental telephone technologies not developed by Bell. Nor did it arise from any fundamental understanding of the need for accuracy and user-friendliness in an increasingly complex system. Its inventor, Almon Brown Strowger, was an "eccentric schoolteacher-turned-undertaker . . . [who] believed calls from his customers were being misdirected to his competitors" (McCarthy 1990, 33). Strowger patented his mechanical switching mechanism in 1889, and the Strowger Automatic Telephone Exchange Company was formed in Chicago in 1891. The first successful installation was in La Porte, Indiana, in 1892.

After a false start using push buttons to control the mechanical switchers, Strowger engineers developed the rotary telephone dial controller in 1896 (McCarthy 1990, 34). After the success of Strowger equipment at the Columbia Exposition of 1893, European countries became interested in the Strowger system. Siemens und Halske licensed Strowger equipment in the German Empire, and the Thomson-Houston Company of Paris brought it to France. Siemens und Halske substantially improved on the Strowger system, and Germany "became the first country to witness the triumph of automatic telephony" (Siemens 1957a, 264). In 1911 the British Post Office decided to replace its London manual exchanges with Strowger exchanges, and the Strowger Company, which had become the Automatic Electric Company in 1908, developed a substantial interest in the Automatic Telephone Manufacturing Company of Liverpool (McCarthy 1990, 35). The success of mechanical switching and the dial telephone made the telephone markedly more user-friendly. Automatic Electric presented it as "The Girlless, Cussless, Waitless Telephone" (30).

Transmission Technology

The most serious problem for telephone technology was the attenuation caused by distance. The farther a signal traveled, the weaker it became.

Fig. 3.3. Projected European Trunk Telephone Cable System Based on the Existing Open-Wire System, 1924 (redrawn from Craemer 1924, 287) *(opposite)*

Until the development of the vacuum tube amplifier, which entered public service on the first line across the continental United States in 1915, distance could only be achieved by increasing the efficiency of the microphone transmitter and the receiver, increasing the power of the batteries used to drive the system, and increasing the efficiency of the transmission lines themselves. Attempts were also made to devise repeaters, the idea being to rejuvenate a weak incoming signal by feeding it into a new transmitter powered by a new set of batteries. Although partially successful, such attempts foundered on the distortion such electromechanical repeaters introduced (Fagen 1975, 253–56). The solution was to use vacuum tube amplifiers at regular intervals to increase the flow of electrons. Since the phone company understood the amplifier as an electronic equivalent of the human repeater in the telegraph system and its own, unsuccessful, electromechanical repeater, the company always referred to it as a repeater. I use the electronic term *amplifier.*

Microphone Transmitters

Early microphones were notably inefficient, and Bell's own microphone was one of the worst. Bell was, however, quick to realize the need for higher efficiency and sought out and purchased other microphone patents. Though interesting, the history of these patents lacks a geography per se. Bell acquired access to the Edison patents with the settlement of the patent suit against Western Union in 1879. Edison's granular-carbon transmitter, patented in 1886 and improved by Anthony C. White, one of Bell's own engineers, in 1890, became the standard transmitter (Fagen 1975, 75–76). It retained this position until the mid-1920s. Thereafter, the switch to amplified systems meant that the efficiency of the transmitter was no longer a crucial concern.

Receivers

An efficient electromagnetic receiver proved significantly more difficult to design than a transmitter. Bell's early "butterstamp" receiver, so named because it resembled the nineteenth-century household device used to mold butter into round pats with a design on top, served from 1877 to 1902. More than half a million were manufactured. The long handle of the butterstamp concealed a long bar magnet. The butterstamp was replaced as the standard receiver in 1902 by a more efficient bipolar design using twin magnets. Five and a half million of these were made (Fagen 1975, 89–98). The last improvements to standard receivers in the pre-amplifier period were made in 1914. Thereafter, the vacuum tube amplifier allowed designers to concentrate on fidelity as well as on efficiency.

The quest for more efficient transmitters and receivers was central to the early history of the Bell organization because of the quest for distance. Measured in decibels (dB), each decibel representing a doubling in volume, thus of efficiency, transmitters enjoyed a gain of 40 dB between 1877 and 1919, and receivers, a gain of 11 dB between 1877 and 1912. Most of these gains, however, came before 1900. Bell defined the model 229 transmitter, of 1895, and the model 122 receiver, of 1902, as their reference units. Transmitters gained only 5 dB from 1895 to 1919, and receivers, 2 dB from 1902 to 1912 (102). Such obsession with efficiency came at a price, of course. As Fagen notes, "The modern telephone user ... would find the peaked characteristics [of the instrument's frequency response] used in the early twentieth century highly unsatisfactory if not almost unusable. The fact remains that many millions of people learned to use this highly characteristic 'telephone speech'" (100 n. 22).

Batteries

All telephones, with the exception of Bell's very earliest experiments, used the sound vibrations striking the diaphragm of the transmitter to generate a variable resistance. This modulated the otherwise constant flow of electrons from a direct-current source, usually a battery (Fagen 1975, 63–64). To begin with, the battery was located close to the instrument and was supposed to be maintained by the subscriber, which introduced considerable variation into the system. Lead acid batteries were messy, dangerous if spilled, and generated limited amounts of energy. Batteries ran down at different rates and were not always conscientiously charged.

Common battery systems, with the batteries at the central-office switchboard, were the solution. Such systems, however, required considerable amounts of energy. Fortunately, commercial electricity generation was becoming normal by the last decade of the century. Large banks of batteries could be continuously recharged from the commercial supply and maintained by properly trained technicians. Because of the unreliability of early commercial electricity-generating systems, most large central offices had backup generators using internal combustion engines that ran on illuminating gas or gasoline (Fagen 1975, 694–99).

Compared with the local battery systems, where one to three cells had provided 1.5–4.5 V for low-resistance circuits (Fagen 1975, 79), central-office systems needed circuits with a higher resistance and ran at 24 V or 48 V. The 24-V systems were introduced in 1897, the 48-V ones shortly thereafter (700). Long-distance service was all at 48 V, which provided considerably more energy for transmission, albeit at some cost. The cost was in crosstalk. As the amount of electrical energy in the line increased, it induced a current in neighboring lines, just as the primary wiring in a transformer induces a current in the secondary wiring. Although to some extent this could be compensated for by simply increasing the distance between wires, that distance was limited by the width of the crossarms on the telephone poles. The current could increase only so much before it was rendered useless by the expense of overcoming crosstalk by increasing the space between the wires. And this was on open wires strung on poles. No such solution was possible in cities once the density of open-wire circuits became too great for poles and the switch to underground cables began. The 48-V circuit represented the effective limit of energy increase for transmission. Common battery systems operating at high voltages also increased the complexity of the system and, along with the problems of transmission, were one of the forces pushing Bell toward a more research-oriented stance. The first task assigned to the first serious theoretician employed by Bell, George Campbell, was to examine the problems of common-battery installations (Wasserman 1985, 66).

Transmission

With transmitter and receiver efficiency as high as practicable and 48-V common-battery systems established, the only other way to improve long-distance performance by the turn of the century was to improve the transmission lines. Transmission technology can be divided rather simplistically into several major periods (fig 3.4). The early metallic circuits of 1880 gave low-capacity service at the urban level and over relatively small areas. Between then and 1900 cable transmission greatly increased urban capacity, and ranges up to 800 mi were achieved for low-capacity systems. Line loading improved capacity at the regional level between 1900 and 1915. Amplifiers made low-capacity national service and high-capacity regional service possible after 1915. Carrier systems improved capacity on long open-wire lines, but not to the same extent as cables had for regional systems. Coaxial cables and microwave transmission allowed a high-capacity system to develop nationwide after 1942. Low-capacity transatlantic service began in 1956 with a coaxial cable powered by submerged amplifiers. High-capacity transoceanic service begin with the Intelsat satellite series in 1965. (Transoceanic wired telephony, satellites, and fiberoptic cables are discussed in chapter 8 because they follow the main, internationalist line of telecommunications development. Similarly, the first attempts at wireless telecommunications are not included in figure 3.4 because their limited capacity and unreliability kept them from being particularly commercial; their import lies in their impact on geopolitics as well as on military technology.) To repeat, telephony is best regarded until very recently as a national enterprise focused on the United States.

The Primitive Period. The earliest telephone line, like the telegraph line on which it was modeled, was a single iron wire with a ground return. Telegraph signals were usually below 30 Hz, and attenuation was very slight at such low frequencies. Because telephone signals had to use frequencies 50 to 100 times higher in order to transmit a reasonable proportion of the range of frequencies used by the human voice, they suffered much more severe attenuation. As long as the ground return was part of the circuit, attenuation made long-distance telephony impossi-

Fig. 3.4. Capacity and Range of Telephone Systems, 1880–1985

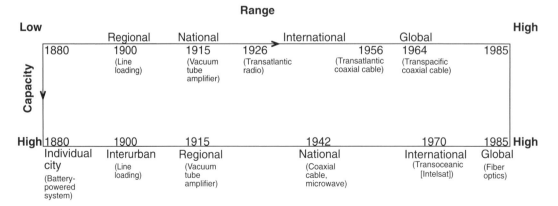

	Range						
Low	Regional	National		International		Global	**High**
1880	1900	1915	1926		1956	1964	1985
	(Line loading)	(Vacuum tube amplifier)	(Transatlantic radio)		(Transatlantic coaxial cable)	(Transpacific coaxial cable)	

Capacity

High 1880	1900	1915		1942		1970	1985 **High**
Individual city	Interurban	Regional		National		International	Global
(Battery-powered system)	(Line loading)	(Vacuum tube amplifier)		(Coaxial cable, microwave)		(Transoceanic [Intelsat])	(Fiber optics)

ble. The accidental discovery that a two-line, all-metallic circuit was much more efficient opened the way for the first long-distance circuits.

Even when the most careful attention was paid to circuit design, transmitters, receivers, and power systems, metallic circuits had a finite range of about 800 mi (Wasserman 1985, 37). New York could just be connected to Chicago, and telephone service was thus possible throughout that area of the United States, variously referred to as New England Extended, or Megalopolis together with the Manufacturing Core. Given the early realization of how useful the telephone was in business, that the region could be provided with telephone service greatly aided the region's emergence in the late nineteenth century as the single most important economic, social, and political unit in the United States. Extensions south and west did not take place until new transmission technologies were developed well into the new century. Bell's origin in New England Bell and the subsequent move of its corporate headquarters to New York City ensured that, for all the company's obsession with long-distance service, it retained a primary focus on New England Extended. Given the state of transmission technology, the parlous business conditions in the South, and the capacity of installed lines, it is important to note here that a high-capacity trunk line system did not reach much beyond New England Extended until after World War II. Although the map of lines for 1925 (see fig. 3.1) shows a nationwide network in accord with Bell and AT&T's stated goal of nationwide service, the lines to the West Coast had very low capacities, as did those to the South.

Line Loading. As Wasserman has clearly demonstrated, understanding how telephone signals were propagated over very long distances required considerably more understanding of theory than the much simpler understanding required for telegraphy. The British physicist William Thomson, later Lord Kelvin, developed an adequate operating theory for the low-frequency signals sent through submarine cables in the aftermath of the failure of the 1858 Atlantic telegraph cable. Thomson's theory explained the exponential rate of decay from the source of the signal as a product of the resistance and capacitance of the line (Wasserman 1985, 25). Another British physicist, Oliver Heaviside, built on both Thomson's foundation and the complex mathematical theories of James Clerk Maxwell to produce a much more accurate theory of transmission. Heaviside included in his analysis the effects of inductance and leakage (25–27, 127–28). He established that "without sufficient inductance, permitting energy to be stored in the magnetic field of the line, efficient transmission would not be possible. Instead, much of the energy of the signal would be transformed into heat" (27).

Although Heaviside's earliest publications on the matter date to 1873 and were republished between 1885 and 1887 in the British weekly the *Electrician* (Wasserman 1985, 134 n. 10), not until the late 1890s was the practical application of his work on inductance picked up by Bell Telephone in its effort to improve long-distance transmission. Few people in the late nineteenth century could understand Heaviside's mathematical insights into the complexities of Maxwell's theories. One of those few was George Campbell.

Campbell joined Bell in 1897 (O'Neill 1985, 243). After receiving a degree in civil engineering from the Massachusetts Institute of Technology, he pursued graduate study in physics at Harvard and in theoretical physics and mathematics at Göttingen, Vienna, and Paris. He was heavily influenced by the work of Heaviside, and he was the first at Bell to understand how to apply it to transmission problems (Wasserman 1985, 30–31). Campbell was recruited by his predecessor, John Stone Stone, "probably the first member of the America telephone community whose interests were almost entirely in the theory of transmission" (Fagen 1975, 243). Stone's groundwork and Campbell's recognition of the centrality of Heaviside's work on inductance allowed Bell to implement much more efficient long-distance service by using *line loading* to increase inductance.

Stone had clearly understood the importance of inductance as early as 1894, and he had proposed increasing it by substituting lines made of high-resistance, high-inductance iron combined with low-resistance, low-inductance copper. The combined resistance of these experimental bimetallic lines caused the advantages of the increase in inductance to be negated by the increase in attenuation caused by increased resistance. Campbell's much more elegant solution was to add inductive devices, for which he used coils, along the line. Although Stone had suggested as much at one point, he probably assumed that the coils also would add resistance (Wasserman 1985, 45). Experimental use of coils in 1893 on the New York–to–Chicago line in fact seemed to provide an empirical rejection of the idea, and Stone may have been influenced by this failure (46). Campbell simply brought to the problem a higher level of mathematical sophistication. He created a much more realistic theory of transmission by concerning himself with the broad range of frequencies telephones had to transmit, not just with average frequencies. As Wasserman notes, "Such a view of the behavior of transmission networks became crucial in the development of the theory of the loading coil, since the required spacing of the coils is determined by the width of the frequency band to be carried over the transmission line" (67).

Campbell's solution was to create coils of very low resistance so that there was virtually no limit to the amount of inductance that could then be applied to a line. Turning this insight into practice called, of course, for much experimentation and cost reduction, but by 1899 Campbell had clearly demonstrated that it could be done (Wasserman 1985, 73–78). Unfortunately, he did not immediately patent the idea, a failure compounded by ineptitude on the part of Bell's legal staff. The result of this delay was that Columbia University physicist Michael Pupin was awarded what became the primary patent on loading in mid-1900. Some indication of how valuable the patent was can be gauged from Bell's buying Pupin's patent for $185,000 plus $15,000 a year over the 17-year life of the patent. The high cost of this buyout was justified by calculations that loading would save approximately $1 million a year in New York City alone (92–97). Perhaps the most significant consequences of this otherwise unfortunate set of events was that it put Bell firmly on a course toward both science-based technology and a more aggressive pursuit of patents (124).

Line loading with coils had three very distinct impacts on Bell's service. The first was clearly geographic in the simplest sense. "With the previous transmission technology limiting long distance, open-wire transmission to several hundred miles, it was possible to design only regional networks. The ability to create a long-distance system with lines spanning two thousand miles represented a large step toward the realization of Theodore Vail's dream of a national telephone network" (Wasserman 1985, 121–22). Figure 3.5, which shows Bell's network as of 1906, demonstrates just how quickly line loading allowed the company to move beyond the roughly New York–Chicago range threshold imposed before loaded circuits became possible. By 1911 Denver, Dallas, and Houston were (just) within range of New York City.

The second impact of line loading was on service in cities. As demand for urban telephone service expanded in the 1880s and 1890s the use of open-wire circuits became problematic. Contemporary photographs show immense crowding along the poles. Weather conditions in the Northeast also caused trouble; sleet storms in particular caused massive service interruptions. The solution in cities was to switch to underground cables in the early 1880s. The use of insulation and twisted pairs of wire for each circuit minimized crosstalk and allowed a large number of lines to be placed in close proximity in underground conduits (Wasserman 1985, 4). This solved the problems of pole crowding and service interruptions from weather, but it also introduced two new problems. The first was much higher attenuation, to the point where the geographic limit of service was about 30 mi. The second was the introduction of what to Bell was an unacceptable dual system. The problems of very high attenuation in the transmission lines plus attenuation at the switchboard meant that calls routed through the underground cable system of a large city could not be connected to the open-wire long-distance lines between cities. Bell had to install special long-distance stations connected directly to the open-wire system to avoid the massive attenuation that would have resulted from switching from an urban cable system to an open-wire long-distance system and back to an urban cable system (37–39).

Fortunately, line loading proved even more beneficial on cables than it had on open wire circuits. Attenuation was so reduced that the problems of combining cable and open-wire systems simply disappeared. Loaded cables had significantly greater range than unloaded ones, yet they could use much lighter wire. Much of the money saved by loading in New York City came from the ability to use lighter wire alone. Preliminary studies revealed savings of 50–83 percent on new lines, depending on the distance cabled. Given that total expenditures might range up to nearly $4.5 million, an 83 percent cut in costs was impressive (Wasserman 1985, 103). In addition, increasing the efficiency of New York City's telephone system ensured the city's continuing centrality in the Bell System.

The third and perhaps the greatest impact of line loading on Bell's service was the switch to loaded cables for long-distance service within Bell's major service region. Bell did not emphasize the full impact of this switch because it went against its model of equalizing service. Line-loaded

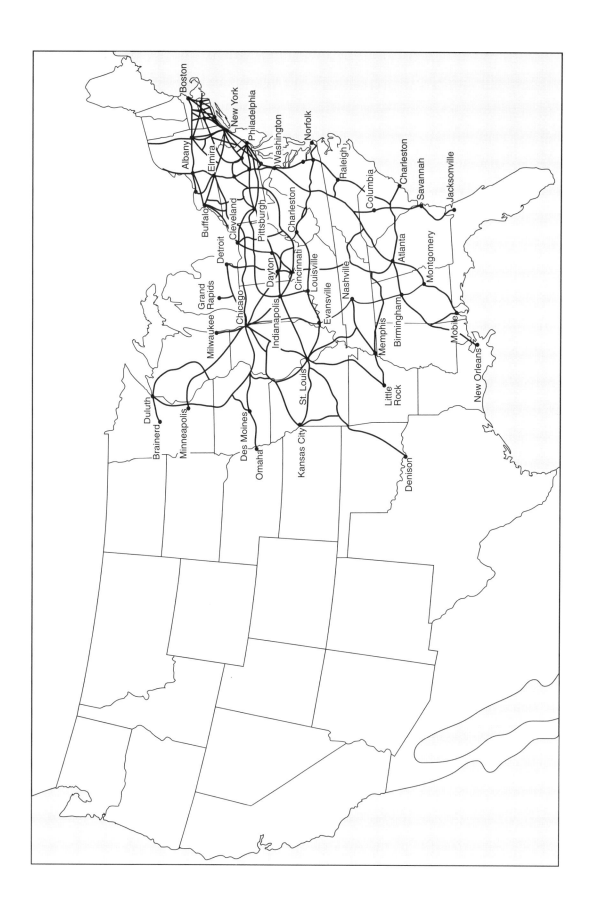

cables provided very substantial increases in capacity but only within the limits of a rather finite geographic region. The first long-distance cables were installed in 1906 between New York and Philadelphia and between New York and New Haven. Following a sleet storm in 1909 that disrupted reporting of President Taft's inauguration, plans were made to cable Megalopolis. The line between Boston and Washington was opened in 1913, although it needed an electromechanical repeater (Fagen 1975, 251).

Loading was, however, not an absolute solution to transmission problems. It slightly more than doubled the range of open-wire circuits, to about 1,800 mi, which was not enough to bridge the continental United States but was enough to connect New York to Denver, Dallas, or Houston (Fagen 1975, 252; Wasserman 1985, 117–18). The increase in range for cables was much more substantial, at least tenfold, but this still did not provide high-capacity service on a more than regional level. Finally, loading brought with it a one-third increase in capacity on long circuits because of greatly improved coil design. Improved coils allowed a third, phantom circuit to be superimposed electrically on two existing metal circuits, so that four wires could thus carry three calls simultaneously (Fagen 1975, 236–38). The standard cable in Bell installations after 1910 was thus the *quad,* "consisting of twisted pairs twisted together" (239).

Amplified Transmission. Telephony in the United States achieved national range only with the development of an effective vacuum tube amplifier, just in time for the completion of the first transcontinental line, from New York to San Francisco, in 1915. Bell had stated the goal of transcontinental service several years earlier, placing its faith in its growing research ability to deliver the appropriate technology. Lee De Forest had patented the audion vacuum tube as a "device for amplifying feeble electric currents" in 1907, introducing a third or control electrode into Alexander Fleming's 1904 thermionic rectifier. From 1907 to 1912 De Forest concentrated on developing the audion to detect modulated radio waves. In 1912, however, John Stone Stone, now working for himself, recognized what the audion might do and persuaded De Forest to demonstrate its ability as an amplifier to Bell officials. Within less than a year workable amplifiers had been developed and tested on the New York–to–Philadelphia line, and by early 1915 they were being used in commercial transcontinental service (O'Neill 1985, 261). This first transcontinental line started as a loaded line to which three, then six, and eventually eight amplifiers were added. Loading, however, reduced the usable bandwidth to about 900 Hz, barely enough for intelligible speech. Steady improvement in amplifiers allowed the line to be unloaded in 1920 and reequipped with twelve amplifiers, and the bandwidth doubled to a more intelligible 1,800 Hz. This remarkable achievement came at considerable cost: in 1915 a 3 min call across the continent cost $20.70 (Fagen 1975, 262).

Despite such costs, the demand for long-distance service accelerated rapidly. By 1925 the toll lines of the Bell system were spread nationwide (see fig. 3.1). In 1900 there had been about 300,000 mi of long-distance circuits; by the end of 1926 there were 3 million (Fagen 1975, 347). Most

Fig. 3.5. Long-Distance Lines of the American Telephone and Telegraph Company, 1906 (redrawn from Fagen 1975, 345) *(opposite)*

of these circuits were, however, in Megalopolis and the Manufacturing Core. Figure 3.1 is misleading inasmuch as it shows routes, not route capacities. The first transcontinental line to San Francisco consisted of two open-wire circuits from which a third, phantom circuit could be derived (O'Neill 1985, 5). Direct service from New York to Los Angeles opened in early 1928. Before that an average of 51 calls per day were routed via San Francisco. The *New York Times* reported that "it is estimated that these circuits . . . will be used for an average of seventy-two conversations a day during the early Spring, which is the busy season for telephone calls between New York and Los Angeles" (12 January 1928, 30).

Despite the need for open-wire lines on such very long circuits, by 1926 some 45 percent of the long-distance lines were in cables (Fagen 1975, 347). The use of vacuum tube amplifiers allowed cable service to reach beyond the confines of Megalopolis to cover the Manufacturing Core with a high-capacity system. Such cables were difficult to design and expensive to manufacture. Because of their high capacitance, they had to be loaded even when they were amplified, and the high losses meant that an amplifier was needed every 50 mi or so instead of only every 250 mi as in open-wire, unloaded circuits. A single 278-pair cable, however, provided a very much higher service capacity than an open-wire system and was weatherproof to boot (O'Neill 1985, 7). The design of the New York–Chicago cable was begun in 1921, and the line opened for service in October 1925, providing 10 times the capacity of the previous fully equipped pole line (7–8). Problems with bidirectional amplification in the cable lines meant a switch to four-wire circuits for long-distance transmission, rather than carrying both conversations on one pair of wires. The 278-pair cable thus carried only 139 voice-frequency calls. Even so, the difference in circuit capacity between New York and Chicago, with 139 direct voice circuits, and New York and San Francisco, with two direct and one phantom circuit, is noticeable.

The final major advance in cable transmission was the development of coaxial cable by Bell Laboratories. Its genesis lay partly in the need for a transmission system with a bandwidth of at least 1,000 kilohertz (kHz) to service the high-definition television systems being forecast all over the electrical world by the early 1930s. Espenscheid and Strieby's first experimental cable could carry one television signal or "more than 200 telephone circuits" (Espenscheid and Strieby 1984, 732). Coaxial cables with an inner diameter of 0.3 inch suffered much less attenuation than did the older quad cables. Coaxial cable with a diameter of 2.5 inches had the same attenuation as open wires (735, fig. 7). Although coaxial cables' high attenuation required frequent amplification of the signal, since the outer tube of the cable served as both conductor and shield, frequent amplification was possible without distortion. A coaxial cable could thus be installed for transcontinental distances, providing the United States with its first high-capacity nationwide system. As figure 3.6 shows, in 1939 the L1 coaxial cable system, using 600 amplifiers, was under way across the southern United States (Brittain 1984, 731). In 1953 this L1 system gave way to the L3 system (fig. 3.7), with a transcontinental capacity of 1,860 voice circuits, or 600 voice circuits and a full-bandwidth 525-line television signal of the type standardized by the National Television Standard Committee (NTSC) (O'Neill 1985, 248).

Carrier. By the 1870s it was appreciated that telegraph lines could be multiplexed, that is, made to carry more than one signal simultaneously, by using different frequency bands for different messages. Alexander Graham Bell was, in fact, experimenting with the wider frequency band needed for multiplexing when he realized that such a wide band made voice transmission possible. It was not, however, until the development of the vacuum tube amplifier that telephone multiplexing became practical. The vacuum tube not only amplified but also provided a stable source of oscillation for the frequencies above voice range needed for multiplexing and the necessary source of modulation and demodulation to enter and retrieve voice signals at beyond voice frequency from the telephone line. Experiments began in 1914, and the first commercial installation of Type A carrier was in 1918. Type A carrier used frequency multiplexing to provide four two-way channels above the voice-frequency channel on each open-wire pair. Type B carrier reduced the number of channels above voice frequency to three, was applicable to longer-distance lines, and used separate frequencies for east-west and west-east transmission. Modified further to improve the somewhat limited frequency band each carrier signal could transmit, thus to improve voice quality, it became Type C carrier. Despite Type C carrier's capacity advantages, its range was only about 2,000 mi (Fagen 1975, 277–90). Type C carrier was first used on the Pittsburgh–St. Louis line in 1924, and it ultimately provided 1.5 million voice-circuit miles on open wire by 1950. It was not fully removed from service until 1980 (O'Neill 1985, 6). As electronics improved, carrier technology became the dominant means of increasing the system's capacity. In 1926 only about 2 percent of Bell's toll-circuit miles were in carrier. That figure had risen to 65 percent by 1950 and to 98 percent by 1976 (Fagen 1975, 347).

Type C carrier was improved several times, but carrier technology came close to reaching its range limits with Type C, hence its long retention in service. Type K carrier, allowing twelve conversations per quad, entered service in 1937 (O'Neill 1985, 80–94). Type J carrier, allowing twelve conversations on open-wire lines, was also developed in the late 1930s. Type J carrier allowed up to 256 circuits to be carried on a normal pole before crosstalk became a problem, compared with 70 circuits using Type C carrier and 30 circuits when transmission was at voice frequency only (94–100). Despite the successes of Type K and Type J carrier, Type C carrier was not really replaced until the advent of the much more sophisticated T1 carrier system in 1962. T1 applied line-division multiplexing and pulse-code modulation to commercial telephony, shifting from analog to digital transmission. It allowed 24 conversations over two ordinary cable pairs. Because transmission was digitized, it could tolerate a very high degree of noise, interference, and distortion, and it could use regenerative amplifiers to almost totally remove such signal decay, thus making it serviceable in systems with large numbers of amplifiers (Hoth 1962, 358). Regenerative amplification did not necessarily mean that it could be used over long distances; in fact, Bell Laboratories

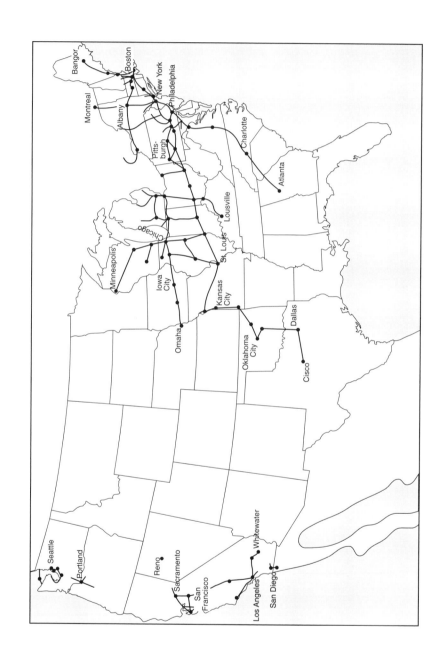

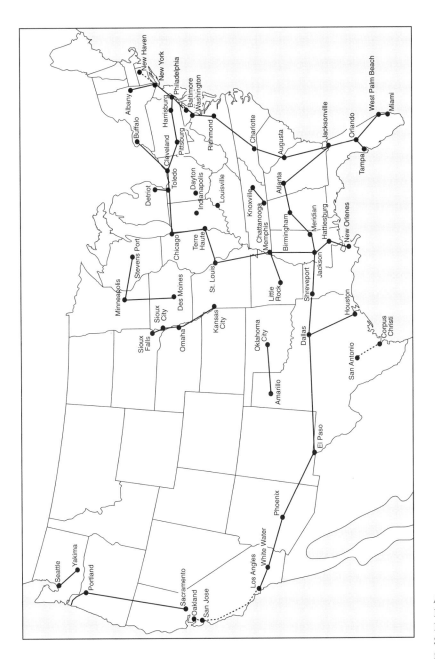

Fig. 3.6. L1 Long-Distance Coaxial Cable Routes of the Bell System, 1939 (redrawn from Jewett 1940, 6) *(opposite)*

Fig. 3.7. L3 Long-Distance Coaxial Cable Routes of the Bell System, 1953 (redrawn from O'Neill 1985, 248)

initially stated that 50 mi represented the limit of its practical use (363). Despite such early limitations, T1 carrier offered a huge capacity increase at the urban level.

Microwave Transmission. Despite many early efforts at wireless telephony and its relative success in transoceanic service using short-wave transmission and beam antennas, wired systems were always cheaper on overland routes until the advent of microwave-transmission systems following World War II. The technical impetus for microwave transmission was the development of radar systems as an integral part of a high-technology air war. Microwaves offered a very high bandwidth at extremely high frequencies for transmitting a large number of simultaneous conversations using frequency multiplexing on carrier, and the magnetron and klystron microwave generators developed for airborne high-frequency radar provided much more powerful sources of microwaves than did the vacuum tubes used in lower-frequency systems. In the aftermath of World War II the Federal Communications Commission (FCC) gave preference to AT&T's expansion of high-capacity cable links over a rapid move to microwave technology. In one author's estimation the adoption of microwave technology was thus seriously slowed (Cantelon 1995). However true this may be of America, it does not explain why microwave technology did not take off in the country that had done most to pioneer microwave radar, Britain, where neither the FCC nor AT&T held sway.

There were, however, pioneering attempts at microwave transmission in the 1930s, using first vacuum tubes and then klystrons. Marconi early realized the potential of high frequencies for line-of-sight communication. He experimented with "a spark system operating on a wavelength of 2 to 3 metres having a directional, or 'beam,' aerial" as early as 1916 (Isted 1991, 115). In 1931 he began experimenting with a 600-MHz system for telephony some 20 km across the Gulf of Genoa in the Ligurian Sea using cylindrical parabolic reflectors (Isted 1991, 119; Olver 1995, 86). The most notable of these pioneering microwave efforts was the construction by Standard Telephones and Cables (STC), the successor company in Britain to IT&T, of a 17.6-cm-waveband microwave (then called "micro-ray") telephone relay across the English Channel in 1931 (STC 1958, 37). Very few hard data on this STC technology were ever published, although there were contemporary accounts in such journals as *Electrical Communication* and the *Electrician*. The vacuum tube that generated these high frequencies was mounted at the center of a parabolic reflector of 3 m in diameter that gave a 3° beam. Power output, at 0.5 W, was very low (O'Neill 1985, 156–57; STC 1958, 37). The contemporary technical descriptions in English may have been sketchy because "the equipment was largely developed by French engineers in the Paris laboratories" of Le Matériel Téléphonique (*Electrical Communication* 10 [1931]: 20). The technology was also clearly embryonic and hard to manage (O'Neill 1985, 157). However, E. G. Bowen, the pioneer of high-frequency AI radar for Britain's Royal Air Force, complained in his autobiography that he was never given access to this microwave technology and that if he had been, the whole course of radar development in Britain would have been different (Bowen 1987, 142). Since he does not elabo-rate on this claim, it is hard to understand on what grounds he made it. A low-power, high-frequency system based on the vacuum tube that worked as well as the German systems of the late 1930s might have slowed the British search for alternate technologies that led to the cavity mag-netron and high-power, high-frequency radar (see chapter 7).

STC proved the possibility of microwave relays, but the technology could not really be implemented until World War II, and it could not be used commercially until after the war. The cavity magnetron, developed in Britain for AI radar, at first produced 1 kW or more of power at a wave-length of about 10 cm. Unlike the klystron, it was not tunable, and it took a great deal of effort to make it so (Smits 1985, 157–63). Tunable mag-netron oscillators operating at very high frequencies made high-band-width microwave transmission practical. The U.S. Army enthusiastically adopted microwave field systems in the last part of the war. Bell Labs had prototypes working by 1943, and from the fall of 1944 high-capacity telecommunications systems were being installed with great speed in the invasion of Europe because there was no need to lay wires. Such sys-tems were hard to sabotage as long as the relays were guarded, hard to tap into for intelligence purposes because of their extremely focused beam, and nearly impossible to decode when they were tapped into be-cause of pulse-code modulation (PCM) of messages (Cantelon 1995, 566–67).

Although it was limited by the need for line-of-sight transmission, and despite the impact of the FCC rulings in favor of AT&T's cable system, a microwave relay network grew rapidly in the United States. The FCC de-cision delayed the growth of microwave transmission by perhaps a decade. When such growth did begin, it was largely controlled by AT&T until a round of court and FCC findings in the mid-1960s ceased to priv-ilege AT&T (Cantelon 1995, 580–82). Microwaves offered what coaxial cable promised, but at much lower cost: the ability to carry either televi-sion signals or very large numbers of telephone calls. Between 1955 and 1965, "microwave radio, with its low start-up cost, short construction in-tervals, and superior ability to carry signals for the rapidly expanding na-tional television network captured the lion's share of the growth: it in-creased its capacity by a factor of ten. Coaxial systems, in contrast, often involved time-consuming acquisition of a continuous right-of-way, con-siderable construction, and higher initial costs" (O'Neill 1985, 256). But microwave relays also had a major flaw: they could be disrupted by nat-ural disasters or nuclear attack. This caused a return to the L4 coaxial sys-tem, with a capacity of 3,600 circuits, in the early 1960s.

Bell's first successful microwave system was the TD2 system. It was in-tended initially to connect New York to Chicago, then Los Angeles to San Francisco, and then Chicago to San Francisco (O'Neill 1985, 295). By 1980 the TD2 route system touched nearly every major city in the Unit-ed States and Canada (fig. 3.8). Because microwave systems operate only on line of sight, the TD2 network towers usually have been situated on the highest available pieces of real estate and are thus generally very no-ticeable in the landscape.

The Telephone Elsewhere

This chapter dwells extensively on the historical geography of continental telephony in the United States. With perhaps the exception of STC's early experiments with microwave transmission and certainly with the development of the magnetron in Britain as part of the radar war effort, all of the fundamental technical problems of telephony were addressed and solved in America, most of them by Bell Laboratories, a few by GTE. All of these technologies, including microwave transmission, were put into practical form and entered service in America.

Yet there is a geography of European telephony to consider, as well as some technical contributions, even if these were mostly refinements. As in the United States, telephony began in Europe as an urban phenomenon. Given the relatively small area covered by most European polities, it was technically possible to construct national networks much earlier than in America, although this was not done. Early technical constraints to telephony between European polities were removed by the vacuum tube amplifier in 1915. Yet telephony between most European polities was not achieved until the early 1920s, before which time no supranational organizing body existed. Even the most developed industrial nations—Germany, Britain, and France—lagged far behind the United States in the adoption of telephony even after all the technical problems had been solved.

Siemens und Halske, in Berlin, had improved on the Strowger automatic exchange. In 1913 they began experimenting with an electron tube amplifier developed by Alexander Meissner and based on tubes patented by Robert von Lieben in Vienna in 1906 (Siemens 1957b, 11, 14). "On the 20th of August, 1914, the German Post Office installed a temporary telephone line with Lieben tubes between the Central Headquarters in Luxembourg and the Staff Headquarters in East Prussia, over which the telephone conversations were held which led to the dispatch from retirement of Gen. Paul von Hindenburg to the Eastern front, and to the Battle of Tannenberg" (14–15). On 21 August Count Helmuth Johannes Ludwig von Moltke, the German chief of general staff, removed Generalleutnant Karl Ludwig von Prittwitz from command of the German forces facing rapidly advancing Russians in East Prussia because of the defeatism evident in his voice in their telephone conversation of the previous day (Nebeker 1996). Prittwitz was immediately replaced by von Hindenburg, and the result was the sweeping German victory against superior Russian numbers at the Battle of Tannenberg on 27–30 August.

Much of the explanation for the geography of European telephony has to do with the ownership by each European state of all forms of communications within the state. Nowhere was the American pattern of separate ownership of the different forms—by the state of the postal service, by Western Union of the telegraph, and by Bell and then AT&T of long-distance telephones—in evidence. European polities ran all three, establishing separate niches for mail, telegraph, and telephone. The mail service was cheap and ubiquitous but relatively slow. The telegraph, more expensive than mail service but also faster, was also ubiquitous because it followed the railroads. Messenger boys, first on bicycles and later on motorcycles, delivered messages speedily to the door of the recipient at a reasonable cost. It was still the predominant form of rapid long-distance communication through the 1950s and 1960s in most European polities. The telephone was a device for the urban middle and upper middle classes, as well as for businesses, although European companies depended much more heavily on the telegraph than did American companies. This was in part a result of discriminatory pricing. Whereas Bell provided unlimited local service as part of the monthly fee and charged tolls only for long-distance calls—hence the Americanism "toll call" for a long-distance call—in Europe even charges for local calls were based on time. Local toll fees, high monthly service fees, and policies that preferred business to domestic users kept telephones out of all but the homes of the better-off.

Even so, a pattern of apparent ubiquity grew up in Europe in the 1920s, once the necessary organization for the exchange of telephone service between different polities had been established. This exchange was proposed as early as the Paris Telegraph and Telephone Conference of 1910, although the real impetus "even among technical people . . . was aroused only when a trans-continental service between New York and San Francisco [8,000 km] had been established" (Craemer 1924, 286). The longest distance in Europe, from Gibraltar to the Ural Mountains, was only 6,250 km. The first step in realizing such an organization was Frank Gill's proposal in his presidential address to the Institution of Electrical Engineers in London in 1922. Gill, who was then European chief engineer of Western Electric, made his proposal based on American experience, suggesting a European equivalent of AT&T with representation from European governments (Craemer 1924, 286). Gill's address sparked a response from the French Telephone Administration, which held an international meeting in the spring of 1923 and established the Comité Consultatif International des Communications Téléphoniques à Grande Distance (CCI).

The CCI established a uniform set of technical standards and a political organization. The International Chamber of Commerce also played a central role in pushing for Europe-wide service. The findings of its 1927 Stockholm congress can, in fact, be taken as a general commentary by the European business community on the rapidly increasing value of long-distance telephony.

a. [It was] an effective instrumentality for removing impediments to international trade, and
b. . . . the ability to secure speedy personal communication between distant cities should be a factor in stabilizing business through a more orderly and economic movement of goods.
c. . . . effective telephone service tends to facilitate all the processes of production and distribution.
d. . . . ability to communicate information quickly tends to minimize the range of price fluctuations and thereby lessen the tendency to speculation.
e. . . . any instrumentality which tends to stabilize business, facilitate its

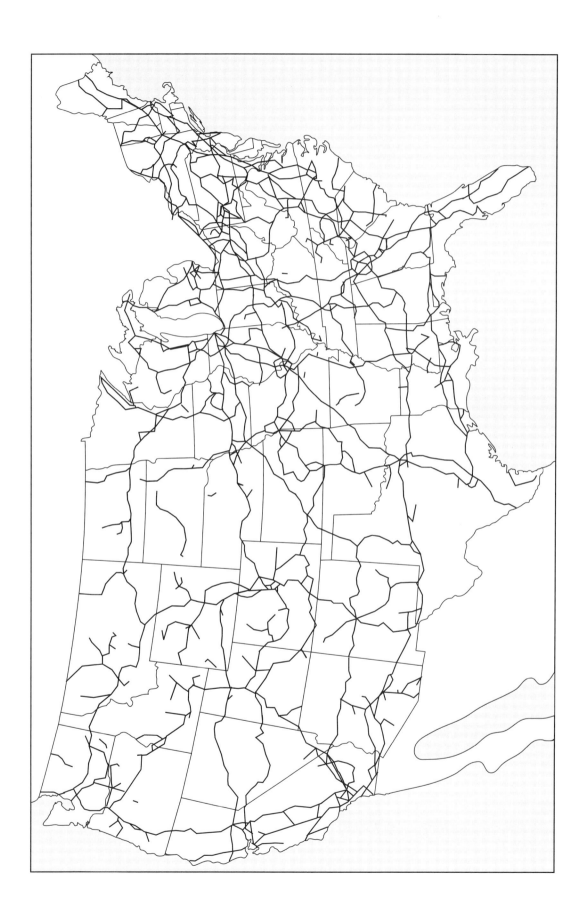

processes, and effectively extend the field of operations with consequent increase in the volume of trade, should lessen the difficulties of international settlements, as after all, all settlements have to be made ultimately in goods or services.

f. ... the ability to communicate voice-to-voice, easily and speedily, whenever desired and needed, cannot but be a means of improving social relationships and developing a common economic and social viewpoint which inevitably must have an effect in promoting a better understanding between nations. (Quoted in Gill 1929, 193)

Items *a* through *d* could easily apply to American interstate trade as well as European international trade, and America had achieved this interstate trade on a regional level in Megalopolis and the Manufacturing Core by the late teens. Items *e* and *f* clearly refer to European conditions in the aftermath of World War I, the Armistice, and the Treaty of Versailles. However true it was that all reparations had to be paid for in goods or services, all the goodwill in the world could not remove the fact that all German governments after World War I were hellbent on avoiding such reparations (Kagan 1995). In the same way, no amount of voice-to-voice contact would diminish the increasing tension between polities for whom, in the absence of an emergent American hegemony, World War I had settled nothing.

The long-distance service that came about as a result of CCI was expensive and slow, especially in comparison with the service of AT&T. Although delays had been significantly reduced by 1927, there was little to boast about. The average delay during busy hours on a Paris-Berlin call fell from 130 min in 1925 to 68 min in 1927; that on a call from London to Amsterdam fell from 61 min to 34 min (Gill 1929, 191). By 1927 AT&T claimed an average toll-call delay of 1.5 min and stated that "about 90 percent of all toll calls are now handled while the subscriber remains at the telephone" (195). Gill also notes that the Bell rate for a person-to-person call, the most expensive service, was 23.4 gold francs for 3 min, whereas the CCI "lightning rate" was 100 gold francs for 3 min. "It is doubtful if the delay would be any less than the delay on the Bell system, and the service would be inferior, in that no Particular Person Service would be afforded and the individuals required would have to be sought after communication was established. . . . The C.C.I. Urgent Rate, with *Préavis* fee [to notify the recipient that a call was coming] would be 33.3 gold francs for a 3-minute call. The delay would almost certainly be considerably more than in the Bell system call" (196).

Gill clearly had an AT&T ax to grind, but his conclusion was that such problems would remain as long as there was no unity of control and that "the individual nations of Europe are far too small to form the economical units for this service, and I cannot escape the conclusion that, sooner or later, there will have to be found some single unit for the administration of the European International Service, unless indeed Europe is to rest content with a service relatively much inferior to the plant which she has provided" (Gill 1929, 199). Inferior service was, of course, to be the case. Other than CCI, no mechanism for European long-distance

Fig. 3.8. TD2 Microwave System Routes, 1980 (redrawn from O'Neill 1985, 307) *(opposite)*

telephony developed to speed ervice, and Europe remained a messy patchwork of national systems with poor intercommunication until the 1970s. This is a classic case of software technology failing to keep up with hardware, inasmuch as European equipment was essentially licensed AT&T equipment and an adequate network of lines had certainly been completed by the late 1920s.

The emergence of the Nazi state after 1934 and its increasingly successful military assaults on its neighbors through 1941 created one possible set of political conditions for Europe-wide telephony. Craemer noted as early as 1924 that "the geographical position of Germany and the geographical structure of Europe make it necessary that most of the chief lines of Europe should pass over the German system" (289). In the rebuilding period after World War I, however, the Reichspost stayed with coil systems even though Telefunken had developed carrier a little earlier than Bell, in 1912 (Schulz 1959, 71). During World War II the Nazis built a network of high-quality coaxial telephone lines that extended throughout the Third Reich, the conquered territories, and the territories of such allies as Italy. Carrier, and the cable systems to carry it, was installed from Russia to Sicily (Petzold 1996; Schulz 1959). The telephone engineers of the Third Reich clearly thought in terms of continental scale (Petzold, personal communication, August 1996). Within Germany proper, telephone lines also carried radio and television signals. By 1945 the line from Berlin to Munich via Nuremberg was carrying the 180-line television system. The line from Berlin to Hamburg and the beginnings of the line from Berlin to Frankfurt am Main were installed to carry the 441-line television system as well as numerous telephone messages (fig. 3.9).

Given the technologies of telephony that developed between its invention by Bell in the 1870s and the achievement of geographic ubiquity combined with high-capacity service by the late 1940s, it was problems of political organization that slowed the spread of telephony around the globe, not problems of technology or economic organization. Bell and, later, AT&T were perfectly viable models for effective economic organization within a capitalist system. Similarly, a powerful centralizing polity, such as Nazi Germany, could have achieved the same end.

By the early 1920s Bell had convincingly demonstrated what could be done at first the regional and then the continental level. Bell cut its teeth on urban systems in the period before 1900, catering to a rapidly rising demand for local service. As the appetite of American consumers for long-distance service increased, Bell moved to satisfy that as well. Between 1900 and 1915 Bell thus created an effective high-capacity system over the major urban and industrial regions of the United States, Megalopolis and the Manufacturing Core, an area at least as large as the largest European polity. After 1915 it extended a low-capacity service throughout the continental United States.

At no time in this period did Bell enjoy a monopoly over telecommunications, although it has always popularly, and erroneously, been called a monopoly. It never merged with the telegraph companies, and its oc-

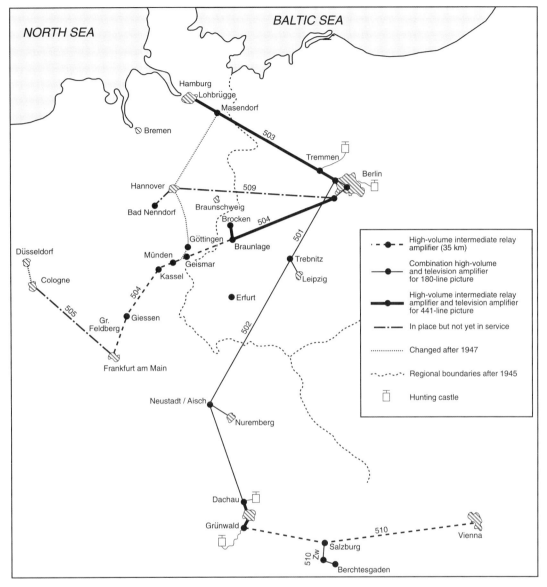

Fig. 3.9. Coaxial Cable Routes in the Third Reich, 1945 (redrawn from Bauer 1996)

casional moves in such a direction were inevitably thwarted by the American commitment to competition. In Europe, on the other hand, in industrial urban societies equally ripe for the growth of telephony if only at the national level of shared language, the emergence of nationalized telecommunications trusts stifled the early success of telephony. The telegraph remained the predominant long-distance telecommunications medium long after Bell had demonstrated that a huge market for telephony existed and produced the technology with which to achieve it.

Whereas telegraphy achieved international range very early in its history, the history and geography of telephony were such that its promoters and supporters were convinced that it was at best a continental-level technology. Early experiments with submarine cables convinced telephone engineers that only the shortest cables would be successful. The first major link was the cable under the English Channel in 1891, linking London to Paris (Gill 1929, 190). The development of the vacuum tube amplifier as the major technology for achieving long distances confirmed this conclusion, at least until the development of extremely compact vacuum tubes and coaxial cables in the late 1930s, when Bell engineers began to conceive of tubular amplifiers integral with an armored cable and having a small enough diameter to allow them to pass through regular ocean cable-laying machinery.

As a result, the history of telephone technology was not only national but concentrated in the one polity that enjoyed the necessary conditions for its easy success: political refusal to nationalize all telecommunications and a sufficient geographical size for Bell to become obsessed with distance. American policy helped even more by making it quite clear to Bell from the earliest opportunity that no merger with the telegraph interests would ever be tolerated. When Bell's initial patents ran out, American policy ensured that Bell would not be allowed to retain total control of local telephone companies. This forced Bell into a series of evolutions, each one concerned with the achievement of longer-distance service, and thus its eventual transformation into a mainly long-distance company. This development of a natural monopoly is the second root of Bell's obsession with long-distance service (the first was the simple geography of the United States).

Bell's obsession with distance produced an obsession with research in order to retain its science-driven lead. As a result, Bell led in the development of nearly all of the most vital telecommunications technologies of the period from 1900 to the early 1980s. It perfected or produced the two major amplifying devices, the triode vacuum tube and the transistor. It developed practical theories of transmission and new types of cables, microwave transmission, and so on. This magnificent set of technical achievements has been more than adequately documented by the several volumes of Bell Laboratories' *History of Engineering and Science in the Bell System.* Bell forced, or was forced into, long-distance technology at a time when most competent engineers would have said the telegraph was both very much simpler and perfectly adequate for the foreseeable future. Bell created the first transcontinental and transatlantic wired telephone systems as well as the first telecommunications satellite. I return to the geopolitical implications of the last two technologies in chapter 8.

CHAPTER FOUR

Radio Telegraphy, Radio Telephony, and Interstate Competition, 1896–1917

Britain's navy, her merchant marine, her foreign trade, her Empire, her dominion over many lands and all the seas have depended in the past upon her world-wide system of communications. What the nerves and sinews are to the body that system has been to British rule. In peace and in war the vast network of cables and wireless, spreading across the oceans' depths and high through the ether has given her the power to see, to hear, to do— if not all, at least to see and hear and do much more than her commercial and naval rivals. If any one factor of her supremacy has been more vital than others it is this.

—Ludwell Denny, *America Conquers Britain*

The struggle to create a wireless alternative to the submarine telegraph cable for international communications began in the late 1890s, intensified in the immediate aftermath of World War I, was settled in the early 1920s with the creation of a system of global telecommunications that was complete in its essentials by 1939 and was not successfully challenged until the completion of the first transatlantic submarine telephone cable in 1956. Wired telegraphy required control of extensive real estate if overland and landing rights at critical locations for repeater stations if submarine. Wireless telegraphy offered to all those who had the technology access to the seemingly unlimited and uncontrollable medium of the upper atmosphere, or ether.

At first the struggle was between Britain and Germany, but by 1911 America had also entered the fray. British interests using Marconi's spark-transmission system were generally dominant through about 1917, slipped badly as America began a technological end run around Britain using continuous waves in the lower end of the frequency spectrum, and were generally restored by Marconi's pioneering shift to high-frequency continuous waves in the mid-1920s. No single nation was to emerge fully victorious from this struggle for the ether, but the struggle had the effect of reinforcing the importance of the long-term British control of the submarine telegraph lines that ran through the ocean depths.

In *America Conquers Britain,* published in 1930, Ludwell Denny stated his belief that American low-frequency continuous wave radio had brought the British telecommunications monopoly to an end. Yet at precisely this moment the British were completing an all-important commercial and imperial chain of high-frequency continuous wave stations using highly directional beam antennas and forcing the competing submarine cable and wireless interests into one crucial monopoly. However,

Denny understood perfectly well the consequences of an American failure that was not then quite yet obvious. Britain's "navy, her commercial fleet, her factories, her international banking system, however great, are tangible things that can be faced and fought in the open by her competitors. But her sway over communications has been almost an unseen thing, penetrating banks, export centres, military staffs, and foreign offices of rival nations, carrying their commercial and diplomatic secrets— secret to everyone but those concerned and the all-hearing ears of the British cables" (367).

In discussions of geopolitics and the hegemonic struggle between polities for control over the capitalist world-system, communications technology has been even more poorly served than technology in general. Without efficient transportation technologies, spatially extensive world-empires and world-systems simply cannot operate, as Innis indicated. If communications operate at the same speed as transportation, such spatial extension will be limited by the time costs of overcoming distance. Britain was the first polity to exceed these limits when it created the first global telecommunications system. In the constant struggle for hegemony between core and would-be core states in the world-system, Britain's success in exceeding such limits was quickly recognized and envied, its technology copied, and alternative technologies sought. In this context, however hyperbolic Denny's statement may sound, it is fundamentally correct. A polity that could not break the effective British monopoly of telecommunications could not hope to succeed Britain as hegemon.

MARCONI AND LOW-FREQUENCY WIRELESS TELEGRAPHY USING SPARK TRANSMITTERS

There is a considerable literature on the early development of radio technology. Two books by Hugh Aitken, *Syntony and Spark* (1985b) and *The Continuous Wave* (1985a), do an excellent job of describing the two main periods of technological developments covered in this chapter and in chapter 5. The first of these periods was marked by the development by Guglielmo Marconi of a viable system of spark radio transmission and by Oliver Lodge of the crucial concept *syntony,* or tuning both transmitter and receiver to the same frequency. The second period, marked by the emergence of continuous wave technology, is far more complex, covered at length by Aitken in *The Continuous Wave.* Aitken is, however, more concerned with the technical problems of getting continuous wave technology to work in the second decade of this century, and his account closes with the emergence of RCA in the early 1920s to control the jumble of conflicting patents that had emerged in the development process and bedeviled progress in America. Susan Douglas, in *Inventing American Broadcasting* (1987), provides a fine account of the institutional and cultural frameworks within which long-distance radio developed but she does not discuss the geopolitical and geostrategic frameworks that drove that development. Like Aitken, Douglas concludes in the early 1920s with RCA, although her interest is in the transformation of RCA from a clearly geopolitical attempt to monopolize global two-way wireless communication to a broadcasting company involved in the provision of both radio receivers and radio programming in the rapidly growing home-entertainment business.

Simply stated, the history of wireless communications technology up to World War II falls into four main periods: theoretical underpinnings; spark technology; the search for continuous wave technology; and high-frequency, short-wave technology.

Theoretical Underpinnings

In 1865 the British mathematician James Clark Maxwell, basing his work on that of Michael Faraday, put forward his mathematical model of electromagnetic fields. Maxwell's model implied that electromagnetic fields could be propagated through space as well as through conductors (Aitken 1985b, 22). Such fields would travel as waves propagated at the speed of light. Indeed, Maxwell's model implied, light *was* extremely high frequency electromagnetic radiation within a narrow range of extremely short wavelengths (fig. 4.1).

In 1888 the German physicist Heinrich Hertz generated electromagnetic waves at the lower end of the radio frequencies, propagated them without conductors, detected them, and measured their speed as that of light. Hertz, for whom the frequency of a wave's oscillation in cycles per second is named, made it possible for others to conceive of electromag-

Fig. 4.1. The Electromagnetic Spectrum (redrawn from Inglis 1990, 8)

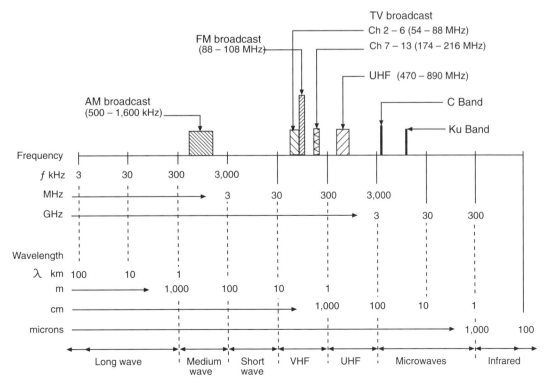

netic radiation as a potential carrier of information. Hertz died in 1894 at the tragically young age of 36 (Appleyard 1930, 140).

Also in 1894 Guglielmo Marconi, son of a Scots-Irish mother and a prosperous Italian landowner and silk merchant, began to experiment with the use of Hertzian waves for "wireless" telegraphy. Marconi was no theoretician, but he had great drive and practical ability, and his early role needs to be clearly understood. Marconi had access to Augusto Righi, at the University of Bologna, one of the few people in Europe who fully understood Hertz's work in the 1890s (Aitken 1985b, 183–84). Marconi's mother clearly understood what he was about, arranged his education with Righi, and paid for much of his experimenting over the objections of her husband (Douglas 1987, 15–17). Although Douglas does not document it, it seems probable that she also imbued her son with a sense of the importance of communications to the British Empire. Innovating based on the experimental work of Hertz and Righi, Marconi was confident enough by 1896 to travel with his mother to London to attempt to sell his system of wireless telegraphy to the British Post Office after the Italian Ministry of Posts and Telegraphs turned it down (17). His mother's relatives in London had no such reservations about the potential of his technology to overturn the profitable monopoly of the submarine cable interests.

One of the few other people in Europe who understood Hertz's work when it was first published in July 1888 was the British physicist Oliver Lodge. As a student Lodge set himself the task of experimentally verifying Maxwell's equations. As a professor of experimental physics at University College, Liverpool, he did so, publishing the results of his experiments in August 1888 (Aitken 1985b, 80). Lodge used long wires as wave guides, so that he did not experimentally prove that electromagnetic radiation could be propagated through empty space, but his experimental proof of Maxwell's theory was as valid as that of Hertz and had just as much implication for wireless telegraphy. Lodge, who was as much an engineer as a theoretical scientist, demonstrated the rudiments of what easily could have been developed into a system of radio telegraphy at Oxford in 1894 (119–23). Although he was a teaching and research professor with a well-funded laboratory in one of the new universities springing up in the industrial and trading cities of late Victorian Britain, he did not promote the practical possibilities of his technology, at least not until Marconi arrived in England in 1896 and filed his first patent application. In 1897 Lodge filed two patent applications of his own for "improvements in syntonized telegraphy." What Lodge described was not merely a system of wireless telegraphy based on Hertzian waves, since Marconi had essentially patented such a system in 1896. Lodge's contribution lay in his thorough understanding of the necessity for syntony, generating electromagnetic waves at "a particular frequency of oscillation" and using a receiver tuned to the same frequency. Without syntony, wireless telecommunication was commercially useless. Marconi's original system used such a prodigious amount of the electromagnetic spectrum that only one transmitter could be worked at any given time. Lodge's work created the basis for accurate division of the spectrum into discrete components, which made wireless commercially viable.

One further theoretical understanding was needed, and this arose after Marconi had empirically demonstrated in 1901 that a wireless signal could be propagated across the Atlantic. All previous theory pointed to radio waves' being propagated only in a straight line. Marconi's signal should have gone straight into space. Two physicists, A. E. Kennelly in America and Oliver Heaviside in England, thus theorized almost simultaneously in 1902 the existence of an ionized layer in the upper atmosphere that would reflect radio waves. In reality, the situation was much more complex. The very low frequency long waves used by Marconi were not reflected by the Kennelly-Heaviside layer but followed the curvature of the earth. Nor was the reflecting layer simple: after considerable exploration it proved to consist of three major ionized reflecting layers, D, E, and F, the last of which sometimes split into two layers, F_1 and F_2, with diurnal and seasonal variations (fig. 4.2). Radio exploration of the upper atmosphere was required before these various layers could be mapped out and a commercially reliable service begun; much of this exploration was carried out in the early 1920s as the first transatlantic wireless telephone service was being planned. In at least some ways the difficulty of propagation ensured that wireless communication would not easily become the possession of any one company or country early in its history. Understanding precisely how radio waves were propagated at the planetary level required a vast international scientific and engineering effort. As researchers moved from the low-frequency long waves to the higher-frequency short waves in the mid-1920s they also laid the groundwork for the development of radar technology in the 1930s.

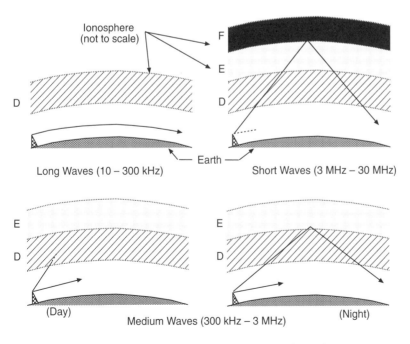

Fig. 4.2. The Three Major Layers of the Ionosphere and Their Effects on Wireless Transmission (redrawn from Inglis 1990, 37)

D, E, and F are the three distinct layers in the ionosphere

The hypothesization of a reflecting layer in the atmosphere by Kennelly and Heaviside in 1902 marks the end of the period of basic theory formulation in radio and its initial empirical testing. Maxwell's equations had been twice verified (by Hertz and Lodge). Marconi had demonstrated that Hertzian waves could be generated by spark transmitters, propagated through empty space, and carry information using Morse code. Lodge had demonstrated that the radio component of the electromagnetic spectrum could be divided into discrete components by tuning for specific wavelengths. And Kennelly and Heaviside had suggested why radio waves seemed to follow the earth's curvature.

There was still, of course, a vast amount of work to be done, much of it in refining the theoretical underpinnings. But the focus of development now moved to the engineers who would create a working technology, the creators of systems of wireless communication, the patent holders, and the geostrategists. From all of these Marconi stands out. From the very beginning, and at the very crudest level of technology, he understood the need for systems, patent rights, and the proper geostrategy. Much of the practical development of the first technology—spark transmission—was also at his hands. The failure to consolidate the lead Marconi gave Britain was due to Britain's failure to understand the geopolitical importance of wireless. The bizarre and destructive internal squabbling over the control of wireless came about because geopolitical theory did not catch up with technological reality for a generation. Mackinder (1904) certainly created a framework within which changes in transport technology could be clearly seen to drive geopolitics, but it would take nearly thirty years and one war's worth of experience with wireless to make politicians and military leaders recognize that the transport of information was at least as important geopolitically as that of people and goods.

Spark Technology

Marconi generated radio waves in the simplest way possible: by generating a stream of sparks of high intensity and regularity (Aitken 1985b, 186). In this he did little more than improve on the work of Hertz and Righi. He detected the sparks with a coherer of the type invented by Edouard Branly in Paris, information about which was easily accessible in the scientific journals of the time. A coherer was a switch that was turned on by receipt of a weak radio signal. It was then returned to its "off" state by tapping or shaking (103). Marconi innovated by interrupting the flow of sparks with a Morse key, thus allowing telegraphic messages to be sent. He innovated further in his design for the first really efficient antenna, which made the sending and receiving of his messages much more efficient.

As Aitken rightly points out, Marconi's driving obsession, and thus the secret of much of his success, was with the range of his apparatus (1985b, 191–92). Unlike Hertz or Lodge, who were satisfied with proving Maxwell's theory by signaling from one room to another, Marconi was trying to create a technology that would allow him to signal to ships at sea or across the great ocean masses. Here, however, technology and obsession interacted to misdirect long-range radio for two decades.

Generating electromagnetic waves with spark technology meant emitting them all across the radio spectrum. Only a tuned circuit avoided this prodigality, and it was not until Marconi's British patent number 7777, in 1900, that he began to focus his signals into narrower parts of the radio spectrum. The "four sevens" patent was, however, not the master patent on syntony that Marconi needed and pretended he had. That honor had already gone to Lodge, and until Marconi bought Lodge's patent in 1910 other companies were more than willing to risk infringing the "four sevens" patent.

In the early days Marconi fed broadband signals to short antennas that effectively emitted only the short-wave components of such signals, thus acting as de facto tuners. Such waves seemed to travel only short distances. In actuality they traveled very much longer distances, being reflected by the Kennelly-Heaviside layers, but they could not be detected close to their point of origin. By 1896 Marconi had found that larger antennas, which sent the long-wave component of this broadband signal, sent them longer distances. Such long waves were actually *ground waves*, which followed the earth's curvature. Since they did not have to be reflected by the ionized layers of the upper atmosphere, they were easily detected even close to their point of origin. According to Aitken, "What Marconi learned was that larger antennas meant longer distances . . . [and] when it seemed that wavelengths could be lengthened no further, there was only one way to achieve greater distance: higher power" (1985b, 196–97). From then until Marconi's "rediscovery" of the short waves in the mid-teens the formula for increasing the distance radio waves would travel was simple: longer waves, taller antennas, higher power.

Marconi was not the only proponent of spark systems. The German Adolph Slaby had been moving in a similar direction. In 1897 he was invited through diplomatic connections to witness Marconi's test transmissions across the Bristol Channel for the British Post Office. Slaby returned to Germany and, with Wilhelm von Arco, developed the Slaby-Arco system of wireless telegraphy. Slaby and Arco, who were associated with the powerful German electrical company A.E.G., merged in 1903 and at German government direction with Carl Braun's company, part of the huge Siemens und Halske engineering company of Berlin. Siemens notes that the Kaiser himself brought pressure to bear on Siemens to complete the merger (Siemens 1957a, 184). The official name of the new company was Gesellschaft für drahtlose Telegraphie M.b.H. Berlin, but it quickly became better known better as Telefunken (*Electrician* 68 [10 November 1911]: 171).

As a typical German quasi monopoly of the period, backed by an efficient higher education system that was strong in science and engineering and having the Imperial German Army as a major client, Telefunken prospered. Marconi was shut out of the German market. Surprisingly, however, Telefunken came up with few innovations. Its main innovation was the "quenched spark" system, which considerably reduced wavelength oscillations in the primary circuit (fig. 4.3), thus increasing the

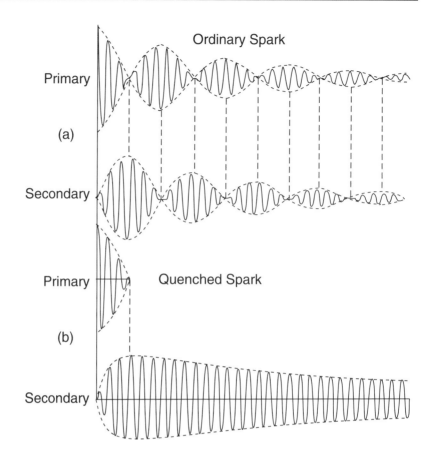

Fig. 4.3. Telefunken's "Quenched Spark" Wireless Transmission (from *Electrician* 68 [10 November 1911]: 171)

amount of power the antennas could transmit. In effect, Telefunken created a variant of spark technology that behaved as much as possible like continuous wave technology. This improvement, plus extremely sound engineering and politically guaranteed sales, ensured their ability to survive in a market otherwise dominated by Marconi.

WIRELESS IN GEOPOLITICAL CONTEXT

The relative German failure to develop and sell wireless telegraphic equipment is surprising. Wilhelmine Germany was far more oriented to scientific and technological development than was turn-of-the-century Britain. Germany had five times as many students in technical colleges and universities and spent twenty times as much on university grants (Barnett 1988). In 1900 "Germany's manufacturers produced twice the output (by value) of Britain's heavy electrical industry" (Pocock 1988, 3). Yet by 1912 Marconi had sold nearly twice as many sets, 953 to Telefunken's 478. In the more open competition for sales outside the British and German empires Marconi was even more dominant: 322 to 114 (4). Pocock suggests a direct link between telecommunications and empire.

British trade slipped away from the industrialised nations to the less-developed territories of the Empire during the last quarter of the nineteenth century. Trade depended upon communications. Submarine cables provided a reliable means of exchanging information and transmitting instructions from central government and business offices to overseas dependencies. Steamships supplied a comparable (albeit slower) service for goods and personnel. Radio telegraphy supplemented these two existing systems allowing for communication from ship to ship and from ship to cable terminal on shore. The existence of the Empire was an essential modifying factor in the nation's social and economic system. It created a need for radio technology and a market for the products of the radio industry. (6)

Technological development in Britain proceeded along precisely the commercial lines Pocock suggests until World War I, with Marconi seeing the possibility of profit from serving ships at sea, offering a cheaper service than the submarine cable companies or serving less heavily trafficked routes where the high capital cost of cables was not justified. But Pocock's analysis leaves out geopolitics. As Britain's effective telecommunications monopoly in submarine cables slipped slowly away, wireless telegraphy seemed to afford a means to restore that monopoly. To the navy, probably the most innovative large organization in Britain at the time, it offered the possibility of centralized communications, command, and control (C^3), so that ships at sea would no longer be effectively lost between telegraph stations but could be redirected by their masters in Whitehall whenever necessary. The British navy sold the British government new technology throughout the Edwardian period. Turbine-powered ships; all-big gun armament for dreadnought battleships; submarines; oil fuel; aerial bombing from naval airplanes; land dreadnoughts, called "tanks"; and wireless were, in modern terms, force multipliers that let fewer taxes do more work. Pocock is right to say that the empire created a need for wireless telegraphy, but it was not just a commercial need, it was also a geostrategic and geopolitical one.

This description oversimplifies what was really a complex process in which four forces more or less continually interacted under the umbrella of geopolitical necessity: technology, geostrategy, control of patent rights, and systems of communication. In geopolitical terms Britain had the clearest vision of any Edwardian polity of the nightmare future made visible by Mackinder, in which the day of sea power polities was over and that of land polities was beginning (Mackinder 1904). The rise of imperial Germany clearly influenced and then reinforced Mackinder's writings about the Heartland (Blouet 1987). The clear implication was that Britain could not allow any unfriendly continental power to dominate Eurasia. A powerful land state could build a much more powerful fleet than Britain's if it came to control the natural and human resources of the Eurasian landmass. The British navy thus had no choice but to seek new technologies as force multipliers. Of the four forces subordinated to the geopolitical necessity that led to wireless, technology was clearly the most significant overall, not least because it was intimately linked to the initial ownership of patent rights and the development of systems of communication. The second most significant force was geostrategy, which must be roughly divided into commercial and political forms.

Geostrategy was linked somewhat to patent rights, more so to systems of communication. It was obvious almost from the start that wireless could be used for global telecommunications and equally obvious that a global network of stations was needed to pass messages along. In the geostrategic struggle for global wireless Britain and America thus did better than France, which in turn did better than Germany. Success came from colonial possessions and influence. In this regard America acted as a classically European colonial power in the late nineteenth and early twentieth centuries, scooping up the remnants of the Spanish Empire as it collapsed and negotiating treaty rights throughout Asia.

PATENT RIGHTS, DOMESTIC POLITICS, AND INTERNATIONAL AGREEMENTS

From the beginning Marconi understood the importance of radio for international communication. Unlike other inventors, he therefore went to considerable lengths to control patents, both his own and those originated by others. His individual competitors were never able to control as many patents as he did. In the geopolitical struggle for international radio three techniques were developed to block Marconi from developing an effective monopoly: the international agreement; manipulation of domestic politics; and patent pooling.

Coming from the weakest patent position, Imperial Germany, which held none of the fundamental radio patents, twice used international agreements to successfully force open the limited amount of the electromagnetic spectrum that could be used with even the best spark technology. British ratification of such agreements was a function of a domestic political struggle against an emerging Marconi monopoly.

In Britain the government controlled Post Office, which had misjudged the value of Marconi's patents in the early days, sought to extend its domestic telecommunications monopoly to an international one by denying Marconi's attempts to create an imperial chain of spark transmitters in the early teens. The situation in Britain was confused by the eruption of the Marconi scandal at the highest levels of government in 1912 (see below). British submarine cable interests also had a stake in Marconi's lack of success, thus further clouding the issue.

Among the earliest group of patents three stand out: Marconi's first radio patent (1896); Oliver Lodge's patent on syntonic circuits (1897), held by British courts to be the basic patent on tuning and extended for seven years in 1911; and Marconi's own tuning patent (1900), British patent 7777, usually called the "four sevens" patent. Giving Marconi's first patent the date 1896 creates certain problems, since only the provisional specification was filed in that year. The complete specification was filed in 1897, and neither was a public document until the patent was issued in July 1897. Before Marconi's patent became public, Lodge's syntony patent was filed. Since Marconi's claim to monopoly rights over maritime radio were based on the 1896 patent, the first radio patent ever issued, the timing is important (Aitken 1985b, 203).

When Marconi first went to Britain he attempted to sell his 1896 patent to the British Post Office, which had finally achieved a monopoly over telegraphy and telephony within British territories in 1896 by the Telegraph Acts (Aitken 1985b, 214). Although the Post Office, in the person of its chief electrician, William Preece, recognized the value of Marconi's technology and patent, its offer to Marconi was a niggardly £10,000. Preece may simply have believed that Marconi had no other possible buyer. In fact, Marconi had his mother's family. Born Annie Jameson, of the Irish whiskey distillers of the same name, his mother was related on her mother's side to the Scottish Haig family, the major distillers of Scotch whisky (Dunlap 1941, 6). Douglas credits Annie with much of her son's success. She took him to Britain and opened the door to the finance of the city of London. Almost from Marconi's arrival his cousin, Henry Jameson-Davis, urged him to set up a company to exploit his technology (17). The poor offer by the Post Office persuaded Marconi to do so. Using the 1896 patent as its principal asset, the Wireless Telegraph and Signal Company was incorporated in 1897 with Jameson-Davis as its first managing director (Aitken 1985b, 224; Douglas 1987, 65).

Given the restrictions imposed by the Post Office's statutory monopoly of internal communications, Marconi's development of his technology as a method of ship-to-shore communications made sense in more than a commercial way. The Telegraph Acts ruled out private revenue-earning service in Britain or in British territorial waters but allowed a private company to send messages for its own use and the use of others as long as no direct charge was made for messages handled (Aitken 1985b, 234). These provisions explain Marconi's refusal to sell its equipment, to allow its use by other than Marconi employees, and to communicate with non-Marconi installations except in emergencies. The refusal to communicate with non-Marconi installations, which became known as the policy of "non-intercommunication," became a severe political problem for Marconi's.

Despite these statutory limitations, Marconi was able to use the 1896 patent to develop an effective company that in 1901 contracted with Lloyd's insurance to monopolize ship-to-shore communication around the British Isles. Yet Marconi's equipment was technologically simple and easily replicated, and although his patents were dubious, they were not tested in the courts until after 1910 (Aitken 1985b, 238). What the Lloyd's contract meant in practice was that "any major shipping line, if it wished to take advantage of the worldwide network of marine intelligence that centered on Lloyd's of London, had no alternative but to see to it that its vessels were equipped with Marconi apparatus. And if the shipowners themselves showed some reluctance to take this step, or to commit themselves to this particular line of equipment, pressure from marine insurance underwriters was likely to tip the scales in favor of integration into the Lloyd's sponsored system" (236).

The contract was supposed to run until 1915, although the pressure brought to bear on it by German and American interests caused it to be voided by the second international radiotelegraph conference in 1906. Without this contract it would be hard to understand why Marconi's was able so quickly to dominate wireless telegraphy. The contract with Lloyd's produced rapid and profitable growth for Marconi since a ship fitted with any other equipment could only communicate with a Marconi-equipped ship or shore station when in distress. In early 1902, Prince Heinrich, brother of the German kaiser, returned from America to Germany on the *Deutschland,* which was not allowed to communicate with Marconi stations because it was equipped with Slaby-Arco equipment. The prince was outraged at Marconi's refusal to carry his messages, and "the incident had immediate political repercussions" (Headrick 1991, 120). The German government ordered German stations to use only German sets, forced the creation of Telefunken, and in August 1903 convened the first international radiotelegraph conference in Berlin. Austria, France, Russia, Spain, the United States and of course Germany used the conference as a forum for attacking the policy of non-intercommunication. Britain and Italy attended the conference but refused to sign the resolutions against Marconi. The Marconi interests contended, accurately, "that the real purpose of the conference was to open the way for promotion of German wireless" (Dunlap, 1941, 148). Conferences were an important element in geopolitical maneuvering from the beginning. Mance (1944) provides a useful summary of the provisions of the main international conferences, agreements, and communications unions.

The Germans called a second international radiotelegraph conference in Berlin in late 1906, with the same agenda as the first. On this occasion American delegates led the opposition. Of the 30 nations in attendance 27 signed the draft treaty and 21 favored free marine intercommunication (Headrick 1991, 120). The British Parliament ratified the treaty by a single vote in 1907, and it went into force in 1908, effectively ending the Marconi monopoly by forcing it to end the policy of non-intercommunication.

The conference results were not all negative, however. Germany and America had proposed reserving the longer waves for government and military use and relegating commercial stations to the shorter, 300–600 m waves. Since only the longer waves were known to be usable for long-range communication, this would have relegated commercial radio to short-range ship-to-shore functions, which would have been devastating to Marconi's long-range plans. British negotiators successfully argued against this reservation of the long waves for military and government use in return for giving up the principle of non-intercommunication (Headrick 1991, 120).

The American role in the 1906 conference requires some explanation. The American navy was particularly hostile to Marconi's company, which it regarded as synonymous with British government interests. Its opposition was cemented by a belief that the tremendous concentration of international submarine telegraph lines through London gave Britain an immense geostrategic as well as economic advantage. Despite repeated British claims to the contrary, the American navy was firmly, and probably accurately, convinced that the British government was reading its mail.

The concern of the American navy was simple. Britain could easily bring the international submarine cable system under its almost total control in the event of war. Even cables that had not been laid by British capital had been made in British factories and laid by British ships on British-surveyed routes. British naval power, surveys of the cable routes, and ownership of nearly all of the ships that laid and repaired cables ensured that the cables of a hostile power could be easily cut and spliced into the British cable network, as German cables were in 1914. If Marconi were able to dominate international wireless telegraphy, Britain would thus continue its effective global monopoly of telecommunications. As Aitken cogently remarks, "This was an unwelcome prospect to those who looked to the new technology of wireless to create an alternative system to the British-controlled cables" (1985a, 253). Aitken further notes that the American navy's opposition to Marconi was Anglophobic, not xenophobic: it was quite willing to buy large quantities of German equipment to the detriment of American entrepreneurs. Only with its support of Federal Telegraph's arc transmitters after 1911 did the navy consistently turn to American suppliers.

This aspect of the navy's role in the development of radio is not evident in Howeth's *History of Communications-Electronics in the United States Navy* (1963). What Aitken describes as Anglophobia, Howeth puts down simply to the refusal of Marconi's American arm to sell its equipment outright. In 1902 the navy considered Marconi equipment the best (45). But field tests that year did not include Marconi equipment, and the navy subsequently chose German Slaby-Arco equipment as its standard equipment, buying 20 sets in early 1903 and another 25 later in the year (52–55). American manufacturers howled. The 1906 report of the secretary of the navy indicated that the navy's subsequent policy would be "to purchase different types of wireless apparatus from the various manufacturers in this country . . . in order to encourage competition. It is believed that this method of procedure, together with the stimulus afforded by prospective commercial profits, has produced a development of the art in this country equal if not superior to that attained abroad" (quoted in Howeth 1963, 106).

The extent to which this statement indicates a promotion of American wireless interests is indicated by the number of sets owned by the navy in 1906. Out of 105 sets 45 were German Slaby-Arco and 6 were Telefunken (Howeth 1963, 106). Yet as Howeth admits, Marconi abandoned its earlier policy in 1906 and subsequently "offered the outright scale of its equipment" (88). The navy never bought Marconi equipment, and Howeth is strident in his condemnation of Marconi, going so far as to make the erroneous statement that "by 1912 their patent position in the United States had become questionable" (147). Howeth concludes that the Atlantic Communication Company, a totally German-owned subsidiary of Telefunken, would, "except for the development of the [continuous wave] Poulsen arc . . . probably have become the Navy's sole source for the procurement of [quenched spark] transmitters following 1912" (148). Marconi may have been willing to sell its equipment to the American navy after 1906, but the navy was certainly not willing to buy.

The third international radiotelegraph conference was convened in London in 1912 in the wake of the *Titanic* disaster. It did little to alter the work of the 1906 conference except express the opinion that "in the general interests of navigation, there should be imposed on certain classes of ship the obligation to carry a radio-telegraphic installation" (Howeth 1963, 165). The practice had already been adopted in America, where in 1910 Congress required wireless telegraphy installations in all ships carrying more than 50 people (Headrick 1991, 121). The principal beneficiary of this legislation was American Marconi, which in the aftermath of the American courts' enforcing the primacy of its patent position in 1911 had bought out the large network of ship and shore installations controlled by United Wireless.

Although it was agreed that a fourth conference would be held in Washington in 1917, World War I caused its postponement until 1927. A fifth conference was held in Madrid in 1932, and a sixth in Cairo 1938. The primary concern of these conferences was not wireless communications but the allocation of frequencies for the new technology of commercial broadcast radio in order to reduce interference between stations. Because of the growth of long-range wireless telephony using the higher frequencies, the 1938 conference did pay attention to the proper allocation of such transmission frequencies and insisted that the use of the noisy, broad-spectrum spark transmitters still common on older merchant ships be significantly reduced (Howeth 1963, 501–12).

Britain did not, however, need foreign manipulation to prevent Marconi from achieving global dominance of international radio communications. Competing interests within Britain were quite capable of doing it on their own. The British Post Office, after flirting with Marconi in 1896 and 1897, became actively hostile to Marconi's as it pursued narrow self-interest rather than what should have been the national interest. The first evidence of this was the Post Office's arranging for Professor Slaby to attend Marconi's Bristol Channel trials in 1897. Slaby quickly copied Marconi's emerging spark technology. The second evidence was the offer by the Post Office of a mere £10,000 for Marconi's patent rights. Marconi's relatives and friends helped him form his own company, capitalized at £100,000. Marconi received £15,000 in cash, a three-year contract as chief engineer at £500 per year, £60,000 of paid-up shares, and working capital to the tune of £25,000 (Aitken 1985b, 224; Douglas 1987, 17). Marconi notified the Post Office of this arrangement when it was in its formative stage, but the Post Office refused to improve its terms.

The bureaucrat who negotiated for the Post Office, and who had been most impressed by Marconi's demonstrations of 1896 and 1897, was William Preece, who had himself been working on the possibility of wireless telegraphy through induction. As late as 1907 Preece was still being scolded by members of the British Parliament on his failure to capture Marconi's technology. The Post Office and, to a lesser extent, Preece thus became Marconi's enemies. After the apparent failure of German interests to force abandonment of the principle of non-intercommunication at the 1903 conference the Post Office stabbed Marconi in the back. As early as March 1904 "Postmaster General the Earl of Stanley recommended to the cabinet that the government license all radiotelegraph stations and bring Great Britain into closer agreement with other governments. . . . [When] the Convention for the Regulation of Wireless Telegraphy [resulting from the 1906 conference] . . . came up for ratification before the House of Commons . . . Postmaster General Sydney Burton . . . supported ratification" (Headrick 1991, 120–21). Ratification, in 1907, was by a single vote. Even so, the straw that broke Marconi's plans for long-range wireless telegraphy before World War I, embodied in his radical proposal to build an imperial chain of spark transmitters, was the Marconi scandal of 1912. The Post Office's part in the scandal certainly is not obvious.

THE IMPERIAL CHAIN

The scheme for an imperial chain grew out of a motion at the imperial conference held in London in June 1911 "that a scheme of wireless telegraphy approved at the Conference held in Melbourne in December, 1909, be extended throughout the Empire, so as to enable the Empire to be to a great extent independent of submarine cables." Herbert Samuel, the Postmaster General, concurred, noting that the British government proposed that "the Post Office and the local administrations in India and the various Dominions" begin by developing six stations. The ones in Britain, Cyprus, and Aden would be paid for by the home government, the one at Bombay by the Indian government, and the one in Western Australia by the Australian government. The costs of the station at Singapore would be shared equally (*Marconigraph,* July 1911, 4). In March 1912 the government signed an agreement with Marconi's to build stations in London, Cairo, Aden, Bangalore, Singapore, and Pretoria (fig. 4.4). The *Electrician* called it "an event of the first importance . . . in fact, the first of many large schemes for the employment of wireless telegraphy on a really comprehensive scale" (68 [15 March 1912]: 905).

With this proposal Marconi's reached the pinnacle of its pre–World War I power. The company's annual report for 1911, published in the *Electrician* in June 1912, emphasized just how much had been achieved in that year (*Electrician* 69 [14 June 1912]: 414–15). Three items of real importance are noted in the report; one is not. The three published items were the acquisitions of the Lodge patent and of virtually complete control of wireless traffic in both Russia and the United States. The unpublished item was a clandestine deal with Telefunken to stay out of each other's markets. The most important item was a single line noting that "the acquisition of the patents of the Lodge-Muirhead Syndicate has further strengthened the company's patent position" (415). Lodge's 1897 patent on syntony had been upheld by the British courts in 1911 as the master patent on tuning and had, as British patent law allowed, been extended for seven years, to 1918, because Lodge had not been able to commercially profit by it. This decision effectively invalidated Marconi's "four sevens" patent of 1900. Marconi finally did the sensible thing and bought Lodge out.

With Lodge's patent Marconi could make a behind-the-scenes deal with Telefunken, the one item not included in the company report. Tele-

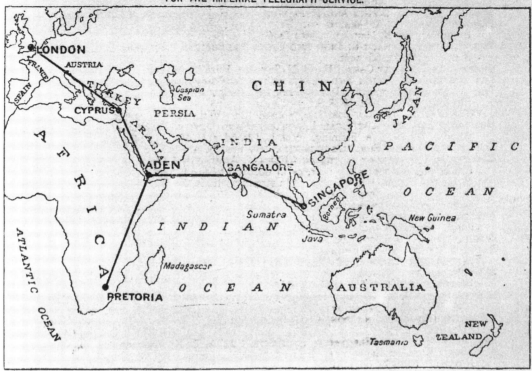

STATIONS TO BE IMMEDIATELY CONSTRUCTED BY MARCONI'S WIRELESS TELEGRAPH CO. LTD., FOR THE IMPERIAL TELEGRAPH SERVICE.

Fig. 4.4. Marconi's Proposed "Imperial Chain," 1912 (from *Marconigraph,* April 1912, 5; photo courtesy of the Library of the Institution of Electrical Engineers)

funken had long accepted that its equipment infringed Lodge's 1897 patent and simply denied the validity of Marconi's "four sevens" patent. As long as Lodge did not seek to enforce his patent rights Telefunken was not concerned. But when Lodge had his patent extended, developed a fighting fund, took on Marconi, and won by Marconi's capitulation, Telefunken had no choice but to make a deal with the patent's new owner. This it did, mostly in secret, in 1911 and 1912, agreeing to end litigation and share patents. Through its Belgian subsidiary Marconi negotiated with Telefunken to set up the Deutsche Betriebgesellschaft für drahtlose Telegraphie (DEBEG) in January 1911. DEBEG "took over the wireless business of the entire German Mercantile Marine, with the resources of the hitherto rival systems for ship-to-shore and ship-to-ship signaling to be pooled and operated on the basis of full intercommunication" (Baker 1972, 132–35). Marconi and its Belgian subsidiary held 45 percent of the shares in DEBEG, Telefunken the remaining 55 percent. A similar agreement was reached with Austrian interests in which Telefunken, Marconi, and a Belgian banking consortium each held a third of the shares. With the German problem resolved, Marconi could thus turn to global expansion.

The second item of importance published in the company report for 1911 was that "in October last the control of the Russian Company of Wireless Telegraphs and Telephones was acquired" (*Electrician* 69 [14

June 1912]: 415). This gave Marconi's control of telecommunications within the Russian Empire and effectively froze the Germans out since the deal with Telefunken had specified Russia as an area of open competition between Marconi's and Telefunken.

The third item concerned the acquisition of the American company United Wireless. Acquisition of the Lodge patent, regarded by American as well as German courts as the master patent on tuning meant that United Wireless could now be taken to court for infringing Marconi's patent rights. As the company report for 1911 put it,

> The greatest progress has been made in the United States of America. For some years past the American Marconi Co. was able to transact but a comparatively small business owing to severe competition by the United Wireless Telegraph Co. Proceedings at law were commenced against that company . . . [which] admitted infringement. . . . The United Co.'s business is now being directed by the Marconi Wireless Co. of America. . . . The satisfactory ending of the litigation should convert the business of the American company into one of considerable importance controlling all the coast stations of importance on both the east and west coasts, besides practically the whole of the American Mercantile Marine at present fitted with wireless telegraphic installations. Considerable impetus to ship-to-shore business in the United States is to be expected from the extension of the scope of the existing law rendering wireless telegraphy compulsory upon practically all vessels. . . . A further agreement of great importance was entered into with the Western Union and the Great North-Western Telegraph Companies. This agreement furnishes the Marconi Co. with some 25,000 telegraph stations for the delivery and despatch of Marconigrams throughout the United States and Canada. . . . The construction of stations, placing this country in direct communication with New York (instead of, as at present, passing though the station at Glace Bay [Nova Scotia]), will be carried out as expeditiously as possible, as will the construction of stations at San Francisco communicating through the Hawaiian Islands to the Philippines, China and Japan, and from New York south to Cuba, Panama and subsequently with each of the South American States. (*Electrician* 69 [14 June 1912]: 415)

In 1913 the managing director of Marconi's, Godfrey Isaacs, elaborated on the decision to expand American Marconi. He noted that in April 1911 he and Marconi had traveled to America and come to an agreement to buy out the shareholders of United Wireless to prevent an expensive Supreme Court action. The tangible assets of United Wireless

> covered practically the whole of the wireless business of America. They had practically 500 installations on board ships and 72 telegraph stations all round the coast of America . . . there was a possibility of very big business in America providing [Marconi's] company had money. . . . But to create a long-distance telegraph service to all parts of the world without the means of collecting and distributing messages in America would be like playing *Hamlet* without the Prince of Denmark. He determined that there were two courses open—either to arrange with a company like the

Western Union Telegraph and Cable Company . . . -[or] to compete with the Western Union by constructing wireless stations all through America . . . He preferred arrangement to competition. (*WW*1 [April 1913]: 130)

The directors of American Marconi agreed to issue new stock to cover the $1.4 million cost of the tangible assets of United Wireless. They also agreed to a further issue of up to $7 million, part of which would be used to construct "high-power stations to communicate with the different parts of the world," the rest if it became necessary to go into direct competition with Western Union on its home ground. Armed with this agreement, Marconi was able to persuade Western Union "that it would be wise to enter into an arrangement, and after long negotiations they came to terms" (131).

In early 1912, therefore, Marconi was on the verge of total domination of global radio communications. A deal had been made with Telefunken, the vast potential market of Russia was under control, and the Americans were brought to heel by the acquisition of United Wireless by American Marconi and Marconi's deal with Western Union. The construction of the stations proposed in the 1911 report, plus those of the proposed imperial chain, would have given Marconi access to the whole world. An article in the October 1912 issue of *Marconigraph* entitled "A Wireless Girdle Round the Earth," by the chief engineer of American Marconi, makes this quite clear (see fig. 4.5). The proposal was to route signals across the Pacific from Singapore via Manila and Hawaii to San Francisco. Australia and New Zealand were generally absent from the first proposals for an imperial chain because their governments were negotiating with Telefunken through its Australian subsidiary, the Australian Wireless Company. Marconi successfully sued both the Australian Wireless Company and the Australian government for the right to inspect their equipment for patent infringement. In the event, the deal with Telefunken settled the issue, and "a new Australian Company . . . formed which would purchase the interests of the Marconi Company and the *Telefunken* Company. . . . Thus, complete control of wireless in Australia passed into Australian and British hands" (Baker 1972, 134). The American route made available by the acquisition of United Wireless was more direct and presumably, since it served the American market, more profitable.

Fig. 4.5. Marconi's Proposed "Wireless Girdle Round the Earth," 1912 (from *Marconigraph,* October 1912, 255; photo courtesy of the Library of the Institution of Electrical Engineers)

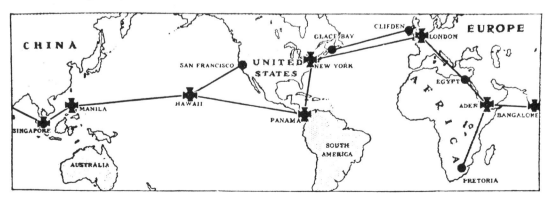

Work on Marconi's stations in the San Francisco area and Honolulu proceeded apace. By December 1913 the transmitting plant at Bolinas Bay, some 18 mi northwest of San Francisco, was nearing completion despite the problems of getting structural steel, wire rope, and the like by sea from the east coast of America, across the Isthmus of Panama by rail, and by sea again to Bolinas Bay (*WW* 1 [December 1913]: 556). Given the lack of any existing settlement and the problems of constructing one, the Bolinas Point site was obviously chosen for its ability to generate a substantial amount of hydroelectric power. The article in *Wireless World* shows a large power house in the course of construction. The receiving station, which required far less power, was located a further 18 mi north, in the town of Marshall on Tomales Bay, a site accessible both by sea and by a narrow-gauge rail line from Sausalito (558).

DERAILING GEOPOLITICAL INTERESTS: THE MARCONI SCANDAL

The stage on which the Marconi scandal was acted out was thus one on which Marconi had achieved an effective near-global monopoly outside the British Empire. The scandal ruined both Marconi's and Britain's chances of regaining effective global dominance of telecommunications before World War I and opened the door for America's rise. The British Post Office threw the first stone, and Sir Henry Norman outlined the two main charges: "On Friday last . . . the Postmaster General [Herbert Samuel] moved . . . 'that a select committee be appointed to investigate the . . . establishment of a chain of Imperial Wireless Stations.' . . . Sir H. Norman said what was known as 'the Marconi agreement' had been criticized on the grounds that it was a bad and imprudent bargain and was a bargain tainted with corruption" (*Electrician* 70 [18 October 1912]: 57).

It was the second of these charges, that of corruption, or what in modern terms would be referred to as insider trading, that led to the scandal. The charge was leveled against the managing director, Godfrey Isaacs; his brother, Attorney General Sir Rufus Isaacs; and several Cabinet members, including the Chancellor of the Exchequer, David Lloyd George. A great deal, amounting eventually to a vicious anti-Semitic attack, was made in the more scurrilous political press of the fact that the Isaacs brothers and Herbert Samuel were also Jews.

The first charge, that it was a "bad and imprudent bargain," clearly arose from second thoughts by the Post Office and perhaps from sour grapes over previous dealings with Marconi's when the Post Office had felt outmaneuvered. As indicated in the parliamentary debate, the agreement that Parliament was being asked to approve had already been signed by the Postmaster General. In Sir Henry Norman's opinion, "The Postmaster General deliberately and to the public disadvantage contracted himself out of . . . the rights to use any invention for the service of the Crown on terms to be agreed upon, or in default of agreement on terms to be prescribed by the Treasury." Norman was referring to the seemingly generous terms of the agreement, which was to pay Marconi's a royalty of 10 percent on the gross receipts from the imperial chain for 28 years. Patents ran for only 14 years, and although Norman did not say so in debate, Marconi's "four sevens" patent was due to expire in 1914 and the seven-year extension of Lodge's patent, now held by Marconi, would expire in 1918. Norman did, however, refer to the provisions of the "unanimous resolution of the Imperial [Defence] Conference to which the Government was supposed to be giving effect . . . 'that a chain of British State-owned wireless stations should be established within the Empire.'" The operative phrase was, of course, "State-owned." The Marconi agreement was a partnership between the state and a private company in which, in Norman's words, "the state would thus forego even the advantage of saving middlemen's profits" (*Electrician* 70 [18 October 1912]: 57).

This rather narrow, penny-pinching point of view was reinforced by a suspicion that Marconi's spark technology was rapidly becoming outdated by the emergence of continuous wave arc technology. This suspicion surfaced in such comments, again by Norman, as, "The ratification of the agreement would inevitably result in the formation of a great monopolistic trust which would restrict the development of wireless telegraphy, double its cost, and set back independent scientific progress for a generation. The case for reconsideration from the scientific side was at least equally strong as that from the commercial side" (*Electrician* 70 [18 October 1912]: 58).

Two replies to these criticisms are worth quoting. First, "the Attorney-General (Sir Rufus Isaacs) intervened to explain his slight connection with the subject matter of the agreement and strongly repudiated the insinuation that he (as the brother of Mr. Godfrey Isaacs, managing director of the Marconi Company) had used his influence to get the contract for the company, or that he had dealt in the company's shares" (*Electrician* 70 [18 October 1912]: 58). Second, Herbert Samuel, the Postmaster General, similarly denied any financial involvement and strongly defended his responsibility for the contract on the grounds that only Marconi had the proven technology to create an imperial chain of stations at 2,000 mi intervals. He referred directly to the unstated claim that continuous wave technology using arc transmitters might be superior. "His experts had advised him that they could not be certain that the Poulsen Syndicate could cover the distance required by the contract. It was true that . . . [the] Syndicate had sent messages 2,000 miles [from San Francisco to Honolulu] by night, but they made no claim to do that by day. The chain of wireless stations must work by day and by night if for no other reason than on the British Empire the sun never set" (59). Although the Poulsen Syndicate, by which Samuel really meant Elwell's Federal Telegraph Company out of Palo Alto, had made a respectable tender for the imperial chain, "the Committee decided that it was impossible to accept the Poulsen offer without a test. [Samuel] found that a complete test could not take place for 12 months, and he had, therefore, accepted the Marconi tender" (59).

In the runup to World War I, when the needs of imperial communications and defense were clearly paramount, Samuel's decision made sense.

For geostrategic reasons an alternative telecommunications system to the submarine cables was needed immediately, and only Marconi could deliver it. Federal Telegraph had a plausible technology, but it had not yet demonstrated it, and in 1912 at least America's position in any future war between the European powers was uncertain. Pro-German sentiment in America was strong, and the American navy had not helped by its obvious pro-German, anti-British bias in acquiring wireless communications technology. Federal was, in any case, a ward of the American navy.

Thus the scandal was not a result of commercial, technological, or geostrategic considerations except inasmuch as the last of these impinged on the growing geopolitical competition between Britain and America. In reality the scandal flew in the face of all three considerations. It erupted in part because Isaacs and Samuel were deliberately evasive in their use of the term Marconi Company. As their lawyer later admitted, they were not denying trading in American Marconi stock, only in shares in "the British, Canadian, or Spanish Marconi Companies" (Baker 1972, 145). Isaacs' published defense was that he had traded openly in American Marconi shares at open-market prices (*WW* 1 [May 1913]: 128–29). The two men's unstated defense was that since such shares had been issued largely to recapitalize the domestic operations of the United Wireless Company, they had not believed that there was any conflict of interest. "The charge of grave and improbable corruption . . . [was] based on the following facts: the managing director of the Marconi Company was Mr. Godfrey Isaacs, while the Attorney-General in the Liberal Government was his brother, Sir Rufus Isaacs. The Postmaster-General was Mr. Herbert Samuel and (an extraordinary weight was put on this) all three were Jews" (Donaldson 1962, 20).

The unstated anti-Semitism behind the charges was a nasty side issue, although it was to result in a successful action for libel by Godfrey Isaacs against Cecil Chesterton and his muckraking *Eye Witness* magazine in June 1913 (Dunlap 1941, 215). In the aftermath of the court decision the London *Daily News* described the motives behind the scandal as "anti-Semitism and antagonism to the government" (216). But when it came to insider trading in the shares of American Marconi, Sir Rufus Isaacs does seem to have been open to exactly that charge. The unstated defense, that the shares were domestic American shares and that the imperial chain was not related to them, is disingenuous at best. Godfrey Isaacs was well aware that the share issue was partly designed to finance long-range stations to link American Marconi to the rest of the world (*WW* 1 [May 1913]: 131). The October 1912 issue of *Marconigraph* was unequivocal evidence of the centrality of the American route across the Pacific to such plans for "A Wireless Girdle Round the Earth."

Upon the successful conclusion of the patent infringement suit against United Wireless in 1912 the stage was set for a considerable expansion of the business of American Marconi. Godfrey Isaacs took personal responsibility for placing 500,000 of the new shares issued to recapitalize American Marconi. Because the American stock market was awash with worthless, usually watered stocks in wireless communications companies, the shares could not be sold in New York. Half of the shares

Isaacs therefore sold on immediately to the London brokerage firm of Heybourn and Croft, and of the remainder the American directors took 150,000, Marconi himself held 10,000, and British Marconi directors and employees took 31,500. Godfrey Isaacs offered his two brothers some of these American shares: Harry, a London stockbroker, took 50,000; Sir Rufus, the attorney general, initially declined but was persuaded a few days later to take 10,000 of the shares bought by his brother Harry. Sir Rufus then sold 2,000 of these to two of his colleagues in the government: 1,000 each to David Lloyd George, then Chancellor of the Exchequer, and Lord Murray, then the chief whip in the House of Lords (Dunlap 1941, 205).

After heated debate for nearly a year the House of Commons Select Committee "was unable to find unanimity, but in May 1913 all agreed in acquitting the Ministers involved of all corruption charges" (Dunlap 1941, 211). Lloyd George admitted, "I acted thoughtlessly, carelessly, mistakenly, but I acted openly, innocently, honestly" (quoted on 213), adding that there was a vast difference between indiscreet private investment and corrupt behavior. The separate investigation of Lord Murray by the House of Lords expressed the opinion that "those who held public office should on no account speculate in stocks and shares" (Baker 1972, 147). A revised contract was duly drawn up, signed, and ratified by the House of Commons on 7 August 1913 (148). When war broke out a year later, the contract was canceled without any stations having been completed. Although the whole issue seems relatively minor in retrospect, the imperial chain was delayed a full year, thus it was not finished before war broke out. Had Marconi's begun work in 1912, the Egyptian and Indian stations might well have been in service by the outbreak of World War I.

Long-distance service across the Pacific also expanded in this period. American Marconi began construction of a San Francisco–Honolulu service in 1913 and opened it a year later. The service was extended to Japan by mid-1915 "at the instigation of the Japanese Government" (Baker 1972, 155). The possible geostrategic value of such service was emphasized when the German surface raider *Emden* disrupted submarine cable services in the Pacific by partially destroying the repeater installation on Fanning Island. It was, however, the German experience itself that spoke most convincingly for wireless as opposed to submarine cables. As soon as World War I broke out, British cable ships cut all the German transatlantic cables within days and hauled the cut ends ashore in Britain so that the British submarine cable network could be expanded. This meant that Telefunken's powerful station at Nauen, near Berlin, became Germany's "sole means of telegraphic communication with the outer world. This station, at the time the most powerful transmitting station in the world, at once opened up traffic with another new German giant at Kamina, in Togoland, with Windhoek in S. W. Africa and with German stations which had been built in the U.S.A., pouring out propaganda on a 24-hour basis" (159).

Propaganda some of it may have been, but it allowed the Germans to at least somewhat counter Britain and France's otherwise total dominance of the information flow from Europe to America. In reality, however, few of these radio stations lasted very long. In August 1914 Telefunken's stations at Dar-es-Salaam, at Samoa, and on Yap in the Caroline Islands were destroyed, the Germans destroyed the station at Kamina to prevent its being taken over by a rapidly approaching British force. In September the Australians took the German radio installations on the Marshall Islands and New Pommern. In November the Japanese took the Kiauchau station. "The last powerful station of what had been intended to be the German Imperial Chain, Windhoek, . . . fell on 12 May 1915, to a military force sent by the Union of South Africa" (Baker 1972, 160).

Because the British had no imperial chain, the British navy quickly contracted with Marconi to install and man 13 long-range stations—at Ascension Island, the Falklands Islands, Bathurst, Ceylon, Durban, Demarara, the Seychelles, Singapore, St. John's, Aden, Hong Kong, Mauritius, and Port Nolloth. All but the Falklands installation were operating by mid-1915. Because of severe supply difficulties the Falklands installation took until late 1916 (Baker 1972, 161).

The Marconi scandal is unusually fascinating since it is not at all clear who benefited by it; it is only clear who did not. Several possibilities exist, one of which was broached after World War I, two by Godfrey's libel suit against Chesterton, and one at the select committee inquiry before the war. After the war it was all too easy to blame anything on the Germans, and the temptation was not resisted. Documents retrieved from Germany after the war show that Sir Henry Norman had had considerable dealings with Telefunken in early 1914 (*Electrician*, 5 March 1920, 279–80); however, they do not implicate him in earlier dealings. In any case, the agreement between Marconi and Telefunken seems to have been amicable. Marconi had profitable investments in DEBEG's marine business and in the Austrian subsidiary. The geostrategic benefit to Germany of British failure to complete the imperial chain would certainly have been tangible if German submarines had been able to cut British submarine cables, but they were never able to do so.

As the London *Daily News* put it, Godfrey Isaacs' successful libel suit against Cecil Chesterton raised the issues of anti-Semitism and antagonism to the government. Both are almost certainly true charges, but Jews were well embedded and generally accepted in British public as well as commercial life by the early 1900s, as the presence of both Isaacs and Samuels in senior government posts clearly demonstrates. And after all, Isaacs easily won his lawsuit. Anti-Semitism looks like a cover, not a cause. The milder charge of British xenophobia against Marconi's Italian name hardly bears up. Marconi's English, upbringing, manners, and social connections were by all accounts impeccable (Douglas 1987, 16). His mother was Scots-Irish, at the time he had an Irish wife, and King George knighted him in 1914. Antagonism to the government is entirely plausible, but whoever was responsible for the Marconi scandal would not be the first or last political actor to stoop to dirty tricks. Since no government fell, it is hard to ascertain what benefits the scandal might have conferred on whom.

The fourth possibility seems the most likely to contain a major element of truth. In his appearance before the select committee Godfrey Isaacs, managing director of Marconi, advanced the notion that the scandal was caused by the London supporters of the Poulsen Syndicate, which had made a rival bid for the imperial chain. Isaacs' published testimony is of interest because the timing implies that it was unlikely that any German interests were involved: "Directly I returned from America [in April] I was informed that there was a very strong attack to be made on the Marconi contract with the Government. It was talked about very freely. I was also told there was a very strong syndicate which was promoting or had been endeavouring to promote the Poulsen Company, and that they were strongly supported by influential persons, among whom were members of Parliament, who would assist them in preventing the Marconi contract being carried through" (*WW*1 [May 1913]: 135).

Godfrey noted that he had paid little attention to these rumors at the time but that he had heard in July that "those who were connected with this syndicate were arranging to make an attack on the ministers to ensure that the contract would not go through Parliament" (*WW*1 [May 1913]: 135). This implies that the Marconi scandal was a purely commercial move on the part of the Poulsen Syndicate. That may have been the case, but the Poulsen Syndicate had all the business it could handle from the American navy. In fact, the American navy benefited from the scandal as much as the Germans did in terms of geostrategy. Further, it was to the geopolitical advantage of America that Marconi's grandiose plans to link America into a Marconi-run "girdle round the earth" come to naught because of the failure to develop the British imperial chain before World War I. By the time plans for the imperial chain were revitalized in the aftermath of the war (*WW*8 [17 April 1920]: 41–43), Federal Telegraph and the American navy had created their own chain of continuous wave Poulsen Arc transmitters around the world (fig. 4.6).

The late nineteenth and early twentieth centuries were a critical period in the world-system. British hegemony was waning in a period of multipolarity, albeit far more in the economic than in the military sphere. Aaron Friedberg (1988) describes Britain in this period as "the weary titan." In Paul Kennedy's reading (1987) Britain was guilty of "Imperial

Fig. 4.6. Poulsen Arc Wireless Services of the United States Navy, 1919 (from *Electrician* 84 [28 May 1920]: 598)

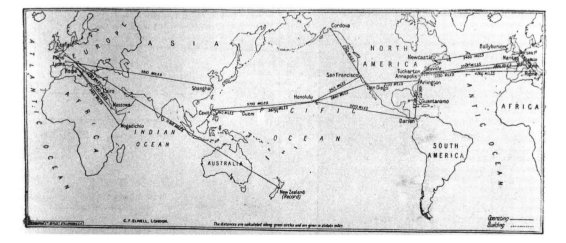

overreach," investing heavily in its military to hang onto its world power but about to find the expense impossible to maintain. Two powerful challengers, would-be hegemons—Germany and the United States—were clearly waiting for the crucial stumble. Yet the overreaching and weary titan did not collapse even under the crushing expense of World War I. Imperial Germany, it is true, was crushed militarily and economically by the war. America, on the other hand, emerged as by far the world's dominant economy, in the process replacing Britain as the world's greatest creditor nation.

It is clear from its actions during the first 21 years of this century that the U.S. Navy believed it could contribute to achieving American hegemony in two main ways. The first and most obvious of these was by the development and management of the era's most geopolitically significant weapon system, the battle fleet. The second was by the relatively clandestine development of long-range wireless as an alternative to the British submarine cable system. Before and during World War I the navy was relatively easily able to fund Federal Telegraph by means of what today would be called a "black budget." In the glare of postwar public scrutiny that was no longer an option. Equally, the war had suspended the normal commercial rights of patent holders, who were forced to allow massive infringement in the interest of the national good. In 1919 normal market forces reasserted themselves, and the navy's only solution was to persuade all the relevant holders of radio patents into the patent pool of RCA, best described at its outset as a fundamentally geopolitical entity. Although RCA seemed a sure-fire success to begin with, a combination of political and technological problems intervened. The corporation thus developed into a much more conventional, profit-oriented company attuned increasingly to the provision of entertainment to the American public. It is to this American failure to develop low-frequency wireless as a viable alternative to British submarine cables and the success of British attempts to develop high-frequency wireless that I turn now.

CHAPTER FIVE

Challenges to British Telecommunications Hegemony
Continuous Wave Wireless

"Now you're the very man," she said,
"Who'll tell me what I want to know.
Who was it laid these silly beams
To places where the Cables go?"

—*Zodiac* 20 [March 1928] 258

CONTINUOUS WAVE TECHNOLOGY

Continuous wave wireless was obviously preferable in theory to the oscillating waves produced by spark technology, however thoroughly these were damped or quenched. Continuous waves offered two presumed advantages. First, they would be more amenable to accurate tuning; thus the amount of power needed to send a signal would be reduced. Second, they offered the enticing possibility of wireless telephony. Given the technology of the first decade of this century, indeed of all telecommunications technology until the 1930s, wired transoceanic telephony seemed impossible. Attempts to send telephone signals through submarine telegraph cables failed dismally because of their low bandwidth. Wireless telephony alone promised to conquer the ocean gaps.

Considerable thought was given to wireless telephony before the end of the period of spark technology. This thought is well summarized in Ernst Ruhmer's *Wireless Telephony in Theory and Practice*, published in Berlin in 1907 and quickly translated into English. Ruhmer effectively dismissed even the most refined spark technology on the basis of his experiments in 1904–5. Although it was possible to generate sparks rapidly enough to transmit speech, "the discharges do not follow one another with anything like uniformity . . . transmitted speech . . . was rough and broken like that of a stammerer" (Ruhmer 1908, 118).

Ruhmer concluded that only the uniform oscillation of continuous wave technology could offer the possibility of wireless telephony (fig. 5.1). Continuous radio waves can be generated in three ways. Two of these—the arc and the alternator—were discussed by Ruhmer, and the third—the triode vacuum tube—was being discovered at about the time he was writing. Vacuum tube technology developed apace because of the demands of World War I and immediately thereafter because of the emergence of the broadcast entertainment industry. Vacuum tubes made possible a move to much higher frequencies. Together with tightly focused

Fig. 5.1. Damped Oscillations from Spark Transmission and Uniform Waves from Continuous Waves (redrawn from Ruhmer 1908, 130)

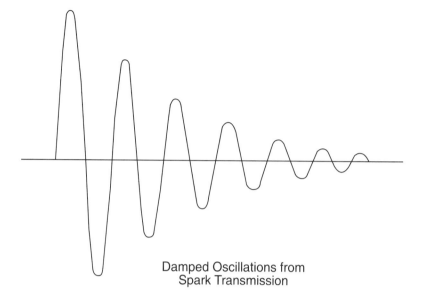

Damped Oscillations from
Spark Transmission

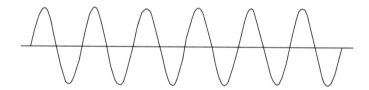

Uniform Oscillations from
Continuous Waves

beam antennas, it drastically lowered the costs of wireless telegraphy to the point where the submarine cables were challenged. It made possible a very low capacity system of wireless telephony around the globe.

ARC TECHNOLOGY

Very high rates of sparking generated oscillations that approached continuous waves. This had been the basis of Ruhmer's experiments of 1904–5. It was a simple leap to conclude that the arc represented a continuous spark and would thus generate continuous waves. This simple idea was not, however, so easily translated into practice until the Danish engineer, Valdemar Poulsen, found that placing the arc in hydrogen rather than air produced the appropriate oscillations (Ruhmer 1908, 154). Poulsen's subsequent 1902 patent provided the basis for the first continuous wave generator, although most of the technical development was done in America by C. F. Elwell and the Federal Telegraph Company. In the event, arc technology was virtually never used for wireless telephony, although it became an important element in wireless telegraphy in the teens.

Elwell was a student in electrical engineering at Stanford University. Asked by a group of local bankers to evaluate their purchase of a patent for wireless telephony that used damped wave sparks, he soon recognized its uselessness. Elwell also recognized that on the rare occasions when he got usable results the spark gap was so small that he really had a small arc discharge (Mann 1946, 378). He became familiar with Poulsen's work through the scientific literature, traveled to Europe in 1909, and returned with an option on Poulsen's technology. Elwell first tried to raise money on Wall Street to finance his development of Poulsen's technology. Unsuccessful, he went back to Denmark to renegotiate his option, returned to California, and demonstrated wireless telephony to the president of Stanford University and several prominent faculty members and citizens of Palo Alto. In 1909 they financed the Poulsen Wireless Telephone and Telegraph Company, split in 1911 into the Poulsen Wireless Company (the holding company) and the Federal Telegraph Company (the operating company) (380, 386–87).

Elwell's first wireless telephone transmissions, between Stockton and Sacramento in early 1910, were interrupted by deliberate jamming by the large United Wireless transmitter in Sacramento, but he found that arc telegraph signals were not so interfered with. With the completion of a third station in San Francisco in mid-1910, Federal Telegraph showed that continuous wave stations offered very accurate tuning. Even when all three of its own stations were transmitting simultaneously, selective reception was possible (Mann 1946, 386). This was not possible using even the best spark transmitters, any one of which obliterated the transmissions of less powerful or more distant units. By 1912 Federal had developed a substantial wireless telegraphy network to compete with the wired system (fig. 5.2). The most spectacular achievement of 1912 was to connect San Francisco to Honolulu, 2,400 mi across the Pacific, albeit only at night because of atmospheric interference by day. Federal severely undercut the submarine cable rates, charging only 25¢ a word to the latter's 35¢. The press rate fell from 16¢ a word to 2¢, and the Honolulu newspaper contracted for a minimum of 1,500 words a day. Before 1912 it had received only 120 words a day (389).

The arc's spectacular success on the Honolulu–San Francisco link caused the U.S. Navy, which had long been seeking an effective American-controlled technology, to install an arc transmitter next to its state-of-the-art spark transmitter at Arlington, Virginia. By night the navy could thus reach the geostrategically vital naval base in Hawaii, signaling 4,500 mi with one repeat at San Francisco. The navy quickly became Federal Telegraph's biggest customer and the first major military user of continuous wave technology. In 1913 the navy ordered a transmitter for Darien, in the Panama Canal Zone, and ten ship-based transmitters. Although the arc required a more skilled operator than a spark transmitter did, it was rugged, compact, reliable, and quiet. Ships had plenty of power to generate electricity for the arcs, and their compactness and quiet was appreciated on board ship. Spark transmitters were notoriously

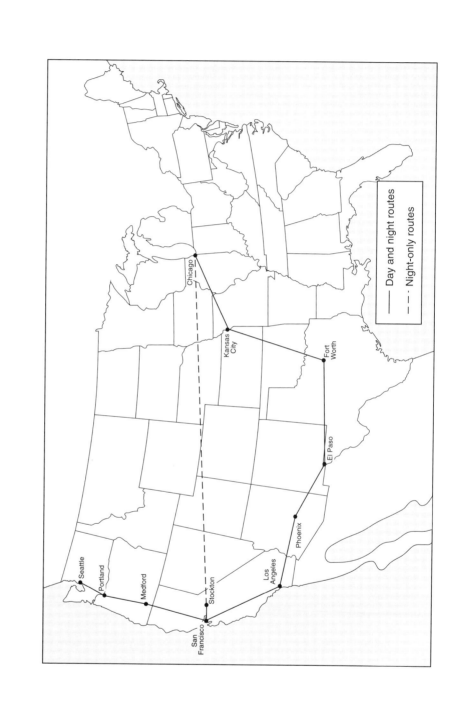

Seattle
Portland
Medford
San Francisco
Stockton
Los Angeles
Phoenix
El Paso
Fort Worth
Kansas City
Chicago

——— Day and night routes
– – – Night-only routes

noisy, requiring soundproof rooms. By 1917 the navy had completed not only the Darien station (in 1915) but also stations at San Diego, Pearl Harbor, and Cavite, in the Philippines. The navy thus had complete control over communications from Arlington across the Pacific (Howeth 1963, 221–24).

Despite the considerable technical successes of arc technology in the first decade and a half of this century, two nontechnical items mitigated against it. One was its development on the American West Coast; although the area around San Francisco was known for its innovativeness, it was far from the acknowledged centers of innovation, manufacturing, and finance in the East. Although Elwell's failure to raise money on Wall Street in 1909 can in part be put down to too much "stock jobbing in damped-wave or spark wireless systems" (Mann 1946, 380), it was probably in part also because Elwell was from California. In a way this eerily predates the amazing reluctance of the American business community to accept other radically new technology from the region. The first really effective and user-friendly personal computer, the Apple, was pioneered by two members of the counterculture, Steve Jobs and Steve Wozniak, working out of garages around Stanford University as the Homebrew Club. Yet the American business community ignored them and their company and accepted the personal computer only when it was presented to them by IBM in a radically inferior and far less user-friendly form (I am indebted to Bob Bednarz for this insight).

A second problem for Elwell was the acquisition of the navy as a major customer. Elwell's improvements on Poulsen's technology became quickly shrouded in secrecy. The American navy had no interest in commercial sales, and Federal Telegraph's improvements on Poulsen's arc went largely unreported in the contemporary engineering literature.

Even though it had published a map of the Poulsen arc services of the U.S. Navy in 1920 (see fig. 4.6), as late as August 1921 the *Electrician* referred to "recent improvements [by Elwell] in the continuous wave system to employ the Poulsen arc as a generator" and to such improvements as an "experiment" (87 [26 August 1921]: 254). As Marconi expanded into continuous wave technology in the late teens, he concentrated on the use of the alternator, although he eventually bought Poulsen's patent to cover his bases. But by 1921 the arc had been a proven technology for more than a decade and was, despite its advantages in cost, ruggedness, and relative ease of use, on the point of obsolescence.

ALTERNATOR TECHNOLOGY

That alternators generated continuous electromagnetic waves was known to every electrician at the turn of the century. The problem was generating them at a high enough frequency for radio transmission. Higher frequencies meant higher speeds, so that the problem of building a radio-frequency alternator became one of mechanical as well as electrical engineering. Ruhmer noted that 15 experimental radio-frequency alternators were built or patented between 1887 and 1905 (1908,

Fig. 5.2. Federal Telegraph's Continuous Wave Wireless Telegraphy Network, 1912 (derived from data in Mann 1946) *(opposite)*

132–34). None achieved the required high frequency and high power output for long-distance wireless telecommunication. Two kilowatts of power output was possible at a frequency of 10 kHz, but at the frequencies of 50 or more kHz needed for effective radio use, power outputs were only a few watts.

A decade later the problems seemed to have been solved, and three basic technologies had emerged to allow frequencies on the order of 50 kHz, wavelengths of around 6,000 m, and power outputs in the 100 kW range (Goldsmith 1918, 104). Both the Goldschmidt and Arco-Telefunken alternators developed in the early teens in Germany used electronic frequency multipliers to generate high enough frequencies from relatively low rotational speeds. The Alexanderson alternator, developed in America by General Electric (GE), used mechanical-engineering techniques of the very highest order to obtain very high rotational speeds and generate the desired frequencies directly. By any standards the Alexanderson technology was a remarkable achievement, so remarkable that it may have blinded its possessors to alternatives. Much more has been written about the Alexanderson than about other alternators. New interest in the history of radio has, however, suggested that the Goldschmidt design, however clever, had reached its technical limit in the 120 kW installation at Tuckerton, New Jersey, before World War I (Freris 1995).

The 200 kW Alexanderson alternator in the New Brunswick, New Jersey, station, built by GE for American Marconi but taken over by the navy when America entered World War I, ran at 2,170 rpm to generate a wavelength of 13,600 m. In order to maintain the set wavelength, the speed of rotation had to be held "as nearly absolutely constant as . . . possible" (Alexanderson 1984, 629). The fundamental problem with the Goldschmidt and Arco-Telefunken alternators was that in multiplying the frequencies they generated, they multiplied irregularities and made tuning difficult (Aitken 1985a, 326). Of the three alternator systems the Alexanderson-GE technology was clearly the best, and by 1917 American Marconi was negotiating to obtain it in large numbers (318–19).

The problems with the alternator were several and ultimately fatal. It was extremely expensive; it could only be installed in large, land-based stations; it needed lots of electrical energy; and it could only operate at the lower end of the radio-frequency spectrum, which limited the number of stations that could use it. In the late teens, however, there seemed to be no reason why the alternator would not be satisfactory. Only with the rediscovery of the advantages of much shorter wave transmission were much higher frequencies needed than those provided by the alternators then available. Higher frequencies than those obtained by GE meant higher rotational speeds, and mechanical engineering, materials technology, bearings, and lubricants were not up to such a challenge in the late teens.

Alexanderson himself recognized the fundamental problem of using low-frequency, long-wavelength transmissions. As he put it in 1918,

Experience has shown that the wave lengths which are most suited for transoceanic communication lie between 12,000 and 17,000 meters [ground waves]. This "space in the ether" has already been taken up by five first class transmitting stations which, during the war period and up to the present time, have been in continuous service for transatlantic communication. Of these stations, two are in the United States, one in England, one in France and one in Germany. By extending the range of wave lengths down to 10,000 and up to 20,000 meters, and following the same system of intervals there would be room for about seven more stations or a total of twelve first class transmitting stations. (1984, 626–27)

The need for vast amounts of power to drive alternators meant that they could only be located close to existing or potential sources of fossil fuel or falling water. The technology of power generation in the first part of this century did not allow for profligate use of electric energy. Alternators were useless in mobile applications.

VACUUM TUBE TECHNOLOGY

As stated in the brief account of the development of vacuum tube technology in chapter 3, Bell Telephone's interest in a repeater led it to develop the vacuum tube into an efficient amplifier by 1915 so that Bell could install the first transcontinental telephone links. In part, of course, it was Bell's problems in producing a long-distance telephone system, as well as the promise of immense profits,that persuaded so many people that wireless telephony was a sensible area to explore and invest in. Many respectable engineers in the first decade of this century saw wireless telephony as a substitute for wired telephony even at the household level (Douglas 1987, 157).

By 1907 Bell was in the throes of major corporate restructuring focused on the problems of refining and extending its problem-plagued long-distance system (Douglas 1987, 159). Attempts by such pioneers of wireless telephony as Reginald Fessenden to sell their technology to Bell floundered. In its search for better long-distance service, however, Bell created the very technology Fessenden dreamed of. Bell acquired the patent rights to Lee De Forest's audion, a simple diode in a gas-filled tube derived from Alexander Fleming's 1904 designs. Fleming was a scientific adviser to Marconi's. De Forest, like Fleming, initially saw the diode as an efficient detector of wireless signals, but he radically improved on the design by adding a third element. With this improvement he was able to patent the triode in early 1907 (169–70). What Bell recognized was that De Forest's audion was also a weak amplifier worthy of development as the repeater it so desperately needed to inaugurate its transcontinental telephone service. Much modified, improved, and evacuated, De Forest's triode became, in the form of the vacuum triode, a powerful and reliable amplifier. By 1915, Bell was able to install the first transcontinental telephone link using a chain of such amplifiers.

Ironically, however, triodes were also ideal oscillators and thus radiofrequency generators as well as detectors and amplifiers. De Forest had noted this oscillation as a weak property of his gas-filled triodes as early as 1912 while working for Federal Telegraph (Maclaurin and Harman 1949, 77; Mann 1946, 391–92). As oscillators, vacuum triodes laid the foundation for the first really practical system of wireless telephony. The first attempt at transatlantic wireless telephony, albeit one-way, was carried out by Bell Telephone and the U.S. Navy in 1915. The complex and troublesome transmitter at Arlington used up to 500 vacuum tubes to generate a high enough power output, but even so, only 2 or 3 kW were available at the antenna (Espenscheid 1937, 1107). The receiver in Paris was installed in the Eiffel Tower. It was barely successful because it was nearly impossible to get that many tubes coupled together, but it was also the wave of the future. High-vacuum tubes, called "hard" tubes in the mid-teens, had replaced gas-filled and low-vacuum, or "soft," tubes by the end of World War I. To begin with, individual "hard" tubes did not amplify as much as "soft" tubes did, but "hard" tubes could be coupled together with much more ease and were both easier to use and far more durable. Outputs rose quickly, and by the early 1920s tubes equaled and then surpassed in output even the most powerful arcs and alternators.

Yet this was not the real significance of vacuum tube technology. Its real significance was that it allowed the easy generation of very much higher frequencies and thus much shorter wavelengths. These made possible very long distance transmission and much more directional antennas than had previously been possible. The search for longer wavelengths, larger antennas, and higher-power outputs was suddenly stopped and reversed. Now that higher-frequency continuous waves could be easily generated, long-distance wireless telephony shifted abruptly in the mid-1920s into the shorter wavelengths and the use of small and very directional antennas, called "beam antennas" at the time. Outputs measured in hundreds of watts could achieve distances previously only dreamed of even when hundreds of kilowatts were available. There were two further advantages: first, far more wavelengths were available at the higher frequencies, so that a vast number of stations could operate simultaneously and Alexanderson's fears in 1918 that only 12 first-class stations were possible were bypassed: second, much lower-power outputs were needed. Low power meant compactness, cheapness, and thus the possibility of geographic ubiquity. A station powered by a simple diesel generator could be placed almost anywhere. Low-frequency transmitters needed so much power that they had to be located on top of a major source of falling water or hydrocarbons for generating electricity. World War I did much to force the technical development of the diesel engine as an electric generator to power German U-boats.

MILITARY USE OF CONTINUOUS WAVE WIRELESS

Much of the development of continuous wave radio and vacuum tube technology, as well as the move to the shorter wave lengths is tied to military developments of World War I, covered in detail in chapter 6. The most significant of these military developments was the demand for a return to more mobile warfare. Wireless telegraphy was part of World War I from the beginning, but spark sets were clumsy. Where they were most needed, in airplanes spotting for artillery, they were hardest to use. Vacuum tube sets promised to be easier to use and more versatile. The first experiments began in Britain in 1915, and a viable technology was available for installation in mobile units by the end of 1918. Prince noted that "after the outbreak of war, while carrying on experiments at Brooklands, mainly with a view to producing a continuous wave valve [vacuum tube] transmitter sufficiently simple enough to be used in the air by unskilled personnel, the author was struck by the great advantage which would be gained by the use of articulate speech instead of Morse" (Prince 1920, 377).

This first experimental set used difficult-to-manage "soft" tubes and weighed 10 lb. The batteries to power it weighed 36 lb, and it used a trailing antenna 250 ft long operating on a wavelength of some 300 m. By war's end much more rugged sets using three, five, and seven "hard" tubes that operated on wavelengths of around 150 m and used wind-powered generators rather than heavy batteries were becoming available (Prince 1920, 383). Wireless telephony enabled the aerial observer to maintain contact with the ground while keeping his hands free to fly and defend the airplane. This not only improved air spotting for artillery but also offered the possibility of much improved command and control of mobile vehicles. After the war much of this new vacuum tube equipment was made available on the surplus market.

HIGHER FREQUENCIES AND SHORTER WAVES

The American government fortuitously had reserved for amateurs the higher frequencies and shorter wavelengths, which seemed useless for long-distance wireless telecommunication up to the early 1920s. Using war surplus equipment and the wavelengths below 200 m reserved to them, amateurs demonstrated conclusively by 1923 that low-power, short-wave wireless telegraphic transmissions could cross the Atlantic: "The maximum wave-length used was 200 m. and the maximum power 1½ kW." (*Electrician* 90 [5 January 1923]: 20). In contrast, the high-power military and commercial stations operated at between 15,000 m and 30,000 m and used 250–350 kW.

In Britain, Marconi had been experimenting from 1916 on with even higher frequencies and shorter wavelengths (2–3 m) and reflecting antennas, but was still using spark transmitters because, in the absence of vacuum tube oscillators, only they could reach the high frequencies needed (Franklin 1922, 593). Marconi was exploring the short waves for the Italian navy, reasoning that they "would limit the signals to quasi-optical ranges, thus preventing any eavesdropping by an enemy well down over the horizon" (Baker 1972, 217). By the late teens vacuum tubes had been developed that produced the required high-frequency oscillations, and British Marconi was experimenting seriously with short-wave wireless telephony (Dunlap 1941, 272).

The first technical changes in vacuum tube technology were linked to the antenna system, not the tubes. Marconi had already discovered that reflecting antennas made short wavelength transmissions highly directional. By the mid-1920s further research produced even more directional and thus more efficient antennas. The more a signal could be focused and directed, the less the energy needed to propagate it. The most significant work was that of Hidetsugu Yagi, a Japanese electrical engineer who had received postgraduate training in Britain, Germany, and America. Yagi was particularly interested in the "ultra short waves" below 2 m, what we now call very high frequency, or VHF, transmissions. Highly directional Yagi antennas mounted on rotors are still commonly used by American television viewers. Yagi's classic paper, "Beam Transmission of Ultra Short Waves," published in 1928, was commented upon at the time by J. H. Dellinger, chief of the radio division of the U. S. Bureau of Standards. The paper reads, in part:

> Before 1920 radio was all wrong. The only use of radio was for communication between two points, and it was always done by broadcasting in every direction. It was not until 1920 that we had the advent of broadcasting as such, transmission intended for reception by large numbers of receivers. . . . 1920 marks not only the rise of broadcasting but also the beginning of directive radio. . . . Since 1920 we have had the gradual and partial evolution of beam systems. . . . When they have been fully developed we shall be a long way on the road toward the possibility of carrying on point-to-point communication by directive radio process. (1984, 644–45)

After World War I Marconi revived the idea of the imperial chain, proposing to the Imperial Wireless Telegraphy Committee a network of "twenty-six main trunk stations (five of them in this country) . . . [employing] continuous waves in connection with the high-frequency alternator and the valve" (*Electrician* 84: [12 March 1920]: 283). The chain was to consist of six routes and branches (*WW* 8 [17 April 1920]: 43 [fig. 2]). In introducing the plan *Wireless World* noted that

> the French have already planned out a scheme of wireless communication to link up all their colonial possessions with the mother country. The equivalent Spanish scheme has been for some time an accomplished fact. The stations which will connect Holland with the Dutch East Indies are already being built and the United States possesses numerous stations of sufficient power to communicate over a radius of 3,000 miles. Twenty-five years has elapsed since Senatore Marconi brought his invention to this country, and yet we have not even begun to use wireless imperially, nor do we boast a single English station which transmits time or "scientific" signals. (41)

A year later nothing had happened and an editorial in the *Electrician* stated, "Having regard to past history we have grave doubts as to whether anything will" (87 [12 August 1921]: 197). Two weeks later, however, the same editorial page noted, "It is pleasant at last to be able to record an achievement by the Post Office towards forging the Imperial Wireless Chain" (87 [26 August 1921]: 253). The same issue contained an extensive description of the Post Office's new Leafield Wireless Station (266–68). In February 1922 the *Electrician* analyzed the report of the Wireless Telegraphy Commission, appointed in 1920 at the suggestion of the Imperial Wireless Telegraphy Committee. Other than recommending a mix of arc and vacuum tube transmitters, with arcs being used in the more remote stations, where the high level of technical skill required by state-of-the art tube technology was not available, the commission did little new. In geographic terms it made no change to the prewar Marconi imperial chain or the revised one issued in *Wireless World* in 1920. The 1920 scheme to use the captured German station at Windhoek to serve South Africa's needs was rejected on the grounds that the cost of the required land links to Johannesburg would exceed that of a new station at Johannesburg (*Electrician* 88 [3 February 1922]: 130–32). One point is nicely illustrated by the commission's report: the high energy cost of running these low-frequency, long-wave transmitters. Had it been built as proposed, the Nairobi station would have had its own hydroelectric power plant near the falls on the Thika River since "for the purpose of the wireless station the water of the Thika is sufficient . . . . the utilization of the Charia [River] might therefore be reserved for any eventual enlargement of the wireless station" (132).

Yet three months later the *Electrician* was moved once more to the editorial opinion that

> nothing is being done to materialise an Imperial scheme except to talk about it. And those who know the history of the attempts to bring worldwide wireless communication into being, and who will recall the many commissions and committees which have reported and passed away and the expert knowledge which has been tapped and spilt, may well begin to wonder whether anything ever will be done. As in many other things, Great Britain was the first country successfully to achieve wireless communication. But now we are falling behind in the race. (*Electrician* 88 (19 May 1922): 581–82)

Once again American wireless geopolitics was the stumbling block to British aspirations. Despite the best efforts of the American navy, there had been few grounds for British concern over American aims in the first two decades of the century, when the patent position in America was highly confused. Once Marconi had bought out United Wireless in 1911, there seemed to be even fewer grounds for such concern. All that changed in the third decade, however, with the creation of the Radio Corporation of America as a national quasi monopoly designed to remove British interests from the American stage and to expand American geopolitical interests on the world stage.

At the Versailles peace talks in 1919 President Wilson concluded

> that there were three dominating factors in international relations—international transportation, international communication, and petroleum—and that the influence which a country exercised in international affairs would be largely dependent upon their position of dominance in these three activities; that Britain obviously had the lead and the experience in international transportation . . . ; in international communica-

tions she had acquired the practical domination of the cable system of the world; but that there was an apparent opportunity for the United States to challenge her in international communications through the use of radio; of course as to petroleum we already held a position of dominance. (Archer 1938, 164)

Wilson's conclusion was based at least partially on "the psychological effect produced by the telegraphic broadcasts of his 'Fourteen Points' and the parts played by the American and German radio stations in the arrangements for the armistice" (Howeth 1963, 354). Wilson's own predisposition and the politicking of the navy persuaded him that British maneuvering to control American wireless telecommunications should be countered. The U.S. Navy, which paid $1.6 million in Liberty bonds to take over Federal Telegraph's stations and patents for security purposes when America entered World War I (Mann 1946, 395), was particularly concerned that Marconi not be allowed to gain "exclusive use of the Alexanderson alternator" produced by GE (Howeth 1963, 353). The Alexanderson alternator was the only technology that seemed able to compete with the arc, and Marconi was keen to acquire it.

Admiral Bullard, who was dispatched from the Versailles peace talks back to America by Wilson to deal with the radio situation, noted that in April 1919, since he was director of naval communications, "the fact that the General Electric Company was negotiating with the English Marconi Company for a sale of a number of Alexanderson machines was brought to my . . . notice. . . . I tried at once to prevent it. I had continually in mind the cable situation, and its control by foreign interests" (quoted in Maclaurin and Harman 1949, 101).

Marconi himself approached GE to buy Alexanderson alternators as early as 1915. Either because he was recalled to Italy for military service (Archer 1938, 129–30) or because of wartime pressure on British foreign exchange (Maclaurin and Harman 1949, 100), the deal fell through. In 1919 British Marconi, which had prospered mightily during the war, "embarked upon the ambitious project of attempting to dominate wireless communications throughout the world." Marconi wanted 24 Alexanderson "alternators at $127,000 a piece—ten for the British Marconi Company and fourteen for the American branch" (Archer 1938, 160). This would have met the needs not only of the revitalized plan for the British imperial chain but also of American Marconi to dominate the Pacific and South America. It would have ensured an effective British monopoly of global telecommunications in the low-frequency long waves then being used. Vacuum tube technology simply was not yet capable of producing the power needed to drive long-wave radio transmissions for transoceanic distances. The U.S. Navy led the charge against Marconi's attempted purchase. In early April 1919 Admiral Bullard convened a conference attended by the president of GE, several directors, and the chairman of the board. Bullard

pleaded for an American radio monopoly and argued that the sale of the alternator to the Marconi interests would ensure the British a monopoly in world communications. The directors of the company, although desirous of selling the alternator to an American-controlled company, pointed out that there was no company capable of making such a purchase and their duty to their stockholders necessitated recovering the monies spent in the development of the device. . . . After additional eloquent appeals to their patriotism, the directors voted to cease their negotiations with the Marconi interests. (Howeth 1963, 356)

GE was left with no customers for a technology it had spent more than $1 million to develop. At a meeting in Washington later in April the navy agreed that if GE would establish a new international communications company to use the alternator, the navy would help it obtain a government charter authorizing a monopoly (Howeth 1963, 356). Not everyone agreed on the wisdom of this move. Secretary of the Navy Josephus Daniels' diary entry for 23 May 1919 asks, "Wouldn't GE Co. with monopoly of patents, go in with cable companies & keep up high prices by wireless & cable?" (Daniels 1963, 416). But then Daniels "strongly believed it should be a government monopoly, preferably directed by the Navy. I caused a bill to be drawn up [in 1919] to give Uncle Sam the same relation to wireless messages that he possessed as to mail. My proposal drew opposition from the companies which had been making experiments and owned some patents. They looked to a bonanza if they could force the retirement of the Navy from the field" (108).

E. J. Nally, the vice president and general manager of American Marconi, was well aware of all this "and the Navy's determination to eliminate foreign influence from American radio operating companies" (Howeth 1963, 357). Captain Stanford C. Hooper, in Howeth's eyes "the 'Father of Naval Radio'" (114), told Nally in late April that he believed the new American company should take all of the personnel of American Marconi. British Marconi correctly read the writing on the wall and in September 1919 sold its American stock to GE. In October, RCA was formed as the holder of the assets and business of the former American Marconi and as the employer of its personnel. Owen Young transferred from GE to become the first chairman of the board of RCA. He then proceeded, with the navy's help, to bring order and stability to a complex industry. "Foreign domination of American radio communications had been effectively eliminated but, since no single company possessed sufficient patents to provide a complete system, there still remained the necessity for considerable cross-licensing and agreement between various corporations to insure the success of the Radio Corp" (360).

The result was a complex set of agreements between the navy, GE, AT&T, Westinghouse, and the United Fruit Company, with RCA acting as a patent pool. AT&T held crucial patents on De Forest's vacuum tube technology, and Westinghouse on Edwin Armstrong's feedback or heterodyne circuit, the only really viable way of designing a wireless telephone receiver. The navy controlled Poulsen and Elwell's arc technology through its control of Federal Telegraph. The United Fruit Company had the largest chain of transmitters in Latin America, installed to communicate between its plantations, ships, and the American market (Howeth 1963, 361–63).

The formation of RCA seemingly had solved the navy's problems by
the early 1920s, but its manipulation of RCA was not a success. First, the
other geopolitical actors were unwilling to go as quietly as the navy had
hoped. Second, the navy had not counted on the rise of broadcast radio
as a profitable employment for RCA's patents. Third, Daniels greatly underestimated the postwar mood of the American public, which turned
strongly against continued governmental control of industries taken over
for war purposes (Douglas 1987, 281–84). And fourth, RCA turned out
to have a mind of its own. The naiveté of the navy's geopolitical maneuvering was thus quickly revealed: "RCA immediately embarked upon a
dual policy of playing the British Monopoly against the Washington Government, of co-operating with the British in one field and fighting them
in another. . . . RCA . . . entered an agreement with British Marconi ap-
portioning . . . the world between the two. . . . Thus competition between
RCA and Marconi was not eliminated in South American and Far Eastern areas in which the Washington Government and American capital
claim special interests" (Denny 1930, 381–82; see also fig. 5.3).

Nor was this the most devastating blow struck by RCA to the navy's attempt to continue its wartime control of American international wireless
telecommunications. When it was being set up in 1919, RCA entered into
a Four Power Pact with Marconi, the French Compagnie Générale de
Télégraphie, and Telefunken (Archer 1938, 237–39), based on agreements that "were to run until January 1, 1945. Each corporation was to
have *exclusive* right to the use of the other company's patents within its
respective territories. . . . Thus was organized the first international radio cartel" (Maclaurin and Harman 1949, 107).

The navy quickly recognized that it had created a monster in RCA and
began to push Federal Telegraph as the only remaining competitor. In
briefings for the Washington Naval Conference of 1922 Captain Hooper noted that RCA had "a strong monopoly everywhere except in the Far
East, and anything this conference does to increase its strength may result in serious harm to this country" (Howeth 1963, 368). Hooper therefore pushed for Federal to control the Pacific and China in particular, an
area Marconi and RCA had agreed would be one of limited competition.
Federal was, however, in a precarious financial position, and the navy was
no longer able to finance it. In September 1922 the company was reorganized as the Federal Telegraph Company of Delaware to construct and
operate the Chinese stations. RCA quickly bought 70 percent of the stock
of the new company, making its monopoly of American wireless telecommunications complete (370). Once RCA had acquired Federal, the Four
Power Pact controlled all worthwhile patents. The navy may have closed
Marconi out of the American market on nationalistic grounds, but it did
not close it out of access to the vital patent pool, which by 1922 was organized on the principles of transnational capitalism. American geopolitical goals and the navy's geostrategic goals both failed at a crucial period, just as America should have been emerging as hegemon.

RCA was not totally at fault in this debacle. The financial wisdom of
the Four Power Pact was clear in 1922, by which time "only one of RCA's
six international radio circuits was producing a profit" (Rossi 1988, 178).

As Rossi notes, RCA was set up as a profit-oriented telecommunications company; to produce a profit it had to communicate with someone, and it was in RCA's interest that the stations with which it communicated be technologically on a par with its own (175).

The rise of broadcast radio finished what transnational capitalism started by distracting RCA from its initial goal of controlling international wireless communications. The genesis of this lay in the order brought to frequency allocations in America by the first national radio conference, organized by Secretary of Commerce Herbert Hoover, held in February and March 1922. Hoover declared in his opening statement that his department estimated that at least 600,000 and possibly as many as 1 million radios were already in service, up from a mere 50,000 a year earlier (Archer 1938, 248). This defined the medium waves accessible by war-surplus vacuum tube radio receivers as the arena of commercial broadcasting, thus confirming them to profitable exploitation. The resulting rapid development of commercial broadcasting took all involved by surprise but made the potential for profit in this area of the radio spectrum very obvious. To bring order to broadcast chaos

> required the invasion of that portion of the spectrum between 185 and 500 kc. [kilocycles per second, or kHz], formerly reserved for military and naval usage by national and international laws. Governmental broadcasting was allotted two frequency bands, 146 to 162 and 200 to 285 kc.; private and toll broadcasting was allotted the band 700 to 965 kc.; public broadcasting was allocated the band between 1053 and 1090 kc.; and the amateurs were allowed exclusive use of the band between 1500 and 2000 kc. and the frequency of 910 kc. plus the shared usage of the band 1090 to 1500 kc. with technical and training schools. (Howeth 1963, 502)

At the time of the first American national radio conference the high-frequency, short-wave bands seemed worthless for long-distance transmission, hence the reservation of the lowest frequencies and longest wavelengths for government use. The conference thus removed a very valuable section of the spectrum from the control of governments, the telephone company (toll "broadcasting"), and the emerging commercial broadcasting interests. The discovery, largely by amateurs in the very early 1920s, that high frequencies could allow very long ranges indeed came just too late for government or commercial broadcast interests to lay claim to that valuable portion of the radio spectrum. Had it not had the public-access short wavelengths to expand into, Marconi's might not have survived as a telecommunications company in the 1920s. Once America, by 1922 indisputably in a position of world leadership in both long-range wireless telecommunications and broadcast radio for entertainment, had allocated seemingly worthless spectrum resources to amateurs, the rest of the world followed suit. Subsequent American governmental and big-business attempts to reclaim the short wavelengths by political action in Congress were defeated by the large number of amateur radio operators who could lobby to preserve their interests. In the American struggle to wrest control of wireless telecommunications from Britain the first national radio conference was a serious and self-induced setback. American representatives had previously used the first and second international radiotelegraph conferences to substantially advance American interests. The Washington, D.C., radio conference of 1927 and the Atlantic City conference of 1947 reduced the allocation of the higher frequencies to amateurs (Kasser 1995), but the real damage was done between 1922 and 1927.

The patent pool needed to control international wireless communication was the same patent pool that allowed radio broadcasting in the medium-wave, amplitude-modulated (AM) band in the early 1920s. RCA's purchase of Armstrong's new super-heterodyne patent in early 1923 gave it control of by far the best broadcast receiver on the market (Archer 1938, 297). RCA already controlled Armstrong's earlier heterodyne patent as a result of its agreement with Westinghouse. Any decent radio receiver had to be built along one of these two lines. Severe patent infringement by the large number of manufacturers that were entering the new and profitable market for broadcast receivers caused RCA to sue. Those being sued called their congressional representatives and complained that RCA was a monopolistic trust. What erupted was at least as diverting from RCA's original geopolitical aim as the Marconi scandal of 1912 had been from Marconi's goal of establishing an imperial chain. The congressional inquiry into the Radio Trust put RCA very much on the defensive. RCA argued that "an international task undertaken at the urgent request of the Federal Government . . . should not subject them to attack from the legislative branch of the same government. The inquisitors retorted that the Federal Government had never contemplated the setting up of a monopoly and accused the founders of RCA of taking advantage of the request from the Navy Department to accomplish purposes of their own" (299). The legitimacy of RCA's position is hardly at issue here. Being dragged into a long and expensive fight with Congress over broadcast radio made it much harder for RCA to accomplish the geopolitical and geostrategic goals of controlling international wireless communication. In the Radio Trust fracas of 1923 the U.S. Congress displayed about the same level of intelligence concerning the geopolitical role of wireless as the British House of Commons had done in the Marconi scandal of 1912. But at least the anti-Semitism was missing.

Marconi's problems in the 1920s were much simpler although by no means absent. His vision, as well as that of his company from the 1890s on, had always been control of international wireless telecommunications. Even if Marconi's had been interested in broadcast radio, it would have been prevented from providing broadcast services by the British government, which placed control of internal communications in the hands of the British Post Office in 1906. Marconi's experimented briefly with broadcast radio from its Chelmsford headquarters in the early 1920s, an experiment that ended with the formation of the British Broadcasting Company as a state-owned monopoly. Marconi's problems were mostly ones of national politics and competition from the well-established and politically well connected submarine cable companies. The Imperial Wireless Telegraphy Commission that reported in 1920 was chaired by the same Sir Henry Norman who led the assault on Marconi in the House of Commons in 1912. The 1920 report "criticized adversely the proposal of the Marconi Company, chiefly on account of its comprehensiveness and that it would confer a monopoly on the Company. . . . The report was a bitter attack on private enterprise and in several of its recommendations it went contrary to the recommendations of previous commissions" (Vyvyan 1974, 73–74).

Under British law Marconi was allowed to operate only as an international wireless communications company. RCA had no such clear corporate vision; international wireless communication was not highly profitable without a global empire, and the company was easily seduced from its initial purposes by the high profits to be made in broadcast radio. RCA's failings, Hoover's mistakes at the first national radio conference in 1922, the Four Power Pact, and the technical change in radio communication to higher frequencies and shorter wavelengths all combined by the mid-1920s to restore Marconi's and Britain to a position of international preeminence. Although Britain was not able to return to a monopoly of global wireless communication, by 1939 it was able to achieve the most powerful global presence. America, which might have been the dominant telecommunications power, merely achieved the strong position in South America guaranteed it by the Four Power Pact and a somewhat weaker position in the Pacific and eastern Asia (fig. 5.3).

HIGH FREQUENCIES AND THE DEVELOPMENT OF GLOBAL WIRELESS TELEPHONY

High-frequency, short-wave transmission offered some disadvantages, as well as several major advantages, over lower-frequency long waves for wireless communication. On the negative side were the need to maintain great accuracy in the frequency being used for transmission and the lack of reliability in transmission. On the positive side were the large number of new channels that higher frequencies made available, extremely low costs, high directionality, and suitability for telephony as well as telegraphy.

The disadvantages were relatively easily dealt with. When arcs and alternators were used to generate radio waves, frequency variation was considerable. At the low frequencies at which these technologies could generate, this variation mattered much less than at higher frequencies. But vacuum tubes not only oscillated at very much higher frequencies, they did so with great accuracy. The problem of reliability in transmission was much more severe given the diurnal and seasonal fluctuations of the ionized layers in the earth's atmosphere. It was solved by considerable research into such fluctuations and by using several frequencies on each short-wave communications circuit, depending on the time of day, the weather, and sunspot activity.

The advantages were considerable. Since high-frequency radio waves could be focused into a tight band by using a carefully designed antenna, much less power was needed to drive a signal. This meant much lower first costs and running costs as well as directionality, usually referred

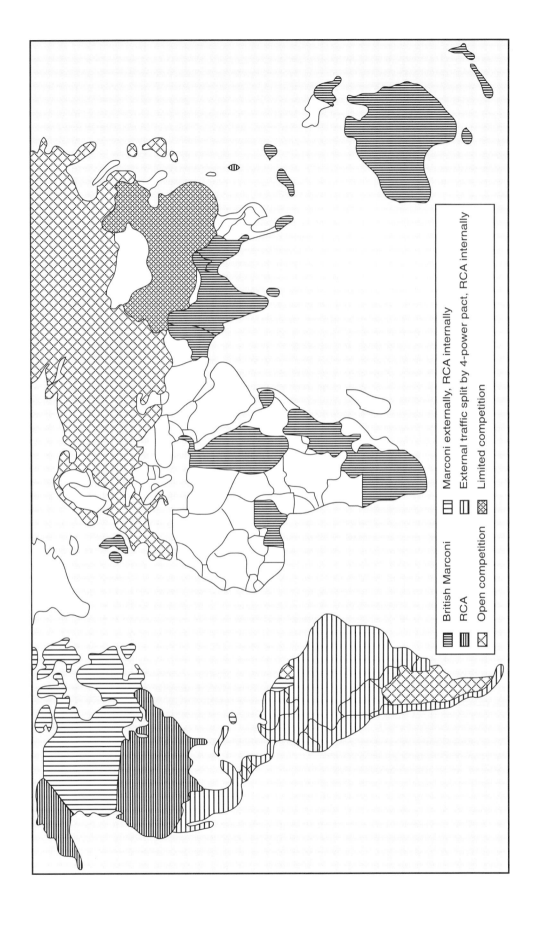

British Marconi

RCA

Open competition

Marconi externally, RCA internally

External traffic split by 4-power pact, RCA internally

Limited competition

to in the 1920s by Marconi's as "beam transmission." Low-frequency transmission required several hundred kilowatts to drive signals long distances. The biggest low-frequency stations, such as those designed to bridge the Atlantic, used 1,000 kW arcs (Mann 1946, 395). It took a great deal of ingenuity and research to approach such power outputs with vacuum tubes. Although the necessary improved high-power tubes for the lower frequencies became available in the early 1920s, the advantage of high-frequency transmission was that cheap, low-power tubes could be used. Power outputs to the antenna could be measured in hundreds of watts rather than hundreds of kilowatts. Eventually they would be measured in watts.

Low-frequency wireless still required high outputs in the early 1920s. Marconi's installed a 100 kW transmitter for transatlantic wireless telegraphy in 1921, and the British Post Office installed a 500 kW transmitter for wireless telephony in 1926, using 10 kW tubes of the type pioneered by RCA and AT&T. By 1931 a 500 kW tube was available for low-frequency transmission (Sturmey 1958, 25). The push to develop such high-power tubes came from the realization of their superior stability for wireless telephony. Alternators, arcs, and even sparks were adequate for telegraphy, but their signals were not stable enough to transmit the human voice. Stability and accurate tuning could only be provided by vacuum tubes.

As long as it stayed in the lower frequencies, international wireless telegraphy had no significant advantages over submarine cable telegraphy except on lightly trafficked routes. First costs were lower, but running costs were higher; only the monopoly profits charged by the submarine cable industry made it seem a viable competitor. The cable companies could have cut their profits, driven wireless out, then restored prices to profitable levels had international wireless developed normally in a market economy. In the first quarter of the twentieth century it was the geopolitical and geostrategic advantages of wireless that seemed significant. By the early 1920s, however, international, transoceanic wireless telephony offered a commercial basis for wireless communication.

In 1923, in conjunction with AT&T, RCA conducted the second test of transatlantic wireless telephony after the American navy's experiments of 1915. This low-frequency system used water-cooled, 10 kW tubes delivering 100 kW to the antenna. AT&T's involvement was crucial technologically since a single sideband transmission technology was used. This came direct from transcontinental telephone practice and had the advantage of reducing the power needed at the antenna by a factor of three (Blackwell 1923, 14). The *Electrician* also noted that because of seasonal differences in transmission "the transmitter should be flexible, as to power output, without requiring investment of capital in peak load apparatus which is idle a great deal of the time." Vacuum tubes met this criterion since the power output of the transmitter could be increased or decreased relatively easily compared with alternators or arcs by adding or removing tubes from the circuit. On the basis of these tests, it "would therefore be possible for any subscriber in the Bell telephone system to speak to Europe" (90 [19 January 1923]: 59).

Fig. 5.3. International Radio Communication, 1922: Marconi and RCA Spheres of Influence (derived from data in Denny 1930, 382) *(opposite)*

The problem was diurnal and seasonal variation in the transmitter strength required to drive telephone signals across the Atlantic. In single sideband transmission "the narrowest band of wave lengths in the ether is used, and all the energy radiated has maximum effectiveness in transmission" (Arnold and Espenscheid 1923, 424). Only about 30 kHz of bandwidth was available at the low frequencies used. About 3 kHz is necessary for speech to be intelligible, and a telephone transmission band had to allow about 1 kHz for clearance between channels. Double sideband transmission needed 7 kHz (3 + 3 + 1), which limited the number of possible channels to four at 7 kHz each in the 30 kHz available. Single sideband transmission allowed some seven possible channels in the available bandwidth (4 kHz × 7). Single sideband transmission also simplified antenna design, since the antenna only had to handle 4 kHz, not 7 (478–79).

The 1923 tests showed that commercial telephony was possible on the 5,000 m band but that higher power was needed to overcome the severe diurnal variations in signal strength caused by part of the Atlantic's being in daylight and part in darkness at certain times of day. During such periods, roughly 6:00 P.M. to 4:00 A.M. in London and 1:00 P.M. to 11:00 P.M. in New York, transmission quality fell off drastically and more than 50 percent of words were misunderstood because of atmospheric interference (Arnold and Espenscheid 1923, 482; see also fig. 5.4). Continued experimentation for the next two years showed that the poorest results occurred "while the United States end of the transmission path is still in daylight, but the London end in darkness" (Espenscheid, Anderson, and Bailey 1925, 175). These experiments culminated on 7 February 1926 in the opening of "reliable, continuous and complete telephone conversation . . . across the Atlantic" (Oswald and Deloraine 1926, 572).

Just as this sustained and expensive research and development effort by AT&T to drive low-frequency, long-wave transmission of speech across the Atlantic was on the verge of success, it was also about to be superseded. From 1916 on Marconi and Franklin had been experimenting with higher frequencies, shorter wavelengths, and focused antennas. As early as 1919 they were achieving wireless telephony in the 15 m band over 78 mi. By 1923 they were transmitting wireless telegraphic signals 2,330 naut mi, from Britain to the Cape Verde Islands, on 97 m using as little as 1 kW of transmitting power. By 30 May 1924 "intelligible speech was transmitted for the first time in history from England to Sydney [Australia]" on 92 m using only 28 kW. These tests proved conclusively that the assumptions embodied in the old Austin formula, that distance was a function merely of longer wavelengths and higher power, were faulty. Marconi expressed the opinion "that by means of these comparatively small stations a far greater number of words per 24 hours could be transmitted between England, India and her distant Dominions than would be possible by means of the previously planned powerful and expensive stations" (*Electrician* 93 [11 July 1924]: 44).

Franklin's radical antenna allowed high-frequency wireless telegraphy along a tightly focused beam (White 1925a). In July 1924 the Post Office contracted with Marconi to build to stringent specifications stations

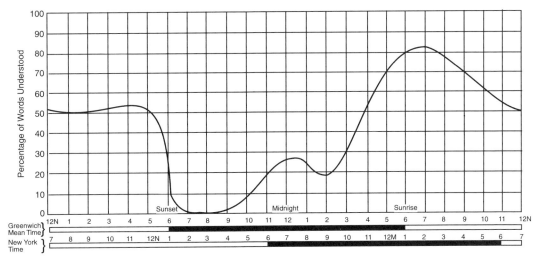

Fig. 5.4. Diurnal Variations in Quality of Transatlantic Wireless Voice Transmission, 1923 (after Arnold and Espenscheid 1923, 482)

for a beam system "capable of communicating with Canada, India, South Africa, and Australia" (Vyvyan 1974, 75). Marconi and Franklin's revolving beam antenna at Poldhu in Cornwall proved this technology by 1925, and "commercial beam stations are now being erected in England, Canada, and South Africa, while within the year it is probable that others will be started in Australia and India" (White 1925b, 430). Compared with the vast fixed antenna needed for long-wave wireless telephony, Marconi and Franklin's beam antenna was small, flexible, and cheap. Vyvyan, who was "entrusted with the carrying out of the Imperial Contract" (Vyvyan 1974, 83), saw "the sixteen years delay in establishing an Imperial Wireless Chain . . . [as] a blessing in disguise. . . . Had the State built wireless stations on the long wave principle their performance would have been indifferent as compared with what can be done with short wave beam communication, whereas had the Marconi Company been permitted to build and operate them on its own account, all the money so expended would have had to be written off" (76–77).

But the British Post Office had in fact erected at Rugby a huge antenna covering "approximately 900 acres" to receive "radio-telegraph communication over a world-wide range without intermediate stations as well as of radio-telephony over long [transatlantic] distances" (*Electrician* 95 [11 December 1925]: 672). Most of the equipment was American or supplied by licensees of American companies, and the station had a transmitter output of up to 500 kW. By 1926 it was a dinosaur, constructed mainly because the Post Office would not risk depending upon Marconi's breakthrough beam technology (Vyvyan 1974, 83). Whereas the Rugby station took electric energy "in bulk at 12,000 V . . . from the mains of the Leicestershire and Warwickshire Power Company" (*Electrician* 95 [11 December 1925]: 673), Marconi's Bodmin station for Canadian and South African service used "three 165 H.P. oil [diesel] engines driving 92 kW generators" to power its transmitters (97 [22 October 1926]: 474). Marconi's Bridgewater receiving station for Canadian and South African traffic needed even less power, which came from "18 HP. two cylinder

Aster engines driving the d.c. generators which supply the station with light and run the motors for charging the receiver batteries." Such self-contained stations were quickly and cheaply erected, and by 1926 Marconi's "had in hand no fewer than 17 short wave transmitting stations" (96 [15 January 1926]: 62). When short waves and beam antennas were used, Marconi said, "20 kW applied to the antenna will give the same results at the receiving end as 200,000 kW under the old system" (97 [22 October 1926]: 474).

Before Marconi demonstrated such performance empirically, there were doubts. "Competitors of the Marconi Company abroad had naturally been watching the progress of short wave development with considerable interest . . . doubt was expressed whether the radiation from a beam aerial would retain directive form when reflected from the Heaviside layer over great distances. The instantaneous success of the beam system was entirely unexpected by them." Marconi's success ensured that its beam system was widely copied. "Two of the main competitors . . . bought beam transmitters, receivers, and aerials, while their research workers devised modifications of the Franklin aerial system, although based in principle on the Franklin model" (Vyvyan 1974, 92). All members of the Four Power Pact of course had access to Marconi's patents, so Marconi's main advantage was that of being the first in the marketplace to sell such technology, not exclusivity through its patent position.

Success also brought problems in spectrum allocation. When Marconi's began to exploit the higher frequencies, it had them almost entirely to itself, but not for long. Some order was brought to this situation by the international radio conference held in Washington in 1927. At the conference the spectrum from 175 m down to 5 m (1.7 MHz to 60 MHz) was divided, and nothing below 5 m was reserved. Since this entire area initially had been reserved for amateurs by the first national radio conference, in 1922, the amateurs were allocated a share of the 785 kHz of spectrum and exclusive rights to a further 700 kHz of usable spectrum, as well as a claim on 6 MHz of as yet unusable spectrum in the highest frequencies under discussion. This area, between 13.1 m and 5 m (23 MHz to 60 MHz), either was not reserved or was allocated to amateurs and experiments because of the inability of available vacuum tubes to oscillate much above 23 MHz and the presumed future ability to move into that area with improvements in such tubes. Contemporary technology also needed 10 kHz of bandwidth for each discrete channel. Of the 58.3 MHz of bandwidth under discussion, 37 MHz thus could not be allocated to commercial use (Vyvyan 1974, 252).

The main impact of the beam system was, however, the extremely negative impact on the revenues of the British submarine cable companies, especially on the routes where there previously had been no competition from wireless. Compared with low-frequency radio, wireless telegraphy using the high-frequency beam stations was three times as fast, used 2 percent of the power, and cost about 5 percent as much (Cable and Wireless n.d., 7). It was significantly cheaper than the submarine cables. The "full" rate for cablegrams was two shillings a word, the "ordinary" rate was six pence. The "full" rate for beam wireless transmission was four pence and the "deferred" rate was three pence, representing reductions, of six to one and two to one, respectively (Barty-King 1979, 203). Geopolitics now combined with commercial interest, and the powerful cable interests represented by the Eastern and Associated Telegraph Companies used their domestic political power to force a merger between their submarine cables and the beam wireless system. This merger came about through the findings of the Imperial Wireless and Cable Conference of 1928 and was a solution to both the competition between cable and beam wireless systems and American expansionism (*Electrician* 100 [23 March 1928]: 321). Marconi ran its own beam stations for communication outside the empire and had built a chain of similar beam stations for the Post Office to serve the empire. Both chains had to be folded into the new company. The forced merger, on terms less than beneficial to Marconi's interests in beam wireless, drove Marconi himself to return to Italy, where he became an apologist for Mussolini.

An extensive and thoughtful article by Roland Belfort, author of several books on telecommunications, in the 23 March 1928 issue of the *Electrician* reflected on the geopolitical implications of the formation of this new company in the light of rapidly building American competition as well as the competition with the Post Office beam stations. The failure of RCA merely threw international competition into a more commercial sphere.

> Probably the most vital factor in the American problem is their determination to secure control of world telegraphic communications. . . . America enjoys the advantage of two vast systems of land telegraphs, the Western Union . . . and the Postal Telegraph. Government interference does not appear likely to restrain their activities. On the contrary, for their policy of world control of telegraphic communications is in consonance with the general policy of the United States—a policy which is developing an Imperialist trend in foreign affairs . . . the rival British companies can no longer escape the realisation of the abnormally competitive attitude of the Americans with their vast organisation—radio, telephones, cables and other future systems, including wireless telephones. (Belfort 1928, 323–24)

The new company was formed in April 1929 as Imperial and International Communications (*Electrician* 101 [3 September 1928]: 127), renamed Cable and Wireless in 1934.

Writing in 1928, Belfort said that "in America it is already being plainly stated that the Marconi-Eastern combination has been formed solely to fight the American radio-cable organisation" (Belfort 1928, 324). In the debate on the merger in the House of Commons on 2 August 1928 the geopolitical implications were evident in the speech by Sir Robert Hutchinson, who described the merger as "a great step towards the defence of the Empire. Had the fight between beam wireless and cables continued, cable would have gone under, and the Government were wise to approach the subject before violent battles were fought between the two systems, in order to preserve the cables, which were absolutely necessary to secrecy for commercial and defence purposes" (*Electrician* 101 [10 September 1928]: 154). In the event, the Depression would intervene to reduce American commercial competition, and the merger between the submarine cable and wireless interests would retain effective global dominance of telegraphic communication for Britain.

The issue of American competition was uppermost in the minds of many of those who commented upon the merger at the time. The *Zodiac,* the house magazine of the Eastern Associated Telegraph Companies, reprinted most of the contemporary comment in London's major newspapers in two articles called "The Big Splice" (20 [April 1928]: 275–81; 20 [May 1928]: 309–12). On 13 and 16 March 1927 the *Daily Mail,* a low Tory and reflexively xenophobic newspaper, printed an article by Sir Robert Donald, chair of the Imperial Wireless Committee for 1924, in which he pressed for a combination of the old cable and the new beam wireless interests: "The imperial mission of such a combination would be to safeguard strategic interests and to establish a complete network of wireless throughout the Empire, to be operated in association with cables. One effect of such combination would be to keep England the center of the world's communications, a pre-eminence which is threatened owing to the divergence of our own interests and the activities of Col. Behn and American bankers who are striving to set up a world monopoly in telephones" (quoted in *Zodiac* 20 [April 1928]: 278). The *Times,* as befitted its status as the unofficial voice of the British government, was more measured. On 23 March 1927 it printed a long editorial making plain that American telecommunications would not be welcome in the British Empire.

> American private interests have already established, with the encouragement of their Government, an extensive and growing wireless-cable system at home and abroad, which will no doubt improve the facilities at the disposal of the business men of America. Competition from this quarter must needs tend to increase in the future, and there should be no question about the capacity of British enterprise to compete, though it is quite unnecessary to regard the question (as it is apparently regarded in some quarters) as primarily one of cut-throat hostility. So long as British control is absolutely secured in the British sphere, there need be no disadvantage in a friendly competition with other systems in other parts of the world. (quoted in *Zodiac* 20 [April 1928]: 280)

The Daily Mail may, however, have had a point. The managing director of Marconi's, F. G. Kellaway, addressing the Ordinary General Meeting of Marconi stockholders in 1928, noted, "How bold some of the American commercial interests are is shown by the fact, of which I am informed on good authority, that within the past year a proposition was seriously put forward that American interests should operate, through a company to be formed in this country, the whole internal and external telegraph and telephone service of the United Kingdom" (quoted in *Zodiac* 20 [May 1928]: 312). Nor did such problems cease with the merger. The earl of Midleton, a member of the court of directors of Imperial and International Communications, complained in a letter of 19 August 1931 to Edward Wilshaw, the company's general manager and secretary, that "comparatively little stress is put on the active foreign competition to which we have been subjected, especially by America. . . . Although within the last year prominent Directors of the Company have visited South America, Canada and Australia for the purpose of safeguarding our interests, visits are not an effectual substitute for the steady diplomatic pressure exercised by all the American representatives in support of their 'nationals'" (Cable and Wireless Archives).

The official position of Imperial and International Communications was summed up in a company pamphlet of 1930, entitled *Imperial Wireless Telephony,* which blamed the British Post Office in no uncertain terms for the American invasion of British telecommunications. This may have been Marconi's point of view, percolating into Imperial and International after the merger. "The Post Office attitude [which led to buying American low-frequency telephone equipment for its Rugby installation] resulted in damage to British prestige, encouragement of American interests in securing influence and business abroad, and tended to discourage British enterprise and research" (Cable and Wireless Archives).

The higher frequencies also allowed wireless telephony over much longer distances than the Atlantic gap, although not 24 hr service of the sort that telegraphy enjoyed. Low-frequency transmitters reached their effective maximum range on the Atlantic service, and then only between America and Britain. "Even Germany is too far from America," reported the *Electrician* (105 [3 October 1930]: 405). Telephone service from America to Europe thus proceeded from the main receivers at Rugby by land lines or local wireless transmitters, thus ensuring that Britain remained at the center of American communications with Europe. Not until December 1936 was an alternative service to the London route opened in the form of a short-wave telephone connection between New York and Paris (Bown 1937, table 1). This was, however, still being reported as "recently completed" as late as December of the following year (*Electrician* 119 [24 December 1937]: 753).

By December 1929 Imperial and International Communications could note that "a direct beam wireless service from England to Japan was opened . . . last August" and that "the feasibility of abandoning certain cables on the East Coast of Africa [in favor of beam wireless] . . . is under consideration" (First Annual Report, Imperial and International Communications, Cable and Wireless Archive, quotes from 17 and 19). From the late 1920s to the outbreak of World War II the development of beam wireless was chronicled extensively in the pages of the *Electrician* as an essentially straightforward process of geographic expansion, with London as the hub. Technical improvements were relatively few, mainly reducing the space taken by the carrier in the single sideband system, which allowed "one transmitter to send two separate telephone conversations simultaneously" (*Electrician* 122 [27 January 1938]: 109). Thorough exploration of the diurnal variations in the Kennelly-Heaviside layer meant that transmitters could be used part of the day on services to one part of the world and part of the day on services to another part. The development of the multiple-unit steerable antenna allowed antennas to be repositioned to direct their beams most accurately (Bown 1937, 1131). By 1937 the map showed a substantial if low-density network of wireless telephone circuits with London as the only hub having direct connections to every major world city (see fig. 3.2). Paris was linked to New York direct but to Tokyo only through its link to Saigon. New York was not linked to Berlin. Out of 96 circuits the Americans operated 29, the British 20, the French 12, and the Germans 9 (Bown 1937, table 1). Although Americans operated the most circuits, 10 of those were short links into the Caribbean from Miami.

Such high-bandwidth wireless telephone links could, of course, also carry a substantial number of wireless telegraph circuits. As indicated in figure 2.5, the combination of these new beam wireless telephone and telegraph services with the older submarine telegraph cables by the creation of Imperial and International Communications, later Cable and Wireless, ensured the continued effectiveness of the British monopoly of global telecommunications.

OTHER COMMERCIAL USES
OF HIGH-FREQUENCY WIRELESS

Although wireless telegraphy and telephony was of central importance to the developing telecommunications industry, radio direction finding (RDF) was also of great commercial value. Ships at sea were able to benefit from RDF by World War I. The development of civil aviation and the move to compact equipment made possible by higher frequencies saw comparable development in RDF for aircraft. Radio "beacons" allowed ships and airplanes fitted with appropriate movable antennas to set a course directly toward or away from such installations.

Marconi was as important to the history of RDF at sea as he was to the history of long-range wireless telegraphy and telephony. Marconi licensed and developed the Bellini-Tosi radiogoniometer to make it suitable for use on merchant ships. The principle was simple: a loop antenna on board the ship was rotated to establish the direction of the beacon; when the loop was turned perpendicular to the beacon, it received the weakest signal; the signal increased in strength as the loop was turned parallel to the direction from which the radio emissions were coming. The problem of determining which direction the signal was coming from was solved after the directional axis was established. The loop was then switched electrically to read the strength of the signal falling on each half of the loop. The half facing the beacon received a very strong signal, the half going away from it virtually none. Determining the distance from the beacon was much harder; a skilled operator was required to read the signal strength.

The adoption of Bellini-Tosi RDF on ships followed the same pattern and logic as had the adoption of wireless. Just as Lloyd's had insisted on the latter for the ships they insured, they began to insist on RDF as it became available. Navies found RDF just as useful as merchant ships did, not only for navigation but also for triangulation on enemy ships transmitting radio signals. This was already of importance in World War I. In the antisubmarine component of World War II, high-frequency direction finding (HF/DF), popularly referred to as "Huff Duff," was an important weapon. The Germans believed that they, and they alone, had the technology to use the highest-frequency radio waves. This proved a serious error, much like their error just before the war, when they believed that only they had radar. U-boats were ordered to broadcast almost continually once they had sighted a convoy so that the German naval command could assemble a wolf pack to do the maximum damage. Allied shore stations quickly triangulated on the transmitting U-boats, and HF/DF allowed the convoy escort to home in on the shadowing targets. The first sets, which entered service in 1941, needed highly trained operators, but by 1942 they were being replaced by sets with a cathode-ray tube display, for which operators needed minimal training. Along with Ultra intelligence from breaking the German Enigma code machine, long-range aircraft, and air-to-surface vessel (ASV) radar, HF/DF allowed the British and Americans to break the German onslaught in the crucial Battle of the Atlantic.

RDF also played a major role in aviation. The British had planned to begin aerial strategic bombing of Berlin in late 1918 using a massive four-engine bomber, the Handley Page V/1500 (Barnes 1987, 131–32). According to *Wireless World,* "Direction Finding sets were ready installed, perfected and indispensable, in these sleuth-hound craft that were on the point of being loosed to nose their deadly way to Berlin when the Armistice rang down the curtain" (*WW* 7 [April 1919]: 40). One presumes that they intended to home on German long-wave radio emissions from Nauen. Direction-finding equipment was also installed at London's Croydon Airport in anticipation of a postwar boom in commercial air transport that never came (Sinclair 1921, 799).

It was in America that RDF came into its own in the commercial airways. A massive network of beacons was installed first for the air mail, then for passenger-carrying airplanes. By 1931 one transcontinental route had been completed and two more were under construction (fig. 5.5). The number of radio beacons expanded from 2 in 1928, the first year they were used, to 281 in 1940, by which time the network of routes flown by American civil aviation was by far the most extensive, not to mention the safest, in the world. Better aircraft, engines, and pilots certainly played their part, but it was RDF that allowed American civil aviation to fly in almost any weather and any season of the year.

Within only a few years the character of wireless communication changed drastically. Despite the technical success of low-frequency, continuous wave wireless by 1920, its value seemed to be purely geopolitical and geostrategic. Relative to the profitable and well-established submarine cables, the commercial value of long-distance wireless telegraphy was questionable because of the high fixed and energy costs of low-frequency transmission. By 1926 wireless was entering a new age. In wireless telephony, even at low frequencies and on circuits of very limited capacity, it had found a unique product that no other technology could pro-

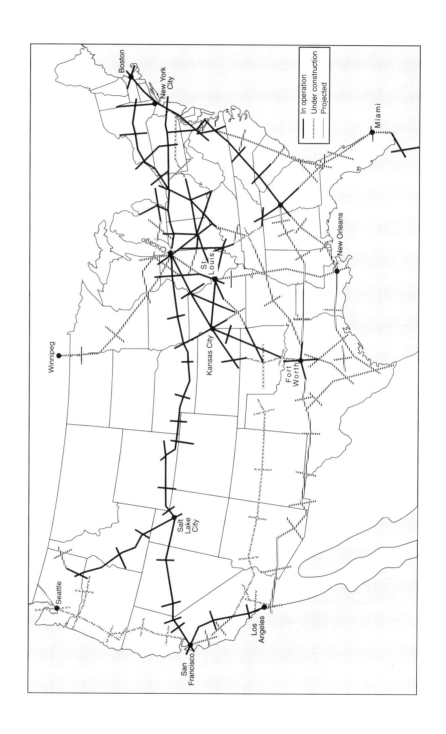

| In operation |
| Under construction |
| Projected |

Boston
New York City
Miami
Chicago
St. Louis
New Orleans
Winnipeg
Kansas City
Fort Worth
Salt Lake City
Seattle
San Francisco
Los Angeles

vide. In high frequencies and beam antennas it had found a transmitting technology with far lower fixed and energy costs that allowed relatively simple installations in almost any location and the use of frequencies not reserved to the government or big business. The improvement offered by short-wave technology was startling.

> With very long waves, a power of 300 kW. is taken from the mains, 150 kW. may be supplied to the antenna, and of this only 17 kW. is radiated, whereas with very short waves, 30 kW. from the mains may give 10 kW. to the antenna of which 7 kW. is radiated. . . . By the use of short waves, an enormously increased number of channels of communication is made available, and this constitutes perhaps the greatest gain that has resulted from their use. If the band of frequencies between 50 and 100 Kc/s. [the long waves] is considered, only 25 channels are available, even though the width of each frequency band is reduced to 2,500 cycles—the minimum required for commercial intelligibility—the carrier and one sideband suppressed, and no spacing allowed between adjacent bands. Against this, the present state of short wave technique permits the use of 200 telephone channels between 6,000 and 12,000 Kc/s., the spacing between adjacent carrier frequencies being 30 Kc/s. This number would be increased to 600 if the spacing were reduced to 10 Kc/s., a figure which should not be considered impossible. (Deloraine 1930, 214–15)

The international telephone network, though not its capacity, grew at an explosive rate once expansion into the high frequencies proved practical. By 1929 "radio telephony . . . provided daily telephone service to more than twenty-six countries. . . . During 1929 and 1930, more than twice as many new international connections were added" (Brooks 1931, 157). Traffic also increased rapidly, albeit within the limits of a system having very little capacity. On the Atlantic route traffic grew "from less than ten messages per day in 1927 to nearly fifty per day in 1929" (Buttner 1931, 251). The number of circuits also increased. By the end of 1929 three high-frequency circuits had been added to the pioneering low-frequency circuit of January 1927: the first in June 1928, the second in June 1929, and the third in December 1929 (Bown 1937, table 1). The high- and low-frequency circuits complemented one another. Diurnal and seasonal variations caused problems but rarely shut down all the circuits at the same time, although it was equally rare that all four circuits could be used simultaneously. Charges were substantial. In 1927 a 3 min call between America and Britain cost $75, and additional minutes were charged at $25 each (Sturmey 1958, 129).

The success of high-frequency technology and the realization of a profitable market for transoceanic wireless telephony had a huge impact on global telecommunications. Despite American attempts to control the new systems, control slipped from the hands of any single country. High-frequency systems were cheap to build and run. Ideal for telephony, they also drastically lowered the costs of telegraphy. The formation of RCA ensured a broad pool of patents that could be shared by all the major telecommunications companies involved in either RCA or the Four Power Pact. Marconi used its beam system to recover much of the world

Fig. 5.5. American Air Mail Beacons, 1931 (redrawn from Duncan and Drew 1931, 859) *(opposite)*

market in the early to mid-1920s but was kept out of the lucrative British Empire trade by the British Post Office, although it did build the beam stations used by that body.

By 1937 a substantial network of wireless telephone, and thus wireless telegraph, links had been installed using high-frequency transmission and beam antennas (Bown 1937, fig. 1; see also fig. 3.2 in this volume). By using their political clout with the British government, the Eastern and Associated Telegraph Companies were able to force a merger between their interests and those of Marconi and the British Post Office. This allowed them to continue to dominate the carrying trade in telegraphy, which the beam stations had seriously threatened. Cable and Wireless thus came to control the lion's share of the world's international cables and many of its high-frequency wireless telephone and telegraph lines. AT&T came next, and the rest were controlled by government-run companies. The Four Power Pact of 1919, consequent upon the setting up of RCA, was still otherwise clearly evident. AT&T had broken from RCA, claiming that its wireless telephone services were a simple extension of its American wired services. Marconi's beam system had been merged into Imperial and International Communications. The French Compagnie Générale de Télégraphie was largely intact. Telefunken had been renamed Transradio and often operated in conjunction with the German Post Office.

The U.S. Navy's dream that in setting up RCA it could achieve American hegemony of global telecommunications thus came to naught. American commercial hegemony was more or less achieved in the Western hemisphere, a sort of Monroe Doctrine of the ether. But it was by no means complete. All American Cables, an RCA subsidiary company, controlled a significant amount of internal traffic in South America. AT&T controlled the Caribbean and services between North and South America (see fig. 5.3). But services between Europe and South America were European controlled, usually in conjunction with South American national companies. Although AT&T controlled services between North American and Asia, Asia was far better served by British, German, French, and Dutch companies. Japanese interests controlled links from Japan to Taiwan, China, Manchuria, and the Philippines (Bown 1937, table 1).

On the basis of their geography, international wireless circuits almost equaled the global submarine cable network by 1937. But capacity was substantially lower, in particular on key routes such as the North Atlantic, in part because there were far fewer links and in part because high-frequency circuits could not reliably operate 24 hr per day. Most significant, no single country controlled the network. RCA failed to do what it was created to do. Instead of creating American hegemony, it merely reduced British monopoly, and without obvious benefit to America. The Four Power Pact meant that international wireless telecommunications were shared through the patents held by Pact members among all the global powers, not controlled by anyone. In this regard the transition from British to American global hegemony was seriously slowed by the failure to control international wireless telecommunications early in the postwar period.

CHAPTER SIX

Military Uses of Radio Communication
The Development of Communications, Command, and Control

Signal cable communication, vastly developed during the static warfare in 1914–18, is unsuitable for moving warfare owing to the time taken in laying lines and the dislocation caused by the lines being broken. The development of radio-telephony enables the staffs to speak directly to their observers. . . .

While developing means of quicker control on our own part, it is no less important to apply our minds to the problem of upsetting the enemy's control, and to apply new means to that purpose.

—Basil Liddell Hart, *Thoughts on War*

Information technology is as important to armed struggle between polities as it is to economic struggle, more so in wartime because the stakes are more immediate. Although wired telegraphy was clearly important to the military, it also had clear limitations. Its main advantage was that it allowed centralized command and control (C^2) of distant forces. This greatly increased the importance of the staff officer relative to the line officer, and the military history of the late nineteenth century is dominated by the emergence of war colleges specializing in the history and theory of war to educate such staff officers. Such centralized C^2 was in line with the emergence of the bureaucratic, national state and the new imperialism as well as the massive increase in industrial bureaucracy attendant upon what has been called the second industrial revolution. The bureaucratic state, the bureaucratic firm, and the bureaucratic military all depended upon better information technology, which until the late nineteenth century meant the telegraph.

But centralized C^2 turned out to be almost as much of a liability as it was an advantage, at least until the proper division of labor was worked out between staff and line officers. Until the experiences of World War I there was a tendency toward excessive centralization of command using systems designed by staff officers at the expense of the initiative of line officers. The particular limitations of wired telegraphy were fourfold: its susceptibility to battle damage; its inability to provide C^2 for mobile warfare; its occasional but disastrous lack of clarity of meaning; and its susceptibility to interception.

In World War I, at the Battle of Neuve Chapelle in early 1915, the British failed to consolidate a major advance because their telegraphic communications had been shattered by artillery bombardment (Nebeker 1996). It became quickly evident that wireless telegraphy was a necessity. Wireless telegraphy using the principle of induction had been around for a while. Using the ground as a conductor had been common commercial practice for short distances (c. 100 m) when lines failed, as had the use of water to conduct signals across rivers (Constable 1995, 14). Early in World War I the French were able to extend the range of ground telegraphy to about 1 km using a vacuum tube amplifier (Nebeker 1996). Such signals were very susceptible to interception.

Second, wired telecommunications were obviously unsuited to mobile warfare. From the inception of the European states system in the 1500s until World War I the form of war practiced by European polities had essentially been mobile and offensive. Fortified houses, towns, and cities simply could not stand against siege guns and sappers. Armies and navies roamed freely over disputed lands and seas. There was no centralized C^2 to guide them, only general objectives. Telecommunications allowed centralized C^2, but as long as the recipient had to be at the end of a wire, mobility was restricted. Historians have attributed the development of immobile trench warfare in World War I to the development of the European railroad network, but it also owes much to telegraphy and telephony. Railroads allowed both rapid national mobilization and the concentration of national and even international industrial production of war matériel along a very tightly defined front line. Guns could quickly pound to dust an opponent's attempt to construct new railroads close to the front line. The front line in World War I thus evolved into what was really several lines. Main-line, standard-gauge, properly built railroads ran parallel to the front just out of gun range. These were connected directly to the ports and the major industrial areas and could rapidly move matériel and armies. Narrow-gauge, cheaply built railroads connected the main lines to the front but moved at much lower speeds. At the front itself shells were moved from narrow-gauge railheads by even narrower-gauge lines, often powered by humans. The speed of movement bogged down to a human walk, and the front line froze in space. In such conditions the immobility imposed by telecommunications merely reinforced the immobility imposed by railroad systems and field guns.

Third was the problem of occasional lack of clarity of meaning. Telegrams were kept short because of expense. Because they were only written words, they lacked the expressive meaning possible in more discursive writing, in speech, and in face-to-face communication. The immediate outbreak of Word War I has been put down in part to a rapid exchange of misunderstood telegrams between Kaiser Wilhelm and Czar Nicholas. The telephone allowed expressive content, although it still suffered from the problems of being a wired system. As indicated in chapter 3, Moltke replaced Prittwitz with Hindenburg just before the Battle of Tannenberg in 1914 because of the defeatism evident in Prittwitz's voice in a telephone conversation.

Fourth was the problem of susceptibility to interception. The global telegraph network centered on London, and it was generally believed that the British were reading everyone's mail. It was obvious that they could, and the events surrounding the Zimmerman telegram indicate that they were certainly reading the important bits. Such telegrams were protected by codes and ciphers, but breaking such codes and ciphers was quickly elevated to major importance in the new field of electronic intelligence. At the local level, opposing armies could eavesdrop on a wired system by the principle of induction provided they could get reasonably near to the line. This was never difficult in the trench warfare of World War I.

All these problems could be solved, at least in part, by wireless, although the fourth problem was never solved. Wireless was as easy to intercept as a wired system, if not easier. In the opening days of World War I the Russians paid a huge price at the Battle of Tannenberg, where, as Liddell Hart notes, "the swift German moves were also made sure through the folly of the Russian commanders in sending out unciphered wireless orders" (1938, 11–12).

THE DEVELOPMENT OF COMMUNICATIONS, COMMAND, AND CONTROL

Once the line system of World War I had emerged, the armies that were bogged down in such conditions became obsessed with breaking out of the trenches. After the horrendous failure of frontal assaults against trenches well dug in and defended by barbed wire and machine guns, with millions dead and the French army in mutiny, it became clear to military strategists that the only solution was a return to the war of mobility. Since, however, centralized C^2 had become necessary to industrialized warfare, the development of mobile communications was vital. Wireless was the obvious way to achieve centralized communications, command, and control (C^3). Given the rapid technological development of wireless between the late 1890s and 1918, from spark telegraphy to continuous wave speech transmission using vacuum tubes, the simplest way to understand the military development of wireless C^3 is by looking at its history. Although the British army and navy both began experiments with wireless in 1899, navies would be the first successful users of wireless C^3, the new air arms would be second, and the armies last. The British army attempted to use wireless in the Boer War, 1899–1902, with singular lack of success, almost certainly because of geological differences. At the frequencies Marconi was using the conductivity of the ground in South Africa was far lower than that in Britain, where tests had earlier taken place. The remarkable success of wireless in the British naval maneuvers of 1899 and in all subsequent marine use was owing to the very high conductivity of the saltwater return, which "markedly enhanced both the performance of the antenna and the propagation of the ground wave" (Austin 1995, 47). Such issues of propagation were, of course, completely opaque to contemporary scientific and engineering skills.

Marconi and his early competitors in Germany and America modulated the radio emissions of an electric spark to achieve wireless telegraphy. Such a technology was cumbersome, used a great deal of electricity, needed a high antenna and a solid ground return, and suffered from the extremely noisy sparks. But it could be installed on ships, where the bulk mattered little, plentiful electricity could be generated from the ship's engines, antennas could be strung the length of the ship at mast height, the ground-return connection between the ship's steel hull and the salt water was excellent, and a sound-insulated room could be set aside for a radio operator, whose nickname quickly became "Sparks." The British navy soon became Marconi's best customer, which may well have led to Marconi's first great commercial success, the requirement by Lloyd's that the ships they insured carry Marconi's equipment. The world's navies were right behind Lloyd's, although national interest prompted the German and American navies to avoid Marconi's equipment.

Neither the Imperial German nor the American navy wanted to depend on Marconi for its radio. In Germany the Slaby and Arco interests, both of whom provided slightly modified Marconi systems, were merged by government edict into Telefunken (see chapter 4). Although Telefunken acquired a near monopoly of radio service to German merchant ships, its real origins lay in naval interests, not commercial ones. The American navy experimented with many suppliers in the first decade of the twentieth century, and although clearly distrustful of Marconi, it was unable to find a reliable American supplier. By 1906 almost 50 percent of the American navy's equipment was German. Only with what amounted to a clandestine takeover of San Francisco–based Federal Electric and its well-developed Poulsen arc technology was the navy able to find a reliable American supplier.

The American navy's problem clearly lay in having to operate under the constraints of a relatively capitalist system. Marconi had achieved an effective commercial monopoly in a capitalist polity by a combination of political astuteness, his commercial alliance with the near monopoly of ship insurance represented by Lloyd's, and judicious research directed to generating appropriate patents or acquiring them from others. Marconi had his detractors, the British General Post Office in particular, but his naval connections outweighed them. Wilhelmine Germany operated sufficiently outside the capitalist system for Telefunken to simply appropriate Marconi's technology, at least to begin with. When the German courts finally recognized Marconi's patents, Marconi agreed not to operate in Germany and Telefunken agreed not to operate in the British Empire, and they came to agreement on various shades of competition elsewhere. The American state could not take such an attitude because it operated as much within the world capitalist system as did Britain. The state would not uphold those who infringed on Marconi's patents, however much the navy may have wanted it to do so. The American navy had thus to buy from Marconi or from his German imitators or find a technology that did not infringe his patents along with a reliable supplier. It could not deal with his American imitators. The early history of wireless telegraphy in the American navy is thus one of using German technology while seeking a reliable American alternative.

The operational history of wireless C^3 is exemplified by the British experience. As with the wired telegraph, centralized C^3 by wireless telegraphy had problems as well as advantages. Those problems first showed themselves in the heat of battle, when it was too late to do too much about them. The most recent historian of "the Great War at sea," Richard Hough (1983), credits wireless C^3 with reducing the initiative of individual ship captains. Reduced initiative and confusion over signaling were certainly the major reasons for failure to win a clear victory when one was in hand at the Battle of the Dogger Bank (1915) and a similar, if less obvious problem at the Battle of Jutland (1916).

ELECTRONIC INTELLIGENCE GATHERING

On the other hand, it was electronic intelligence gathering that led to those and several other engagements between the British and German fleets in World War I as both fleets sought the ultimate Mahanian confrontation between their battle fleets. Early in the war both the British and the Germans decided that they could gain more by reading each other's radio traffic than by jamming it. This decision elevated the making and breaking of codes and ciphers to a major military skill. The British, pursuing a classic Mahanian war of blockade, quickly discovered that a sudden increase in radio traffic on known German naval frequencies indicated that the German fleet was about to leave port in strength. Radio direction finding (RDF) could identify the bearing and approximate range of such traffic. The constant flow of signals to and from ships at sea that centralized C^3 implied was thus an excellent source of information to an enemy even if the signals were in secure codes. British warships merely used RDF to home on German signals.

The British navy also demonstrated the vital importance of code breaking, establishing Room 40 at the British Admiralty to deal with the German naval codes. It was Room 40's ability to break even the most secure German naval codes that resulted in the deciphering of the Zimmerman telegram. Although the British came quickly to recognize that most codes were insecure and that German ships could use RDF to locate British ships just as easily as the British could use it to locate German ships, the lessons sank in slowly. The Germans were able to gather considerable intelligence about British merchant ship movements in both world wars merely by examining ship sailing dates and cargo manifests posted in supposedly neutral Switzerland for insurance purposes. This made their own war of blockade using U-boats much easier: insurance payouts were high. In February 1942 the German cryptanalytic organization known as the B-Dienst broke the British-American-Canadian Naval Cipher Three, which was used to control North Atlantic convoys (Smith 1993, 118).

German U-boats pursuing two wars of blockade twice brought Britain near to collapse from inadequate supplies of raw materials and food. In World War II the Germans were convinced that they had solved the problem of electronic security with Enigma, a mechanical encrypting device of fiendish complexity that was developed from a German commercial coder marketed in the 1920s. As before, what could be encrypted could be decrypted, although decrypting German signals was to take some of the best mathematical minds in Britain, banks of mechanical computers called "bombes," and eventually the first genuine, though non-programmable electronic computer, "Colossus." Reading Enigma wireless traffic, the decrypted form of which the British called Ultra, gave senior British commanders a clear picture of German intent through much of the war, allowing the British to concentrate their much smaller military forces where they would do the most good. In this sense, Ultra information was a classic force multiplier, as was radar in the air war.

The most spectacular use of Ultra, if not the most important, was in the U-boat war. Ultra decrypts or the lack of them account for much of the seesaw activity of the long-drawn-out Battle of the Atlantic. Victories were easy in what U-boat captains call the first "happy time," but they dropped away as the three-wheel Enigma was broken. When the Germans added a fourth wheel to their Enigma machines in February 1942, Enigma became an enigma again. Because it coincided with the breaking of Allied Naval Cipher Three by the B-Dienst, the addition of the fourth wheel could easily have meant German victory in the Battle of the Atlantic. The British decryption effort, located at Bletchley Park, went into high gear and eventually decrypted the more complex four-wheel Enigma (Garlinski 1979, 137–39). The British attached more importance to Ultra than to any other aspect of World War II. Although they shared the technologies of centimetric radar and the jet engine with America during the 1940 Tizard Mission, well before official American entry into the war (see chapter 7), it was May 1943 before they fully shared Ultra (Smith 1993, 151).

America also made substantial code-breaking efforts, although these were initially hampered by cuts in spending on such activities in the aftermath of the Depression. Japanese codes were both complex and subtle, and decryption was hampered by the ideographic subtleties of written Japanese. Ideographs do not translate well into romaji, the romanized version of kanji. One Japanese ideograph may have up to eight meanings in romaji transliteration. American-born nisei were crucial in the effort to understand Japanese codes. The Japanese Foreign Ministry introduced Purple, an early form of Enigma, in 1939. A concentrated American effort broke this code before the American entry into World War II. The Japanese army and navy continued to use conventional code books. The Japanese army used a four-digit operational code that was not broken until 1943; its three-digit tactical code was never broken. The main Japanese naval operational code, the five-digit JN-25, was penetrated in late 1940. The diplomatic traffic was referred to as Magic. After 1944 all this material was presented to American commanders in the Pacific as a form of Ultra (Drea 1992, x–xv, 1–13).

American code breaking produced a massive and useful information flow. For example, Admiral Spruance was able, at the Battle of Midway in 1942, to get accurate intelligence on the disposition of the Japanese fleet. Combined with naval radar to provide accurate data on the day of battle and Spruance's own skill at managing his ships and planes, it produced one of the classic victories of naval warfare, comparable with the battles of Trafalgar and the Tsushima Straits, where superior ability and tactics more than made up for severe discrepancies in numbers and firepower. American electronic intelligence was also responsible for the successful interception and destruction of Admiral Yamamoto's transport plane by Lockheed P-38 Lightnings flying at the very limits of their range, a classic use of a long-range strategic fighter. Yamamoto, who had spent much of his life in America and recognized the inevitability of American victory, was the most brilliant of Japanese naval strategists, and his loss sorely hampered the Japanese navy. Even so, Drea's nuanced interpretation of the use of Ultra in the Pacific makes it plain that the information available to General MacArthur was not as complete as that available from Enigma decrypts in the European theater: "One searches in vain for that 'Ultra state of mind' commanders fighting the war against Germany are alleged to have possessed" (1992, 229).

Smith's recent account sheds a great deal of light on code breaking on both sides of the Atlantic, although he makes clear how far ahead of America Britain actually was at the start of World War II (Smith 1993). But then, Britain's cryptanalytic effort was never hampered by the bizarre cutting off of funds that occurred in America after 1929. "Had Bletchley Park been forced to start truly from scratch, like Room 40 in 1914, Ultra intelligence would scarcely have been possible. Bletchley depended on the expertise painstakingly built up between the wars on minimal resources" (Andrew 1985, 38).

The definitive account of British intelligence is contained in five volumes edited by F. H. Hinsley. The efforts to break Enigma and the considerable assistance of the Polish and French in doing so are detailed in appendix 30 of volume 3 (Hinsley et al. 1988, 945–60). Once the British understood how Enigma worked, they altered the operation of their Typex code machines along similar lines to facilitate decoding (952). The Typex was made available to America when America entered the war, and the Americans produced a version of their own Electrical Cypher Machine to work with the Typex that was called the Combined Cypher Machine. Both the Combined Cypher Machine and the Typex proved totally secure: "indeed the Germans made no serious attempt to solve either system" (Hinsley et al. 1981, 639). In constructing the electro-mechanical bombes that would break the Enigma machine the work of mathematicians such as Alan Turing at Bletchley Park was essential. Turing realized that the Polish approach to decrypting Enigma traffic would not be flexible enough should the Germans change settings rapidly and altered the specifications for the bombes accordingly (Hinsley et al. 1988, 952).

Britain was reluctant to share intelligence information with America in the early part of World War II because of its well-founded reservations about American diplomatic and military security. Ultra was released only to American navy and air force personnel in Britain. A significant number of Rommel's victories against the British in North Africa resulted from the fact that the American military attaché in Cairo, Colonel Fellers, was sending full information about British plans to Washington. These messages were intercepted by Italian and German cryptanalysts and relayed to Rommel (Smith 1993, 111–12).

RDF AND RADAR IN THE BATTLE
OF THE ATLANTIC

Merely decrypting Enigma traffic did not, on its own, assure British and, later, Allied victory in Europe. The Battle of the Atlantic, by many accounts the most crucial of the first half of the war, required one of the most sustained technological efforts of the war. Since not all German orders moved through Enigma, only part of the total picture could be gained by code breaking. Many other factors were also involved, including the availability and adequate supply of long-range, radar-equipped patrol bombers that could police the Atlantic as well as destroyers, corvettes, and other antisubmarine patrol vessels.

Air-to-surface vessel (ASV) radar was also important. The first such, derived from airborne intercept Mark IV radar designed for night fighters, worked in the 1.5 m band and could easily locate ships at sea. It has recently been suggested, on the basis of the flight path indicated by the flight logs, that the American Consolidated PBY Catalina long-range flying boat that located the German battleship *Bismarck,* after she had sunk the British battlecruiser *Hood,* was fitted with British ASV radar (Devereux 1994). The 1.5 m ASV could also locate a U-boat on the surface at night, and British coastal command Sunderland flying boats destroyed many U-boats caught in this manner. When centimetric radar operating on the 10 cm wavelength entered service, British and American patrol aircraft were able to detect the periscopes of German U-boats. Even so, it was usually Ultra decrypts that narrowed the search area.

RDF also came into its own as a major electronic weapon in the Battle of the Atlantic. The second "happy time" for the U-boats was partly a result of their change to wolf-pack tactics. Once a patrolling U-boat located a convoy, it reported back to German naval headquarters and shadowed the convoy until C^3 could assemble all the U-boats within range of the convoy into a wolf pack. Such tactics required considerable radio activity, and the Germans naturally feared RDF. They believed the solution was to use only very high frequency radio, something at which they had been indisputably more adept than the British before the war. As detailed in chapter 5, the British successfully detected German U-boats using shore-based RDF and ship-based high-frequency direction finding (Huff Duff). Mass production of such sets, mostly in American factories, made it possible for every convoy escort to have one by the later part of the war.

In both world wars the side with superior electronic intelligence won. Part of that intelligence involved the information provided by the great volume of enemy radio activity. With accurate RDF this included such dangerous information as the range and bearing of enemy ships. For all the advantages of centralized wireless C^3 to bureaucrats and staff officers at home, it caused the loss of many officers and ratings at sea. Understanding such radio traffic was unnecessary to get results, but electronic intelligence that decrypted enemy messages paid huge dividends. Room 40 and Ultra demonstrated that a false belief in the security of one's encrypted transmissions was perhaps the most fatal error made by either side in both world wars.

THE GREAT WAR IN THE AIR

Origins of the War in the Air

In any history of World War I in the air the absence of the term *wireless* or *radio* is conspicuous. The same is true of the many histories of individual airplanes and airplane manufacturers written for and usually by aviation buffs. In part this can be put down to the romanticization of the Great War in the air: men and tactics, rather than machines and technology, were stressed. From almost the beginning of aerial combat the most successful fighter pilots became lionized as "knights of the sky." When Germany's favorite and perhaps greatest ace, Oswald Boelcke, died in combat in October 1916 he was given a funeral worthy of a kaiser (Fritzsche 1992, 74–81). Such public adulation has been understood as a natural reaction to the horrors of trench war. But most fighter pilots were effective cold-blooded killers who did their best to shoot their victims in the back before they ever saw them. The beau ideal of a medieval-style knightly combat between two well-matched equals was rarely achieved. The job of the fighter pilot was simply to shoot down poorly armed, usually slow and unmaneuverable two-seat reconnaissance aircraft whose particular job was "spotting" for the artillery by reporting the fall of shells by wireless telegraphy.

If we take British production figures as typical, about half the planes built in World War I were designed for artillery reconnaissance; the other half were fighters. Although production figures for wireless sets are lacking or at least hard to access, nearly all artillery spotters were fitted with wireless by the last year of the war. This simple fact has been obscured by romanticization of the air war, omission of any information about wireless in aviation histories, and the classification of data on wireless as secret by the British government. The last is particularly odd since the data classified were technically obsolete by 1920 and a whole series of articles on British airborne wireless was published in *Wireless World* in 1919 and 1920.

Airplanes and Pilots

Out of a total British production of some 55,093 airframes (Penrose 1969, 606), six designs accounted for 23,303 (Bruce 1957). Of these the B.E.2 (3,241), the R.E.8 (4,077), the F.E.2 (2,189), and the Bristol Fighter (3,101) were either dedicated reconnaissance airplanes (the first two) or usually used in that role. The S.E.5 (5,205) and the Sopwith Camel (5,490) were fighters per se. Three other airplanes achieved huge production numbers: the Avro 504 trainer (8,340); the de Havilland D.H.6 trainer (2,282); and the D.H.9 bomber (3,204). No other British airplane exceeded 2,000 in number. German pilots treated the B.E.2 and its little-changed descendant, the R.E.8, with the contempt they deserved. The B.E.2 earned the name "Fokker Fodder" during the "Fokker Scourge" through the early summer of 1916, when the Fokker EI monoplane was the only fighter to have a machine gun synchronized to fire forward through the propeller. They were more circumspect about the F.E.2, one of which once shot down the great Manfred von Richtofen. As its name implies, the Bristol was also a two-seat fighter. It was one of the great designs of World War I, produced numerous aces, was treated by German pilots with the greatest respect, and remained in British service until 1932 (Mason 1992, 86).

Baron Manfred von Richtofen, the highest scoring ace of World War I, shot down 80 airplanes, 46 of them two-seaters (Gibbons [1927] 1959, 211–15). Unlike Boelcke, whose dictums still guide fighter pilots to this day (Hallion 1984, 162–63), von Richtofen left only one enduring word of advice, that when attacking two-seaters it was important to kill the observer first. Although von Richtofen doubled the score of his mentor, Boelcke, he "never inspired the affection or verse of the German public" as Boelcke did (Fritzsche 1992, 82).

Even with an armed observer the B.E.2 and the R.E.8 were easy to shoot down because in addition to being slow and poorly protected, they were designed to be inherently stable. The B.E.2 was designed in 1912 for one specific purpose: to allow the pilot to set his course and then concentrate on reconnaissance. In June 1914, flying the first B.E.2c, Major W. S. Brancker, who "was not an experienced pilot, recorded that, after climbing to 2,000 ft and setting course, he was able to make the forty mile journey without placing his hands on the controls until he was preparing to land. He spent his time writing a report on the country side passing below" (Hare 1990, 148). As the skies above Flanders became filled with hostile fighters such a design philosophy ceased to make any sense, yet the B.E.2 and its successor, the R.E.8, had to soldier on until the end of the war because they were available in large numbers and could, if given adequate fighter protection, do the job that had to be done, namely, wireless spotting for the artillery.

As early as March 1916, Noel Pemberton Billing, founder of the famous Supermarine Company and member of Parliament for East Hertfordshire, made a lasting name for himself criticizing War Office aviation policy. In a sensational speech to the House of Commons he referred to the B.E.2 as "Fokker Fodder" and suggested that its continued deployment meant "that quite a number of our gallant officers in the Royal Flying Corps have been rather murdered than killed" (quoted in Hare 1990, 91). Pemberton Billings' use of the word *murder* was later dismissed as an "abuse of language," and his sensational approach lost him more friends in Parliament than it won him, but it brought the air war very much into the public eye.

Wireless and the Air War

Spark Telegraphy and the Artillery

The B.E.2 was actually intended to carry wireless from its inception. The *Marconigraph,* monthly predecessor of the important weekly journal *Wireless World,* first described "wireless telegraphy and aeroplanes" in its August 1911 issue. The article referred to a lecture delivered to the Royal Institution and "satisfactory results obtained by Mr. Farman . . . using 2 trailing aerials, each consisting of rather thin wire about 100 meters in length" (14). Similar apparatus was shown on line drawings of a pusher biplane in the issue of December 1911, by which time "the feasibility of telegraphing by wireless from aeroplanes has already been amply demonstrated" (4). Much higher frequency apparatus using wing-mounted rather than trailing antennas was shown on a Flanders tractor monoplane by mid-1912, although the plane in question crashed fatally (*Marconigraph,* June 1912, 48–49).

The prototype B.E.2 first flew in February 1912, and "towards the end of March a wireless set which had been constructed at the [Royal Aircraft] Factory was installed" (Hare 1990, 138). The B.E.2 was flown from the rear cockpit, and the pilot, in addition to flying the plane, was expected to key the wireless Morse sender, which was fitted outside his cockpit for his right hand, as a surviving film clip in the British Imperial War Museum clearly shows. Regrettably, the B.E.2c on display in the same museum was not equipped with wireless in February 1995. Hare does not comment on wireless for the R.E.8, in which the observer was at least moved to the rear cockpit, and the rather run-down surviving example at the Imperial War Museum's aviation collection at Duxford did not have wireless equipment in February 1995. Barnwell designed the Bristol Fighter so that the observer in the rear cockpit had "dual controls, wireless, camera and message-launching tube," as well as a defensive Lewis gun (Barnes 1988, 104). The very finely preserved late war example in the Royal Air Force Museum at Hendon on display in February 1995 had the wind-powered generator for the wireless clearly mounted on the lower wing, which photographs and line drawings usually do not show.

By the end of World War I, in fact, the British had evolved sophisticated two-way, continuous wave wireless sets for aircraft use, the vacuum tubes of which needed the substantial generator shown on the Bristol Fighter at Hendon. British wireless use, which was well ahead of German use, breaks down into two periods. The earliest airborne wireless sets, such as that used in the prototype B.E.2, were simple spark sets that could only transmit. The guns for which such wireless spotted communicated to the planes in a "clock" code, laying lengths of cloth in a radial clock-like pattern around their emplacements. Such spark sets operated at relatively high frequencies for the time, in the medium-wave band. Their range was relatively short. Although they were easily jammed, jammers operating on the same wavelengths had equally poor range and had to be moved to whatever part of the front the wireless-equipped plane was flying over. It was easier to send fighters to shoot down such airplanes than to jam their wireless transmissions. The most common set was the Sterling, subject of a lengthy description in a postwar issue of *Wireless World.* The Sterling was derived from the Royal Naval Air Service (RNAS) Wireless Transmitter #1, one of which is displayed at the RAF Museum at Hendon. The Sterling was rugged, simple, compact at $8'' \times 8'' \times 5''$ (*WW* 7 [May 1919]: 101–3) and, at 9 lb, 12 oz, light in weight (Erskine-Murray 1921, 695). The 6 V accumulator to power it weighed perhaps another 10 lb.

One of the few British aviation histories to mention the Sterling by name is that of the usually reliable Penrose (1969). Penrose describes the pilot's task accurately as imagining arbitrary target circles of varying radii denoted by alphabetic letters "and, from the draughty cockpits of the handful of Maurice Farmans and B.E.2cs, [spelling] out in amateur Morse the hour angle and radial reference, using primitive Sterling one-way radios weighing a colossal 220 lbs, yet incapable of transmitting more than eight miles" (42–43). The source of these inaccuracies is unclear. Lifting pilot, observer, and a 220 lb radio would have strained a B.E.2c to the utmost. Nor was the Sterling primitive. It was certainly as good a radio as existed for the purpose in 1915. Its slightly more tunable and powerful successors, the Model 52A and the Model 54A, were not that much better (Erskine-Murray 1921, 694–95). Only once continuous wave sets began to appear in large numbers, from March 1918 on, could the Sterling be plausibly called primitive. As for its range limits, *Wireless World* commented that "with a crystal detector and fair atmospheric conditions it proved efficient up to 20 miles between plane and earth" (7 [May 1919]: 102). Twenty miles was the range limit of all but the very biggest World War I artillery, so it was hardly realistic to describe the Sterling's range as limited in actual use.

Given the scattered nature of the data, the relative lack of interest in military wireless by historians and buffs alike, and the British classification of records pertaining to such use as secret, piecing together even the British use of wireless is difficult. Although it is clear that the RNAS played a major role in the development of airborne wireless in World War I, that role is recorded in the Public Records Office in one handwritten document. The document notes that when the RNAS became part of the RAF in April 1918, "83% of two seater machines and 100% of airships were fitted with wireless. . . . The standard [spark] transmitting set weighed 90 lbs and had a range of 120 miles. A continuous wave system with a range of 400 miles was just being supplied and this had a special attachment enabling wireless telephoning to be carried out" (PROAIR 1/109/15/29). The development of continuous wave systems is well covered by Gossling (1920), but the engineering journals contain nothing on the earlier spark systems. Much the same can be said of the history of the radio in the RAF and its predecessors, the RNAS and the Royal Flying Corps (RFC). *Wireless World* published brief accounts of the spark transmitters when the war was over but clearly was far more interested in the joys of continuous wave technology.

In its April 1919 issue *Wireless World* began an "Aircraft Wireless Section," edited by J. J. Honan, formerly a lieutenant and instructor in the RAF (35–40). In May 1919 the section described the Sterling transmitter (101–3). In November the section described the Model 52A spark transmitter, essentially a Sterling with an air-driven alternator (470–72). In December it described the Model 54A, a more powerful, battery-powered spark transmitter with a range of up to 50 mi capable of operating on two different sets of wavelengths, 200–335 m and 500–600 m (512–14). I have not been able to find production figures for such sets in the PRO, but Honan noted in January 1920 that although crystal receivers were being rapidly phased out in favor of modern continuous wave vacuum tube sets by the war's end, "as a land instrument for the reception of messages sent from the air, the crystal 'Short wave tuner' proved most success-ful. . . . In fact, the Armistice found some thousands of these sets still on active service in and about the front line area. They were mostly em-ployed in the reception of messages from aircraft engaged on patrol or short reconnaissance duties, and more particularly for recording the ob-servation of planes working on 'target spotting' in co-operation with the artillery" (605).

An official British report on the state of wireless telegraphy at the Armistice indicated that each RAF brigade had an average of two to four corps squadrons cooperating on an army front. Each of the roughly 200 gun batteries had a receiver, and each of 50 to 70 airplanes had a trans-mitter. In battle each machine covered about 400 m of the front (PROAIR 1/727/154/3). If the ratio of transmitters to receivers was roughly 1 to 3, then almost all of the 857 corps reconnaissance machines recorded by Bruce as being at the front on 31 October 1918 must have had a transmitter. Bruce lists 1,407 machines at the front on that date that could have been used for reconnaissance: 182 Armstrong Whitworth F.K.8's, 328 Bristol Fighters, 1 lone B.E.2, 222 F.E.2's, and 674 R.E.8's (Bruce 1957). By late 1918, however, the F.E.2's were being used for night bombing, for which they did not need radio, and the Bristols were mov-ing toward continuous wave radio and being deployed increasingly for fighting. Most corps reconnaissance fell to the F.K.8's and R.E.8's. Such a conclusion is reinforced by the statement in *Wireless World* in July 1918 that "to-day all pilots and observers alike are trained in wireless work. It forms part of their professional qualifications, and the radio equipment of a modern aeroplane makes a wonderful advance upon the primitive outfit formerly used. . . . Radiotelephony, when it comes into more ex-tended active practical employment, will . . . still further increase the po-tentialities of the Air Service" (215).

Had the war continued into 1919, continuous wave wireless telepho-ny might well have had an impact on World War I akin to radar's impact on World War II. At a meeting at RAF Headquarters on 18 July 1918, the minutes for which had a fifty-year secrecy notice, it was agreed that "all fighter reconnaissance, bombing, and single-seater squadrons ought eventually to be equipped with telephones" (PROAIR 1/1996/204/273/235). The corps squadrons that spotted for the artillery were not mentioned. Of the 18 squadrons affected by this decision, 2 squadrons of Bristol Fighters were already so equipped. The report concluded that the 800 officers of the remaining 16 squadrons could be retrained in wire-less operation by the end of January 1919 and established the appropri-ate schedule. A much more detailed summary of progress through the Armistice noted five main uses of wireless telephony: (1) in-flight in-struction in formation flying behind the lines; (2) in-flight communica-tion during offensive patrols; (3) reporting reconnaissance results to Army Headquarters; (4) in contact patrol; and (5) communication with tanks (PROAIR 1/727/154/3). No. 22 Squadron worked to develop the use of wireless in offensive patrols at the front. No. 8 Squadron was a train-ing squadron back in Britain working on liaison with tanks. Both squadrons were equipped with Bristol Fighters (Bruce 1957, 141). The other Bristol Fighter squadron at the front to use wireless telegraphy was No. 88, which was experimenting with it for use in both offensive patrols and liaison with No. 103 Squadron, equipped with de Havilland D.H.9 bombers (PROAIR 1/727/154/3; Bruce 1957, 197).

The radical use was, of course, in tanks. In late 1918 the problem was that the use of what were then referred to as short wavelengths, of 200–600 m, required aircraft to trail antennas 30–50 m in length. It was not possible to mount a 30 m antenna on a tank. Even on aircraft the long antenna was a problem. One of the few popular writers to pay attention to wireless was former RFC Captain W. E. Johns, who started his career writing adventure stories for British teenage boys with tales of the Great War in the air based on his own experiences. Johns' hero, Captain Bigglesworth, Biggles for short, started his RFC career flying F.E.2's, moved to Bristol Fighters, and ended in Sopwith Camels. In a story called "Eyes of the Guns," Biggles, flying an F.E.2, disables a German Albatross fighter for reasons he cannot initially fathom: "He remembered his aerial and took hold of the handle of the reel to wind in the long length of copper wire with its lead plummet on the end to keep it extended. The reel was in place, but there was no aerial, and he guessed what had happened. . . . When he had swung around after his loop, the wire . . . must have swished around like a flail and struck the Boche Machine, smashing its propeller!" (Johns [1935] 1992, 137).

Continuous Wave Vacuum Tube Wireless

The possibility of centralized C^3 of combined air and ground operations revolutionized the theory of warfare. Spark telegraphy was a one-way system that could be crudely applied to spot for the artillery. Continuous wave wireless telephony between headquarters and mobile air and ground units meant a return to mobile warfare without losing the bureaucratic advantages of centralization. Continuous wave radio using vacuum tubes to transmit, receive, and amplify voice signals made this possible. The extreme urgency of World War I, in which telecommunications became of paramount importance, caused normal patent agreements to be suspended and, on the Allied side, encouraged massive technology transfers between countries as well as between firms. Because the vacuum tube was, from a manufacturing point of view, a fairly straightforward development of electric light bulb technology, a rapid move to mass production was possible. The effect was standardization at a much earlier period than is usual in a new technology (Nebeker 1996).

Vacuum tube amplifiers emerged from Bell Telephone's need to develop repeaters for the American transcontinental telephone service opened in 1915 (see chapter 3). Lee De Forest, the greatest of all vacuum tube pioneers, was already selling "the Radio Telephone for use wherever wire cannot be employed" at the Panama-Pacific Exposition in San Francisco in 1915 (Tyne 1977, 123). By 1917 De Forest was offering his "Type A Aeroplane Transmitter" for sale (117). Once the decision was made to equip large numbers of airplanes, American companies produced as many as a quarter of a million vacuum tubes for continuous wave wireless telephony. Patent disputes between AT&T, De Forest, Marconi, and others were suspended by the war. American progress toward large-scale production of standardized designs was rapid (131, 146). By 1920 the AT&T engineer Henrich Van der Bijl could publish a definitive and practical work on the production of electrons in the cathode (Van der Bijl 1920).

Despite this immense American research and production effort and considerable British use of American tubes, the history of the military development of continuous wave mobile radio is British. As early as 1914 Marconi was making a short-distance wireless telephone transmitter and receiver using "soft" tubes designed by Captain H. J. Round of Marconi (Tyne 1977, 207–14). "Soft" tubes amplified well but were hard to manufacture consistently and even harder to use. Once American manufacturers demonstrated that it was possible and advantageous to evacuate tubes more thoroughly, they were replaced by "hard," high-vacuum tubes. These were much easier to manufacture and use, although they did not amplify as much. One "soft" tube often had to be replaced by three "hard" tubes. The most common "hard" tube used by the British had its origins in 1916 in the French TM-type tube, for *télégraphie militaire*. This was modified to become the British R-type tube (Gosling 1920), which spawned numerous descendants through the late 1930s (Tyne 1977, 215–24).

A considerable number of continuous wave sets were developed. Most of these are described by Erskine-Murray (1921). The exceptions are the Marconi Types 50 and 55. Type 50 was a four-tube receiver operating on 600 m. The Type 55 was a seven-tube set that could both send and receive (*WW* 6 [February 1919]: 598). Erskine-Murray, "the officer in charge of the wireless experimental section of the Royal Air Force," described three aircraft transmitters for wireless telegraphy, five wireless telephone transmitters, and five vacuum tube receivers. The single biggest problem was developing duplex rather than switched wireless telephony. In a switched system simultaneous transmission and reception, obviously vital during air combat, was impossible. This problem was essentially solved by the Wireless Experimental Establishment of the RAF at Biggin Hill in late 1918 (Eckersley 1920).

Data on the other combatants are singularly lacking. The French were obviously large users and producers of continuous wave equipment, as were the Americans, although little but vacuum tubes got across the Atlantic. The Germans seem to have lagged somewhat. Writing in 1933, Vyvyan, the Marconi engineer most responsible for beam radio, commented that "the enemy up to the time of the Armistice had not been able to design a continuous wave transmitter for use in the air. This was borne out by the capture of a document stating that the English had been able to construct a continuous wave transmitter for use in aeroplanes, and offering large rewards for the salving of any portions of the wireless apparatus from a night bombing machine brought down behind the lines" (Vyvyan 1974, 139).

The only other evidence in the English-language engineering or secondary literature seems to be one report in the July 1918 issue of *Wireless World,* picked up from the French magazine *La Nature:* "Lt. Jean-Abel

Lefrance states that for some little time [German aircraft] have been occasionally equipped with continuous wave receivers of the valve type. Describing the transmitting apparatus, he says that the generator produces alternating current (270v 3a) and continuous current (50v 4a). The machine is driven either by a small airscrew rotating at 4,500 rpm or by the motor" (210). Even if the Germans did have such sets, their vacuum tubes were four to five times as expensive as British ones and "appeared to be mainly hand-made, whereas ours were very largely machine made" (Cusins 1921, 769). Nebeker has, however, recently noted that by late 1918 German production was up to 1,000 RE-16 vacuum tubes per day, about the same as French production of the TM (Nebeker 1996). Germany certainly became a major producer of advanced vacuum tubes after World War I with the publication in 1923 of Heinrich Barkhausen's seminal treatise on the behavior of electrons (Blumtritt 1996).

Wireless in the Ground War

Despite the failure of Marconi sets in the Boer War, wireless telegraphy's importance to ground troops was recognized before World War I, although it quickly came to be used mainly for air-to-ground communication spotting for the artillery. As early as April 1913, *Wireless World* noted that the American delegates to the International Radio Telegraphic Convention in London in 1912, convened in the wake of the loss of the *Titanic,* had reported to the U.S. Congress that Britain and Germany regarded wireless telegraphy as an important weapon, and that America needed to keep up. *Wireless World* further noted that "it is not sufficient to perfect the machine if there is no person sufficiently experienced to take charge of it" and pushed for the Boy Scouts and the Church Brigade "to make themselves proficient in this branch of national service" (1). In the same issue Lord Robert Baden-Powell, the founder of the British Boy Scouts and the international Boy Scouts movement, noted that "wireless has become a favorite hobby with boys of the right sort" (xxi), and in the May issue he welcomed a new monthly series of published lectures to educate such boys (iv).

The limitations of wireless telegraphy in field use were, however, quickly realized. Because of the free use of unciphered wireless telegraphy by his Russian opponents, Hindenburg was easily able to follow their intentions and defeat them at the Battle of Tannenberg in 1914. Wireless telegraphy was not a good mechanism for strategic C^3 at the staff level because it was easily intercepted, decoded, and use by opponents. Its real use lay in wireless telephony for tactical C^3 on the battlefield.

Despite the problems of fitting continuous wave sets in tanks and other ground vehicles, the experience with wireless telephony in the final 100 days of World War I indicated its remarkable potential for breaking the deadlock of trench warfare. The most effective user of this technology was the Canadian Independent Force.

> A most formidable collection of armoured cars, machine guns and pom-poms mounted on cars and lorries, this force would "sail out into the blue"

once the fighting became open enough, engaging machine gun nests and automatic guns with great success. One car carried a continuous wave set which kept the force in touch with a special set at a convenient headquarters behind our line. . . . That it was possible to keep such a detached force in close touch with the commander of the main body of attacking troops is a remarkable demonstration of the value of wireless in modern warfare. (*WW* 7 [August 1919]: 267)

This Canadian force, whose equipment was not specified, seems to have been a "model of efficiency and successful organization" (*WW* 7 [September 1919]: 396). At the very end of the war, they certainly demonstrated the immense future possibilities of tactical battlefield C³ and thus the way out of trench warfare. Many military strategists would comment on the use of C³ in the interwar years (Bidwell and Graham 1982, 136).

In the aftermath of the Armistice there was a hiatus in wireless development. Not until October 1929 was a new design commissioned by the British army to replace the obsolete sets left over from the war. This set, W.S.3, entered trials in 1933 but was difficult to produce in quantity and did not take advantage of the short waves then becoming common in commercial use. Nevertheless, it was the only medium-range control set available in quantity in 1939, and it was not effectively replaced (by W.S.23) until 1940. Wireless sets W.S.2, of 1934, and W.S.7, of 1935, were intended for tank use but were never satisfactory. They were replaced by W.S.9 in 1939. W.S.9 weighed 192 lb, as much as the earlier sets despite its much better performance, and was still difficult to mount in medium and small tanks, although it required only a 6 to 10 ft antenna. The first short-range, intertank set—W.S.14, of 1940—was modified from a civil aircraft set and weighed 81 lb. It was not really until the introduction of the 70 lb W.S.19 of 1941 that a really effective set was deployed, for the first time operating in the Very High Frequency rather than the High Frequency ranges (Meulstee 1995, W.S.3–2, W.S.23-1, W.S.9-1, W.S.14-1, W.S.19-1).

From a geopolitical viewpoint, wireless communications provided an important technology with which to direct the ceaseless economic struggle between competing states in the capitalist world economy. But there were other uses for the radio spectrum when the economic struggle turned into a military struggle. Some of those uses were pioneered in the civil world, but others were to be created out of a perceived need for a particular military purpose. McNeill has written persuasively about the emergence of a "command technology" among the naval forces of the world in the late nineteenth and early twentieth centuries (McNeill 1982). This technology scaled new heights in the war for control of the radio spectrum that began alongside the naval arms race in the early twentieth century, accelerated with the demands of combined land and air operations at the end of World War I, and dominated the entire course of World War II. The polity or set of polities that did best in these struggles for the radio spectrum inevitably triumphed in the major wars of the twentieth century.

Mobile radio meant a return to the war of mobility that had characterized wars of the past and an escape from the trench warfare of World War I. The origins of trench warfare were several. The railroad and the machine gun have been given their share of the blame, the railroad because it allowed rapid mobilization to the rear but not in the front lines, and the machine gun because it allowed a few machine gunners to tie down huge numbers of infantry and cavalry. The telegraph deserves similar credit. It was an imprecise communications device: written messages of the maximum brevity were devoid of the expressive meaning available in face-to-face interaction or the more discursive form of meaning available in letters. Telegrams were open to considerable misinterpretation. In addition, the telegraph suffered many of the same problems as the railroad. Extensive fixed infrastructures representing considerable capital investment allowed extremely rapid military and economic mobilization of the population of the increasingly bureaucratic nation-states that flourished in the late nineteenth century. These bureaucracies owed their expansion largely to the state's ability to control telegraphic communications. Extensive fixed infrastructures were not, however, well adapted to a war of mobility. Only in a fixed zone of conflict could the necessary telegraph lines be laid to carry the messages necessary for managing machine-age warfare, and even there fixed lines were fairly easily destroyed by enemy artillery bombardment.

Wireless revolutionized warfare by allowing the retention of the bureaucratic advantages of centralized control and a return to mobility. It was quickly and enthusiastically embraced by the navies of the world, to which centralized control had come late and in which the tradition of the independent commander died slowly. The air forces, lacking clear traditions, tended to follow naval models but had to work hard to miniaturize wireless technology. The armies, by far the most conservative arm, were eventually changed most by wireless.

World War I produced the six major innovations that identify modern land war. Two were in software: the scientific control of artillery fire and the revolution in infantry tactics; and four were in hardware: tanks, aircraft, machine guns, and wireless (Bidwell and Graham 1982, 172). The last, however, was the common technology that bound the other five together in the C^3 of combined operations. It was not spark wireless that made C^3 possible but lightweight wireless telephone sets using continuous waves generated by vacuum tubes.

The need for scientific control of artillery fire raised its head early in the war. Spotting from tethered balloons allowed telephone lines to be run down to the ground, but hydrogen balloons lacked mobility and were easy targets for fighter planes using incendiary bullets. The little-known but quite remarkably advanced electric helicopters developed in Austria-Hungary by Theodore von Karman, powered by ground generators through their tether lines, would have been much more difficult targets but seem to have been ignored in Germany and came, in any case, too late for the set-piece artillery battles. Observation airplanes proved their value with aerial photography early in the war, and a one-way wireless telegraphic link to the ground allowed immediate artillery spotting. Two-way wireless telephony offered obvious advantages over one-way telegraphy and was in sight by late 1918.

It is ironic but understandable that the tank was a product of thinking about the problems posed by trench warfare by the British navy rather than the army. The navy had long been used to a war of mobility, and so the concept of the tank as a "land dreadnought" came easily. The technology came more slowly. The mechanical problem presented relatively few difficulties. The San Francisco–based American agricultural-machinery companies Best and Holt, later merged as Caterpillar, provided the track-laying technology. Adding guns and armor did not present much of a problem, and even the doctrine of how best to use these new machines in conjunction with air power was in place by 1918 (Hugill 1993, 244–45).

Those who Carver calls "the apostles of mobility" (1979) pushed hard in Britain for new theories of war so that future wars might escape the deadly stalemate of the trenches. By mid-1918 Colonel J. F. C. Fuller, drawing on his experiences in the tank battles that had already occurred and on his position paper for Churchill, "Tanks as Man and Time Savers," had articulated the tactical doctrine of combined air and ground assault directed by wireless (Macksey 1983, 32).

As early as 1923 Fuller proposed that the war of the future would be won by strategic assault from the air as well as by aerial units in tactical cooperation with tank armies on land. He drew a gloomy picture of what he saw as the rapid and certain collapse of civil authority in London under terror bombing with mustard gas (Carver 1979, 35). Reflecting on the success of the German air raids in the Spanish Civil War in 1937, Basil Liddell Hart noted that although the material results of the Gotha bombing raids on London late in World War I were relatively slight, they made a profound psychological impression (Liddell Hart 1944 [1937], 32). In the period immediately after World War I the conclusion that a strategic air offensive was likely to prove irresistible was elevated to the level of doctrine in the writings of the Italian General Giulio Douhet and the American Brigadier-General William Mitchell. The evolution and testing of these theories and doctrines is, however, best left to the next chapter, since the needs of the strategic air offensive led to the development of a system of electronic communication that was much more sophisticated than the simple continuous wave wireless telephone system used in World War I.

POSTSCRIPT

One further consequence of World War I deserves note. The war caused the training of huge numbers of men to use wireless as well as the mass production of high-quality vacuum tubes and sets designed to work in the 200–600 m wavebands. At the war's end these tubes were made available cheaply as war surplus and created a huge boom in wireless. Since some of this material was intended for wireless telephony, it is perhaps unsurprising that people began to experiment with broadcast entertainment, to be picked up by wireless receivers converted from air force sets or built following the plans found in publications such as *Wireless World* and a host of other, more practically oriented magazines. In 1923, just after the first experiments with broadcast radio, *Wireless World,* now a weekly, published instructions for adapting the RAF Type 10 receiver, a five-tube wireless telephone receiver for aircraft (11 [12 December 1923]: 349–51). By 1923 *Wireless World* was filled with advertisements for proprietary chassis that readers could buy and build as well as for war surplus vacuum tubes. These chassis kits were never supplied with tubes because surplus tubes were far cheaper than new tubes possibly could be. New tubes also carried very high taxes and license fees. A four-tube set using war surplus components in the early 1920s cost less than £10; a new one of equal performance would have cost nearly £45 (Bussey 1990, 5).

More to the point, however, the war created both a larger market for trained wireless engineers and a cadre of engineers who created a domestic radio market in the aftermath of World War I. This market in turn would generate the nucleus of electronics engineers for World War II. Flowerdown, the military training center that had trained many of the wireless operators for the British air effort, turned its attention to "commercial courses" once the war was over in order to help demobilized wireless operators adapt their skills to the civilian market (Burch 1980, 6). In his 1923 presidential address to the Radio Society of Great Britain W. H. Eccles was specific about this: "When the war came, the amateurs penetrated in their hosts into the armies, and turned their wireless experience and their talents to the design, construction, operation and improvement of apparatus for use in war" (1923, 51).

When the war ended, "the valve had arrived, and the stay-at-homes envied the service men their knowledge of it" (*WW* 11 [25 November 1922]: 260). Historians commonly write of the advantages to the British and Americans in both world wars of more efficient, better-organized production driven by better mobilization of natural resources. But it is important to remember that modern wars need large numbers of highly skilled technical personnel and that developing and mobilizing such skills is a critical task. In World War I, Britain, for reasons perhaps captured best by Pocock (1988, 6), and commented upon in chapter 4, did much better in the new field of wireless than did Germany. Clearly, Britain substantially outperformed Germany in the use of radar in World War II. This was at least in part because of the ready availability of war surplus equipment, the early onset of broadcast radio, and the general "radio-mindedness" created by World War I.

CHAPTER SEVEN

Communications, Command, and Control in the War in the Air
Radar, World War II, and the Slow Transition to American Power

Sleep on, pale Bruges, beneath the waning moon,
For I must desecrate your silence soon,
And with my bombs' fierce roar and fiercer fire,
Grim terror in your tirèd heart inspire; . . .
Sleep on no more! The bombs are screaming down,
And sleep is murdered in the little town!

—Paul Bewsher, *The Bombing of Bruges*

The development of mobile forms of two-way radio changed the nature of war. Radar would change it far more. The centrality of war for the capitalist polities enmeshed in the modern world-system, in particular wars of hegemonic transition, ensures that such change cannot be taken lightly by any polity with pretensions to hegemony. The relative position of the major polities involved in the most recent hegemonic transition with regard to information technology, or communications, command, and control, is thus of considerable interest. By 1939 four polities were most involved: the declining hegemon, Britain; the two challengers from World War I, Germany and America; and one new challenger, Japan.

Because mobile radio offered a return to wars of mobility as well as the possibility of the centralized command and control so important to the bureaucratic national state, it was a technology that favored offensive strategies above all others. Britain's "moat defensive" geography (as noted earlier the term is Shakespeare's, from John of Gaunt's deathbed speech in *Richard II* 2.2) meant that it rarely favored a purely offensive geostrategy. As Mahan pointed out in the 1880s, Britain controlled the world by controlling its sea trade through wars of blockade. In this sense the navy was a means of both protecting the British Isles from invasion and projecting British power abroad.

But navies cannot control continental interiors, and Britain historically refused to pay the high taxation costs of a standing army. In this respect British army involvement in World War I was an aberration brought on by Mackinder's 1904 restatement of Britain's geopolitical position as being at the mercy of a powerful land state coming to control the Eurasian landmass with railroads and telegraphs and thus able to command the resources of the heartland to build an overpowering fleet.

Adding to the problems identified by Mackinder were the problems posed by aviation. British writers such as H. G. Wells recognized the offensive implications of air power from its very beginnings. From the publication of Wells' *The War in the Air* ([1908] 1967) on, it was clear that the "moat defensive" was drained: Britain was "no longer an island" (Gollin 1984). The loss of the "moat defensive" meant that Britain was the first polity to develop an independent air force, in 1918, to parallel its navy and army. That air force originated as a strategic offensive arm of the navy. This early realization of the need for an offensive air arm seems to be well understood by modern commentators such as Gollin (1989) and Robertson (1995, 25).

The first genuine strategic bomber ordered by the British armed forces, the Handley Page 0/100, was ordered by the Royal Naval Air Service at almost the same time that de Havilland's prototype D.H.3 bomber (of slightly smaller dimensions) was being turned down by the army's Royal Flying Corps on the grounds that "large twin-engine bombers were impractical" (Mason 1994, 14, 44, 48; quote from 48). The link between strategic bombing and the literary tradition of Wells is, however, generally ignored, as is the implication for defensive technology. It was quickly understood that the navy blockaded ports to reduce imports of raw materials and food and the air force bombed cities to destroy industrial infrastructure and civilian morale. The link between this and Mackinder's reformulation of geopolitics is, however, not made by modern commentators. The air force might strike back directly at the heartland in a way that the navy never could. John Maynard says it well: "Without this [aerial] offensive [Nazi] Germany could have prospered, immensely strong and secure behind the defensive insulation of the occupied lands which surrounded her. Her citizens could have enjoyed the benefits of virtually undisturbed peace, their belief in the demoniac [*sic*] regime that governed them unshaken" (12).

In Mackinder's reformulation of geopolitics Germany stood to gain the most from the ability of railroads and telegraphs to make possible continental-scale polities. Although World War I had proved that the combined power of France, Britain, and America could bring Germany to its knees, it had been, in Wellington's words about an earlier attempt to force European integration, "the nearest run thing." The return to the offensive strategy offered by mobile radio further favored Germany, which, with powerful states to its east and west, had always required a tactical offensive geostrategy. The return to the offensive offered by air power was recognized in the centrality of the Luftwaffe as a tactical air force in the emergence of Nazi Germany.

The nineteenth-century history of America stood as a shining example to Wilhelmine Germany of how railroads and telegraphs could help a polity govern a continental mass. After 1919, however, America generally paid inadequate attention to the rest of the world or to its geopolitical position in that world except in the crucial area of naval appropriations, which were not viciously reduced after World War I. American political isolationism, too often seen by American historians as a merely domestic response, was an implicit recognition that America's geopolitical position was that of Britain exaggerated, with the Atlantic and the Pacific a much more efficient "moat defensive" than the English Channel. Mahanian doctrines percolated deep into the American navy as a result, and navalism ruled American military appropriations. As in Britain, the air power of others was obviously a threat to both the navy and the "moat defensive." And as in Britain, it became obvious to persons such as Brigadier General William Mitchell that strategic air power could be a complement to or even a substitute for naval power. Lack of serious American involvement in the aerial component of World War I regrettably delayed the emergence of an independent strategic air force, the need for which was clearly dictated by American geostrategy.

Japan trod a somewhat different road. The Japanese paid lip service to Mahanian naval doctrines, most crucially in terms of seeking the ultimate clash of battle fleets, but without an adequate understanding of the war of blockade that Mahanian doctrine required. Japan never built enough convoy escorts to protect their merchant fleet from American submarine attacks. The Japanese further recognized the need for a strong land offensive in China if they were to acquire the raw materials needed to accelerate industrialization and build a strong battle fleet, a strategy clearly informed by Mackinderian thinking. Japan tied air power to its navy and army, using its army air force as a tactical air force and its naval air force for the strategic offensive.

But whereas the development of mobile radio in the teens and twenties encouraged all forms of offense, the development of radar in the thirties returned the advantage to the defense, most notably against an air offensive. It is therefore hardly surprising that Britain, despite its manifold ineptitude in most matters of Neotechnic technology, was ultimately to have the most success in developing radar as a defensive system. What is surprising is how well Germany did in developing the technical side of radar while still on the offensive and how inadequately Germans turned their considerable skills to the defensive after 1943, success in the nighttime Battle of Berlin notwithstanding. Allied reports after 1945 tended to paint the German war effort as incompetent, but there is more to it than that. The Germans, perhaps driven by the egalitarianism of the Nazi Party, did not maintain lists of skilled technicians and specialists who were not allowed to enter the military as did, for example, the British. The consequence of this was that many skilled radio engineers were allowed to volunteer or were drafted. By late 1943 the Luftwaffe was thus reduced to advertising in the magazine *Funkschau* (loosely translated, "radio show") for "technicians having some skills in electronics" (*Funkschau* 16 [August–September 1943]: 1) The substance of this and other advertisements was that electronic science was of the utmost importance for German electronic warfare. Attempting to establish new electronics research and development units when the human capital had been spent down in this way was extremely difficult (Arthur Bauer, personal communication, December 1995).

The role of America in the development of electronics for the military is ambiguous. America had one of the best electronics industries in the world in 1939, yet the United States Army Air Corps (USAAC) fought

most of World War II using British or British-designed electronics information technology. So strongly did Britain mark air war that British
World War II terminology is still in everyday use in the United States Air
Force (e.g., *bandits* for enemy aircraft, *angels* for height in thousands of
feet, *tally-ho* for enemy in sight). The failure of the USAAC to develop
radar cannot be put down simply to naive isolationism, since a true defensive obsession should have encouraged such technology. Rather, it
must be put down to an extreme case of the "moat defensive" mentality,
in which it was simply assumed that the Atlantic and Pacific were impass-
able and that the job of the USAAC was to project power. Regrettably, the
American navy kept to itself its development of an effective gun-laying
radar by the late 1930s, presumably believing itself to be the only true defender of the "moat defensive." Americans did not lack for an offensive
strategic doctrine by which to apply air power, nor had they neglected to
produce the air weapons to implement it. A whole series of long-range
strategic bombers, starting with the Boeing B-15 and culminating in the
Boeing B-29, were developed from the mid-1930s on. The three best of
these—the Boeing B-17, the Consolidated B-24, and the Boeing B-29—
saw extensive service in World War II.

If America's electronic-warfare capabilities were lacking on the eve of
World War II, Japan's were almost completely absent. This was not for
lack of interest in radar on the part of the Japanese navy or lack of technical ability. Japanese scientists had developed a high-frequency magnetron capable of a continuous output of 500 W by 1939 (Nakajima 1988,
246). But Japan, secure in the vastness of the Pacific behind the walls of
its fleet, was even more oriented toward a "moat defensive" doctrine than
Britain or America. In addition, "Japan had very few scientists and engineers and a shortage of materials as compared with the US and European
countries before the war period . . . there was no comparison between
Japan and the US and England concerning their research programmes
and production capabilities" (258).

Thus, the initial research, development, and application of radar in
air warfare was almost exclusively a product of the Anglo-German component of World War II. American naval radar in the meter band was sophisticated and almost as advanced as that of the Germans, but it was developed in foolish isolation and secrecy (Page 1988). Americans did
sterling work refining and then producing British or British-designed
electronic equipment, much of which the British presented to them
through the Tizard Mission of 1940. Sir Henry Tizard, whose chairing of
the Committee for the Scientific Study of Air Defence, in reality a sub-
committee of the Committee for Imperial Defence, had done so much
to promote radar, proposed early in 1940 that Britain disclose her scientific secrets "to the USA in return for help on technical and production
matters" (Bowen 1987, 3, 150). This became known as the Tizard Mission,
and in the fall of 1940 it transferred to the United States the most recent
British developments: "the jet engine, then in embryonic form, rockets,
predictors and, most important of all, radar in its many forms" (152).

Without American superiority in production most such advanced
British equipment and ideas as centimetric radar and the proximity fuse

would never have reached the front lines in adequate numbers, if at all.

Americans also emulated the structure of the British research effort when they set up the Radiation Laboratory at the Massachusetts Institute of Technology, after Britain's Telecommunications Research Establishment (TRE). Although the Radiation Laboratory did a great deal of development work and set the model for postwar research laboratories, it contributed very little that was original in time to be used in the war itself.

The reasons for the relative success of the British and Germans in developing radar and the failure of all others lies in four areas: (1) geopolitical theory; (2) an obsession with air war after 1918; (3) the social and political organization of research; and (4) the development of commercial electronic television. The radar war of 1936–45 was an epic struggle between two main players, Britain and Germany. Like all great epics, its coherence depends upon a descriptive account. Much of the geopolitical theory involved is covered in chapter 1. The geostrategic component of that geopolitics, as well as the other three others named above, is covered within this chapter. The narrative that results is that of the defining hegemonic challenge and response of the twentieth century, thus of the destiny of our current version of the world-system.

THE DEVELOPMENT OF AIR DOCTRINE IN THE AFTERMATH OF WORLD WAR I

In 1938 the British prime minister, Neville Chamberlain, returned from a visit to Munich proclaiming "peace in our time." The 1938 Munich agreement between Britain, France, Germany, and Italy ceding Czechoslovakia to an aggressively expanding and rearming Germany has been portrayed as the classic example of appeasement, and Chamberlain as a spineless toady. Yet Chamberlain had good reason to sign the Munich agreement. German performance in the air war over Spain, exceptionally well-directed propaganda, and the unfortunate pro-Nazi pronouncements of the politically naive American aerial expert Charles Lindbergh had combined to convince many observers that the reborn Luftwaffe was capable of doing what proponents of strategic air doctrine had been claiming for nearly 20 years, namely, reducing enemy cities to rubble and forcing their civilian populations to surrender. After its experience in the Spanish Civil War, notably at Guernica, the German High Command came to believe "that it would be possible for the fast medium bomber to lay waste any country without opposing fighter defences having an opportunity to interfere" (Green 1959, 9). At the time of the Munich agreement there had been no clear demonstration of a defense against the obvious German threat, although the basis of that defense, an effective system of identifying incoming bombers linked to a coordinated system of fighter control from the ground, was in place in Britain by the fall of 1937 (Bowen 1987, 27). Until the Battle of Britain in September 1940, the thinking in most military and all political circles was, however, that offensive strategic bombing would pose insurmountable problems for would-be defenders. Only in Britain did a defensive doctrine based on the possibility of electronic identification of incoming bombers take root deeply enough in the mid-1930s to flower in time for that key defensive air battle of World War II.

Historians have tended to underestimate British power and commitment to such unconventional forms of power as air power after the turn of the century. The conventional interpretation is neatly summarized in the title of such books as *The Weary Titan* (Friedberg 1988), although Friedberg's concern is only with the years 1895–1905. This underestimation has been made even though Britain produced more airplanes than Germany in 1940 and its production figures continued to exceed Germany's throughout World War II. Edgerton rightly argues that Britain was a "militant and technological nation," a "warfare state" rather than a welfare state (1991, 83). Because he believes that the strategic bombing campaign against Germany was not a success, Edgerton concludes that the British severely misallocated resources to the construction of strategic bombers. He suggests that the commitment to the strategic offensive caused "massive investments in machines which could not be used for fear of retaliation in 1939, did not work between 1940 and 1943, and pointlessly destroyed people and buildings thereafter" (82). Inasmuch as he turns upside down conventional British history, which concludes that the British state is "incapable of planning, of investing in science and technology, or of appreciating scientists and engineers," I agree with him, at least with respect to the period after 1934, when the clear and present danger of Nazi rearmament was recognized. But I cannot accept his assessment about misallocation of resources since he pays little attention to the flexibility of British air doctrine, its switch to defensive science and technology when it mattered, in the late 1930s, and its return to the offensive when the Germans invaded Russia in 1941. I would also suggest that the American view that the air war over Germany was a war of attrition with the Luftwaffe, with the bomber as bait for German fighters, and that without it the invasion of Germany would have been considerably delayed is essentially correct (see McFarland and Newton 1991, ch. 6).

The real flaws—the kinds that lose wars rather than delay victory—in both air doctrine and the development and application of technology appropriate to that doctrine were German, not British or American. The Germans failed to develop or apply a proper strategic offensive doctrine; they confused the twin-engine bombers, which were ideal for tactical offensive, with the much heavier four-engine bombers they really needed for the strategic offensive; they failed to capitalize on the technical superiority in both electronics and fighters that they had in 1939 when their opponents began overhauling both in 1941; they failed to secure an adequate supply of aviation fuel; and they took far too long to understand the need for proper defensive doctrine against a well-pressed home British and American strategic offensive after 1943. They thus failed to win the crucial defensive air battle for Germany in the last year of the war.

Homze notes that German air doctrine as developed by Hitler and Göring in the late 1930s was akin to Imperial German naval doctrine under Tirpitz in the first decade of this century (1976, 55). The air fleet the Luftwaffe brought into being was a "risk fleet," one intended to make any opponent think twice before entering a war with Nazi Germany. It was cheap, with 400 bombers at the cost of two battleships. It was flexible inasmuch as it could be used on the kind of two-front war against Poland and France that Nazi geostrategists took for granted in their irredentist push for Lebensraum. It was, in short—and this was its ultimate Achilles' heel—a "fleet that had to look like a strategic air force abroad while it functioned like a tactical one at home" (56).

In the four years of World War I the development of aerial warfare from its origins in scouting for the army's guns or to locate the enemy battle fleet was immense. Air arms grew in explosive fashion. Whereas in early 1915 Britain had "less than 200 landplanes and 50 seaplanes," of which 60 were experimental, by June 1918 there were 2,630 British airplanes in front-line service. Total production amounted to 55,093 airframes and 41,034 aircraft engines, the latter being the main bottleneck. Germany, America, and Italy produced slightly fewer engines, France more than double what the British produced (Penrose 1969, 9, 606). Britain's RFC and RNAS merged to form the world's first independent arm, the Royal Air Force (Kennett 1991, 92). The main uses of aerial weapons became clear, as did the consequent requirements for specific practical types of airplanes: reconnaissance, ground attack, bombing, and interception. Because of the specific needs of the war, some of these groupings were better elaborated than others. Hallion, for example, traces five distinct generations of fighter airplanes during the war (1984, 115). In the years immediately following the war, air doctrine would come to comprise three basic positions: the tactical offense, the strategic offense, and the defense. The United States Strategic Bombing Survey at the end of World War II graphically summed up the two different styles of offense: "Strategic bombing bears the same relationship to tactical bombing as the cow does to the pail of milk. To deny immediate aid and comfort to the enemy, tactical considerations dictate upsetting the bucket. To ensure eventual starvation the strategic move is to kill the cow" (quoted in Price 1979, 117).

The European combatants of World War I experimented with all three positions, in particular with the two offensive doctrines. The ground attack aircraft evolved from the fighter as a contributor to tactical offense and the combined advance of air and ground forces. The strategic bomber was created to attack the enemy's home territory in order either to damage the infrastructure that allowed the war to be waged or to terrorize the civilian population. Defensive doctrine was poorly developed during and immediately after the war in part because the fighter plane had initially evolved as a way of denying enemy reconnaissance airplanes the ability to operate, in part because by the end of the war the combatants perceived the principal problem to be how to use the offensive to escape from the deadly war of stalemate being fought in the trenches. This is not to say that defensive practice was totally neglected. The German Air Service fought a far more defensive war than its opponents because the British and the French aggressively sought the offense and because the persistent westerly winds over the front pushed airplanes over German-held territory rather over the territory of the Allies. Although the

Germans never developed a true defensive doctrine based on the dictums of Boelcke, one of Germany's greatest aces, they certainly had plentiful experience in defense. The irony is that they construed their own actions as essentially offensive and came to define the fighter pilot as a *Jäger,* a hunter.

After the false start in the first years of the war, when they used slow, cumbersome, and highly flammable hydrogen-lifted airships in a well conceived but poorly executed strategic offensive against British targets, the Germans turned to airplanes. Far too many airships were lost to bad weather, slowness, and the ease with which they burned when attacked with incendiary bullets. A small number of German multiengine strategic bombers were eventually used in the first Battle of Britain, in 1917–18. These caused immense problems for the British by striking "bold and hard and in broad daylight at the very center of the biggest city in the world" (Cross 1987, 42). The *Englandgeschwader* comprised six *staffels* of twin-engine Gothas, with six planes to the *staffel.* The first successful raids on London by the *Englandgeschwader* resulted in three squadrons of front-line fighters being diverted to home defense, and three more soon after. By 1918 the Gothas were backed up by a small number of much larger but highly experimental Zeppelin-Staaken R-planes. By late 1918, 145 Sopwith Camel single-seat fighters and 55 two-seat Bristol Fighters were "defending" the British civilian population. The defense was not that credible. Of the 60 Gothas lost, "twenty-four were shot down or disappeared over the sea. The remaining 36 were written off in landing or approach crashes" (52), the particular weakness of the Gotha being its tendency to turn over on landing because its center of gravity had been changed as a result of a reduction in its fuel supply or its bombload. The combination of combat losses and accidents coming on top of the total defeat of the airships convinced the German Air Service that its strategic offensive was not justified. The giant German four- and five-engine R-planes came just too late and were too few in number to pose a serious threat, although had the war continued, they almost certainly would have. They carried a large number of mobile machine guns and could keep flying with one engine out, which the Gotha could not do. They proved very hard to shoot down. Losses were far lighter for those that made it to combat than was the case with the Gothas.

The geostrategic implication of the German raids was not lost on the British. Cross notes that the raids of 1917 "finally propelled the British War Cabinet into the arms of the strategic bombing enthusiasts" (1987, 53). The result was a refocusing of the British air war effort from reconnaissance and tactical activities toward attacks on enemy cities and infrastructure. By the end of the war in late 1918 this refocusing was, however, still minor. The British strategic bombing force had only 122 airplanes at its height (65). The needs of the last few months of the war forced this strategic bombing force to concentrate on tactical support for advancing ground troops, and its losses were consistently heavy. The refusal of the RFC and the RNAS to detach aircraft for home defense from the western front produced the major change, which was the formation of the RAF as an independent arm of the British military (Robertson 1995, 17).

Yet the apparent failure of the British strategic offensive against Germany did not stop Britain from continuing to develop a strategic doctrine after the war ended. On the other hand, the apparent failure of the German offensive against London persuaded Germany that strategic doctrines were inappropriate and that the best use of air power was tactical. This seeming discrepancy can best be considered in the light of geopolitics, contemporary opinions about air war, and the bombing campaigns themselves. In Britain, the German strategic offensive using airplanes was to hit at the heart of British civil society, at London itself, the seat of government, business, and the monarchy. This last point should not be ignored. To Britons, the monarchy symbolized both the continuity of the British state and the power of the old aristocracy. In World War II the accidental bombing of Buckingham Palace, the refusal of king and queen to leave London, and the frequent appearance of the royal family in bombed areas of the city to sympathize with the victims did much to avoid the morale problems that could have arisen had it been believed that only the homes of the working class were being destroyed by German attacks.

The Germans, however, went almost untouched by the strategic air offensive of World War I. British and French attacks were mostly on German-held French and Belgian cities within 50–75 mi of the lines. Of the German cities that were attacked, only Mannheim experienced a large number of attacks (21–30) on its industries, although aerodromes at Cologne, Coblenz, Frankfurt am Main, and Stuttgart all experienced 16–20 attacks. The remainder of the German cities attacked experienced fewer than 10 raids on their industries and railroads combined (Cross 1987, 64). No air assault was made on Berlin during World War I, although in the very last days of the war the British were preparing for just such an attack using the huge, four-engine Handley Page V/1500 bombers built expressly for the purpose (Barnes 1987, 23). In his study of the German Reichswehr, the military force that developed between the time of the Treaty of Versailles and the rise of Hitler, Corum notes that the "German Air Service, painfully aware of the failure of its 1917–18 strategic bombing campaign to inflict serious damage upon the enemy, had clearly decided in 1920–21 against strategic bombing of the enemy homeland as a serious military option" (1992, 146). The Germans were also aware of the failure of the British strategic offensive of 1918: "The loss of 352 RAF bombers in the 1918 campaign probably cost the British more money than the damage that the whole British campaign inflicted on the Germans" (155).

But the development of Luftwaffe doctrine and technology under Hitler also owed something to the Luftwaffe leadership structure. The German Air Service of World War I inflicted damage on its British and French opponents out of all proportion to its relative size because it fought an essentially defensive battle over its own lines against an enemy committed to the offense. It was the extremely successful fighter pilots of this period who took the leadership roles in the later Luftwaffe. Herman Göring, who as Hitler's righthand man came to command the Luftwaffe, was himself an ace with 22 kills and the last wartime commander of Baron Manfred von Richtofen's famous "circus," *Jagdgeschwader* 1. As Corum notes, "A strategic bombing theory that asserted that 'the bomber always gets through' could not sit well with fighter pilots. In fact the German Air Service's World War One experience had taught the men that the bombers often did not get through" (1992, 168).

Such experiences as those of the Germans over England in 1918 convinced others that long-range bombers posed an essentially unstoppable threat. The "apostles of mobility" pushed hard for new theories of war to replace the deadly stalemate of World War I's trenches (Carver 1979). Although the writings of Gifford Martel, J. F. C. Fuller, and Basil Liddell Hart, the "apostles" in question, have been construed as being about mobility on land, in reality they are about much more. As early as November 1916 Martel outlined the concept "of tank armies operating like fleets at sea" over anything but "very wooded or mountainous country" (quoted in Carver 1979, 22).

Fuller followed a very similar line in his famous article of 1920, "The Application of Recent Developments in Mechanics and Other Scientific Knowledge to Preparation and Training for Future War on Land." In *The Reformation of War* (1923) he noted quite clearly that the war of the future would be won by both tactical and strategic air offensives, the former in cooperation with tank armies on land. He drew a gloomy picture of what he saw as the rapid and certain collapse of civil authority in London under strategic terror bombing with mustard gas (Carver 1979, 35). On the battlefield he foresaw that "wireless reports will be sent back from the air fleet and telephoned from the flag tank to the squadron leaders, who will manoeuvre for position, for light and wind." By 1929 he was convinced that, in the tactical offensive, "co-operation between tank and aeroplane is likely to prove far more important than between tank and infantry" (quoted in Carver 1979, 36 and 40).

In 1925 Liddell Hart published *Paris, or the Future of War.* In it he concluded that "speed, on land as in the air, will dominate the next war, transforming the battlefields of the future from squalid trench labyrinths into arenas where surprise and manoeuvre will reign again" (82–83). As Carver notes, Liddell Hart shared with Fuller the "emphasis on the strategic mistake of making the destruction of the enemy's army the main aim rather than the bending of his will; on the purpose of war being to produce a prosperous and secure peace, and on the major part to be played by the combination of air-power and gas" (1979, 44). Liddell Hart was also a successful popularizer of his own ideas, publishing many of them in the 1920s in articles written for such well-respected British newspapers as the *Daily Telegraph* (45).

In the years immediately following World War I the idea of the strategic air offensive was elevated to the level of doctrine. In *Command of the Air* (1921) Italian General Giulio Douhet argued that terror bombing of cities by fast, unescorted bombers would cause civilian populations to surrender, that fighter opposition would never amount to anything, and that the strategic offensive form of air war would be the decisive element in any future struggle. In *Winged Warfare* (1930) American Brigadier General William Mitchell argued a similar case for the fragility of civilian morale under terror bombing, although he did accept the need for fighter escorts to protect the bombers from defending fighters. British geostrategic thinkers came to accept as doctrine a combination of Douhet's and Mitchell's thought. Most American thinkers preferred to follow a line of reasoning that was more acceptable to the American voter, namely, that the enemy's capacity to wage war could be interdicted by precision air attacks on infrastructure, in particular the transportation and communications systems, energy sources, and key industries producing war materiel. They came to believe that unescorted, self-defending bombers flying at high altitudes could use the extremely accurate Norden bombsight to ensure that attacks on the enemy's infrastructure would not result in significant civilian casualties, what we now euphemistically call "collateral damage." No other polity fully adopted the doctrine of strategic air offensive, although all tinkered with it and all developed prototype strategic bombers. In the event, Germany, the Soviet Union, Japan, and France elected to remain with tactical offensive applications. Although Mussolini was a disciple of Douhet and placed an "uncritical faith in heavy bombers," he had no real understanding of the technical and administrative needs of a modern air force (Overy 1980, 166). The Italian aviation industry, like that of Germany, turned out tactical attack bombers in the mid-1930s that were misused as heavy bombers. The much larger strategic bombers that were needed by the early 1940s were, in any case, well beyond the technical reach of Italian industry by that time.

In immediately post–World War I Britain purely pragmatic considerations persuaded Chief of Air Staff Hugh Trenchard that if the RAF was to be preserved as an independent arm, independent air power had to be cost-effective. He was able to demonstrate in the Mesopotamian crisis of 1922 the practicality of "control without occupation," which greatly reduced the costs of imperial defense by replacing large numbers of ground troops with a small number of skilled pilots and machines. In the austere financial climate that prevailed in Britain in the early 1920s such a use of air power was very successful, and the RAF was able to retain its independence as an offensive arm of imperial power. Trenchard also came to be a disciple of Douhet, arguing that "the nation that could stand being bombed the longest would win in the end" (quoted in Cross 1987, 73). As the RAF developed as a fully independent service, it established an officer-training program at Cranwell and its own set of doctrines, most of them borrowed from Douhet. This meant that in the interwar period defensive doctrine was slighted. Spending on long-range strategic bombers was always much greater than that on short-range tactical bombers or on fighters, even allowing for the far greater cost of bombers, and Bomber Command received the cream of the trainees. Even in the Battle of Britain Fighter Command had to make do with a large number of relatively inexperienced pilots, many of whom came from the university air squadrons established in the mid-1930s. British fighter pilots were well trained but sorely lacking in flight time and combat. They were opposed by equally well trained pilots who had been seasoned by the Spanish Civil War and successful campaigns against Poland and France.

Strategic bombing was also attractive to the British because of its apparently low cost. The human as well as financial costs of fighting the trench war that World War I had quickly degenerated into had been high, and another such war was reckoned by the British government as unacceptable to the British public. Before World War I, Britain had never fought a land-based war in Europe involving vast numbers of British troops. Britain's usual method of waging war with European powers had been to retreat behind the "moat defensive," pursue a Mahanian war of naval blockade against its continental opponent(s), and fund other continental polities to do most of the land fighting. The strategic air offensive offered the possibility of carrying war into the enemy's heartland and substituting capital, machines, and a few well-trained airmen for large numbers of ground troops. Given the appalling memories and costs of World War I, strategic air power thus made perfect sense in British military thinking between the wars. Given the experiences of the first Battle of Britain, the notion of terror bombing civilians also made sense since the need to defend against such bombing had imposed significant costs. What is problematic about British air policy in the period up to 1935 is that it paid almost no attention to the defensive role of air power despite the implications of a policy that stressed terror bombing of the opponent's cities. Such a lack can only be understood in terms of a limited budget, a history of offensive air doctrine, and a belief that bombers would be remarkably hard to intercept, as had proven to be the case in 1917.

Two British prime ministers summed up Britain's obsession with strategic bombing. Stanley Baldwin, prime minister from 1923 to 1929 and again from 1935 to 1937, believed that "the bomber will always get through." Winston Churchill, First Lord of the Admiralty at the beginning of World War I and prime minister throughout World War II, described London as a great fat cow tethered conveniently for the Germans to slaughter at their leisure by aerial assault. In *The World Crisis* (1922) Churchill wrote of the general British conviction in 1914 that Zeppelins flown by the Imperial German Navy would savage British naval facilities as well as London. The origins of such beliefs lay in H. G. Wells' bleak and much imitated novel, *The War in the Air*, in which German Zeppelins pounded Britain into submission, then moved on to America and repeated the performance. This assault, plus the counterassault, primarily by Japanese forces that had mastered heavier-than-air flight, brought on the collapse of the monetary system, famine, plague, and finally a new Dark Age. Reproductive failure—"a large proportion of the babies . . . died in a few day's time of inexplicable maladies" (245)—ensured that any recovery would be slow indeed. Finally, Wells described the survivors as "an urban population sunken back into the state of a barbaric peasantry, and so without any the simple arts a barbaric peasantry would possess. In many ways they were curiously degenerate and incompetent" (245). In later years Wells modified this bleak outlook in the light of the failure of the two German strategic bombing campaigns against Britain, by Zeppelins from 1915 to 1917 and by heavier-than-air craft in the first Battle of Britain, in 1917–18. These failures resulted, however, less from a particularly successful defense than from the failure of the Germans to develop successful aircraft with which to attack. Despite Wells' modifications, the first two of the five sections that make up *The Shape of Things to Come* (1933) are modeled on *The War in the Air*. Near destruction of planetary civilization is followed by the imposition of a technocratic socialist government by an air police, an idea clearly derived from Trenchard's "control without occupation." Many lesser writers agreed with Wells, and many even predated him. Paris paints a convincing picture of Britain as a society well prepared for the negative features of aerial warfare by the time World War I broke out (1992, ch. 2). Other writers of Edwardian science fiction, such as Harry Collingwood in *With Airship and Submarine* (1908) and F. S. Brereton in *The Great Aeroplane* (1911) and *The Great Airship* (1913), were less bleak in their outlook but found aerial vehicles no less irresistible than Wells did.

It is in the light of such thinking as Wells' and the realities of the Spanish Civil War that we must understand the British response to threats of aerial assault as late as 1938. Without accurate knowledge of the direction, altitude, and speed of an approaching attacker, aerial defense was a hit-or-miss affair. Once the defense was notified that attackers were approaching, the fighters had to take off and climb to the altitude of the attackers. Bombers, in contrast, had ample time to climb to their attack altitude and carried sufficient fuel to travel long distances. The power that enabled them to carry large bombloads also enabled them to attain respectable speeds. In the continuous technological struggle that has marked air war more than any other form of modern war, most bombers of the mid-1930s were quite fast relative to the fighters that would oppose them.

The British and the Americans were more committed than any of the other polities that fought World War II to strategic air offense. They differed only with regard to which were the appropriate targets. In America, by the late 1920s the Air Corps Tactical School at Maxwell Air Force Base, in Alabama, had concluded, as Mitchell had, that precision bombing of the enemy's infrastructure was the appropriate strategy. The US-AAC, which had been able to separate its budget from that of the army after 1925 and could thus pursue its strategic aims, recognized that such bombing would have to be by day and that precision-bombing aids, such as the Norden optical bombsight, would be needed. The Norden bombsight thus became the most important secret project of the American military in the interwar years; its costs eventually amounted to 65 percent of the costs of the Manhattan Project (McFarland 1995, 6). A strategic doctrine tied to daylight bombing also implied that bombers would need to fly high and fast and and carry heavy defensive armament. The USAAC argued that a heavy armament of .50 caliber (12.7 mm) machine guns, most mounted in power turrets, would allow a formation of such planes to defend itself against opposing fighters. The fallacy of the self-defending bomber was effectively demonstrated by the German experience in the Battle of Britain as well as by early British attempts to fly daylight raids over Germany. By 1941 both the Germans and the British had clearly demonstrated that even when the self-defending bomber was well armed, it was easy prey to fighters by day, and so both had retreated to the cover of darkness. The German experience in the Battle of Britain demonstrated that even closely escorted bombers were likely to suffer heavy losses against a determined fighter defense.

The main American bomber in the air war over Germany, the Boeing B-17, exemplified the strengths and weaknesses of the self-defending bomber required by strategic doctrine in the mid-1930s. It was relatively fast, flew high, and was well armed with heavy machine guns in power turrets. But it could not fly fast enough or high enough to evade enemy fighters, it was a big target, and it was not easy to maneuver. Fighters, which were highly maneuverable fixed-gun platforms, could mount heavy cannon firing explosive shells, only a few of which were needed to down a bomber. The appalling operational losses self-defending bombers suffered against a determined German defense brought their use to an end. The long-run success of the American strategic offensive thus came to depend on the development of long-range escort fighters. Fortunately, the first of these, the Lockheed P-38 Lightning and the Republic P-47 Thunderbolt, were beginning to enter service in reasonable numbers in 1943. The best of them, the North American P-51 Mustang, which had an adequate range to reach Berlin, performed far better than the later Messerschmitt Bf 109G, and even had an edge on the Focke-Wulf Fw 190, was available by early 1944. Although the American fighters had .50 caliber machine guns, all of them would have benefited from the cannon armament the Germans had at the beginning of the war and British fighters had by 1942. Only with a considerable fighter escort, often numbering close to one fighter per bomber, could American daylight bombers continue to penetrate German air space in order to precision bomb infrastructure.

The British had abandoned precision daylight bombing much earlier, preferring the more Douhetian strategy of night terror bombing of cities in order to reduce their inhabitants to collapse, known politely as "area bombing." The conclusion was that only a night offensive would keep operational losses to acceptable levels and that if night bombing was to be used, the minimum target would have to be a city. Several factors contributed to this change: (1) the British had been on the receiving end of the (comparatively mild) German Gotha and R-plane attacks on London in 1917 and 1918; (2) the British literary tradition emphasized the likelihood of such a collapse of civilian morale; (3) Trenchard had ensured the survival of an independent RAF against a hostile navy and army and an indifferent Parliament in the same way; (4) it offered the substitution of the lives of relatively few airmen and some relatively cheap airplanes for a slogging land war and the huge casualties and vast expenditure on field guns and ammunition suffered in World War I; and (5) the realization, based on the German experience in both world wars and the British experience in the opening months of World War II, that daylight bombing was suicidal.

If precision was unobtainable, then the weight of bombs dropped would obviously become more important. British bombers were thus designed to carry much larger bombloads than American bombers. The best British heavy bomber of the war, the Avro Lancaster, could carry

14,000 lb of bombs 1,660 mi, in contrast with the Boeing B-17G's load of 4,000 lb for 1,850 mi (Green 1959, 134, 36). The Lancaster was designed to lift heavy loads in the thicker air lower down, using darkness as its main defense. The main reason why the B-17's load was so much poorer was that it was designed to fly high in order to stay above enemy flak and fighters. But the B-17's load was also lighter because the proponents of the self-defending bomber insisted that it carry a large number of 12.7 mm machine guns (13 on the B-17G) and plenty of ammunition, which weighed a lot more than the eight 9 mm guns carried by the Lancaster. The British had started down the road to the Lancaster in the early 1930s. British specification B.9/32 (the ninth bomber specification issued in 1932) was for a long-range strategic bomber with a top speed of 190 mph and a service ceiling of 22,000 ft. The most successful contender, the Vickers Wellington Mk I, had a top speed of more than 250 mph and entered service in 1938. With 4,500 lb of bombs, it had a range of 3,200 mi cruising at 180 mph at 15,000 ft (Andrews and Morgan 1988, 340).

Contemporary German bombers performed similarly. The Heinkel He 111, which was so successful in the Spanish Civil War, had a top speed of 258 mph in the 111H-6 variant deployed in the Battle of Britain in September 1940. More important, it cruised at 224 mph at 16,400 ft, a height it took some 20 min to reach. With 3,300 lb of bombs, it had a range of 760 mi (Green 1959, 16). The significantly better cruise performance of the Heinkel resulted in part from the superior aerodynamics common to most German designs and in part from its lack of the drag-inducing defensive armament of British designs, needed because a long-range strategic bomber would be over enemy territory for a long period. The Heinkel was a short-range tactical bomber designed to clear a path for advancing armored ground forces; its rudimentary defensive armament was a debilitating weakness in the Battle of Britain.

British designers quickly moved on to bigger things. Whereas the Luftwaffe canceled the Junkers Ju 89 and Dornier Do 19 prototype four-engine strategic bombers, British specification B.12/36, issued "after Adolf Hitler's statement . . . in March 1935 that the German Air Force already equalled the Royal Air Force in striking power," called explicitly for the heaviest bombers in the world (Andrews and Morgan 1981, 319). Contracts for bombers to this specification went to the Supermarine Company for the Type 317, never built because of the pressure to turn out Spitfire fighters, and to the Short Company for the Stirling. According to specification B.12/36, the largest bomb carried would weigh 2,000 lb and the wingspan should not exceed 100 ft so the bomber could fit existing hangars. Bomb size increased rapidly in World War II, but the Stirling had compartmentalized bomb bays and its maximum load was seven 2,000 lb bombs, which it could carry only 740 mi. Its short wingspan limited its altitude performance when altitude was one of the best defenses (Barnes 1989, 371; Green 1969, 43, 48, 51).

Specification B.13/36 was for high-performance, twin-engine medium bombers to carry not only bombs but also two torpedoes 21 inches in diameter. Designs thus began with unobstructed bomb bays. In the end they would be developed to carry the heaviest loads of the war. Two designs were accepted, the Handley Page Halifax and the Avro Manchester. The Halifax design was changed to include four Merlin engines in the design process (Barnes 1987, 385), but the Manchester entered service using two unreliable Rolls-Royce Vultures (Jackson 1990, 353–55). Rolls-Royce has long been properly lauded for its famous Merlin, but getting the Merlin right in time for the war resulted in the neglect of several other promising engines. The jewel-like Peregrine, two of which powered the impressive and heavily armed Westland Whirlwind fighter, was abandoned because only the Whirlwind used it (James 1991, 266–68). The Vulture, two Peregrines on a common crankcase, was never developed into a trouble-free unit, and the Exe was abandoned in development. The Manchester, which performed excellently when its two Vultures ran properly, was saved by the same expedient that saved the Halifax, namely, converting its design to use four Merlins, effectively bringing an already excellent airframe up to 1940 standards. Renamed the Lancaster, it soon became the mainstay of Bomber Command and carried the heaviest loads of the war, including the massive 22,000 lb "Grand Slam" bombs, designed to smash submarine pens by creating localized earthquakes.

PROBLEMS OF THE DEFENSE

Even though fighter designers strove to provide both high speed and high rate of climb to intercept such bombers, nothing ever took the place of early warning of an attacker's approach. Before radar, all attempts to provide early warning of aerial attack were futile, and in fact most were tried and rejected by the end of World War I (Guerlac 1987, 29–31). By the time an enemy air fleet had entered a defender's territory, it was too late for ground observers to report the fleet's position to defending fighters by telephone or radio in time for the fighters to take off, climb to altitude, and attack. Sound location devices could locate planes by engine noise but had poor range, were rendered useless in windy situations, and became ever more inaccurate as aircraft speeds rose. A plane flying toward the locator at a speed of more than 200 mph was closing in on it at about one-third of the speed of sound. By the early 1930s efficient detectors were vast structures built of concrete and thus totally immobile. Infrared detectors could sometimes detect engine heat from a reasonable distance but were easily fooled by variations in air temperature caused by clouds. Detecting the radio-frequency electrical impulses generated by the aircraft's engines was easily negated by shielding the electrical system, as the famous General Motors engineer Charles "Boss" Kettering recognized before the end of World War I.

To make any fighter defense system work at all, standing patrols of fighters were thus necessary. Given that fighter designers almost always traded endurance for speed and climbing performance, large numbers of planes and pilots were needed. Specifications for fighter airplanes usually included speed and rate of climb, not endurance. British specification F.7/30 required a 25 percent improvement in fighter performance over specification F.20/27, set the design criteria for British fighter planes entering service as late as 1937, and was not really superseded until F.37/34, which was written to embrace the privately developed Hawker Hurricane and Supermarine Spitfire. F.7/30 required a climb to 10,000 ft, the likely attack altitude for bombers in 1930, in less than 4 min and a top speed of 250 mph at 15,000 ft. No plane designed to meet specification F.7/30 ever fully met these requirements, not even the Gloster Gladiator biplane, which entered service as late as 1937 (Mason 1992, 244). Such planes would have been very hard pressed to intercept bombers entering service alongside them—the Heinkel He 111, the Handley Page Hampden, and the Vickers Wellington. The endurance of most of the designs to F.7/30 was less than 2 hr, less at full throttle. In wartime operation at full throttle was very likely.

The designs for which F.37/34 was evolved, which entered service in 1937 and 1938, were capable of much higher speeds and rates of climb than those to F.7/30: speeds of 324 mph in the case of the Hurricane and 364 mph in the case of the Spitfire and a rate of climb of about 8 min to 20,000 ft for both (Mason 1992, 255, 260). But their endurance was no better, nor was that of their opponents. The excellent Messerschmitt Bf 109, which served the Luftwaffe through both the Spanish Civil War and World War II, carried fuel for a flight of only 2 hr. The German fighter ace Adolf Galland, promoted to commander of *Jagdgeschwader* 26 during the Battle of Britain and later the Luftwaffe general in charge of fighters, complained that in one sortie late in the battle "my group lost 12 fighter planes, not by enemy action but simply because after two hours flying time the bombers had not yet reached the mainland on their return journey" (quoted in Cross 1987, 104). Too many German fighter pilots simply drowned in the "moat defensive."

In the economic climate of the 1930s the massive spending that would have been needed to mount standing patrols of airplanes with such short endurance was almost unthinkable. When Prime Minister Baldwin said that "the bomber will always get through" he was speaking what seemed to most people the simple truth, and when Chamberlain capitulated at Munich he did so with the very real threat of the proven German bomber force on his mind. The only solution was an early-warning system that allowed fighters to take off and reach the cruising altitude of the attacking bombers without wasting fuel.

If the geostrategic problems and realities of air war were obvious by the mid-1930s, so was a possible solution. As previously indicated, radio technology had developed at an extremely rapid pace during the 1920s, in particular in the short wavebands. Considerable success had been enjoyed in expanding long-distance communications in the short bands, using cheap, vacuum tube oscillators to generate continuous waves of radio frequency. Without such oscillators the medium (600–200 m wavelength) and short (200–20 m wavelength) components of the radio-frequency spectrum could never have been entered, since arc and alternator technologies could not generate waves much below 600 m. Commercial use of short waves for communication also required a substantial research effort aimed at understanding the interplay of radio waves and the upper layers of the atmosphere. This research focused on waves down into the 20 m band, which could carry signals almost anywhere on the planet when ionospheric conditions for reflecting such waves were right. Radio waves below the 20 m band usually passed through all layers of the atmosphere at all times of the day or night and proceeded straight out into space.

The area below 20 m was seen as valuable, however, because it could carry much more information than the area above 20 m. As demand for long-distance radio telephony increased in the 1930s, the available short bands down to 20 m were quickly filled up. The development of commercially viable high-definition electronic television systems in the mid-1930s led to a vastly increased demand for bandwidth, one television signal consuming as much space in the radio spectrum as several hundred telephone conversations. This required moving below 20 m, thus into the very high frequency (VHF) portion of the spectrum. Line-of-sight communication between stations transmitting radio telephone signals on 18 m also seemed very promising by the early 1930s. Generating such very short wavelengths pushed the very limit of vacuum tube technology. Special, ultracompact tubes were needed to generate waves in the 10 m band.

In recent years there has been a great deal of scholarly interest in the history of radar not only by historians of technology but also by engineers. One recent book goes so far as to describe radar as "the invention that changed the world" (Buderi 1996). Certainly Buderi's claim that radar was the weapon that won World War II is not unreasonable, although his book focuses on the American end of radar development. Given that most of the developers of the new electronics technologies were young when they began their careers in the late 1930s, many are still active and able to contribute to the scholarly discussion on its origins. It is now clear that the geostrategic origins of radar lie in naval activity and interests, not in aviation. It is also clear that the argument for civilian origins must not be totally put aside in favor of well-founded arguments that the main interests were military. Both navies and civilian shipping interests saw radio as the solution to many problems in command and control, if one extends the military concept to include location of obstacles to navigation by commercial ships. From almost the earliest days of radio it was obvious that one could locate any radio transmitter fairly easily, in particular if one could triangulate a receiver upon a series of such transmitters. Here lay the principle of radio direction finding. But it was also obvious that radio waves could be projected toward possible obstacles and received when reflected back from them, the basic principle of radar. Such was the intent of Christian Hülsmeyer's patent of 1904, but even though he demonstrated that his system worked, the newly founded Telefunken Company rejected it (Burns 1988, 6–9). Despite various other suggestions, commercial proposals languished until 1935, when the Société Française Radio-Électrique fitted the French liner *Normandie* with equipment operating on a wavelength of 16 cm and intended to detect obstacles. This was not very successful because of the low power output of the vacuum tubes available (Molyneux-Berry 1988, 46–47).

The other line of development was naval, possibly based on Marconi's restatement of Hülsmeyer's ideas before the American Institute of Electrical Engineers and the Institute of Radio Engineers in 1922. A. Hoyt Taylor and L. C. Young, then working in the aircraft radio group of the Bureau of Naval Engineering at the Naval Air Station in Anacostia, probably heard Marconi's address. They certainly demonstrated that high-frequency radio waves were reflected by ships passing along the Potomac. The Bureau of Naval Engineering did not respond to Taylor when informed of these results in 1922, and Taylor therefore did not return to radio detection until 1930 (Burns 1988, 14–15). In any case, navies' problems were simpler than those of air forces: they needed only range and bearing to locate obstacles and targets. The American, British, German, French, Italian, Dutch, and Japanese navies all invested in radar development in the 1930s, usually behind a veil of considerable secrecy (see, respectively, Page 1988; Coales and Rawlinson 1988; Kummritz 1988; Molyneux-Berry 1988; Calamia and Palandri 1988; Staal and Weiller 1988; and Nakajima 1988). Only the Soviet Union, with no navy to speak of, from the beginning focused its radar effort on aerial defense (Erickson 1988).

THE BATTLE OF BRITAIN, SEPTEMBER 1940

The story of the development in Britain of radar as part of a defensive system against a strategic air offensive has been told often enough that it does not need repetition here except to contrast it with efforts in other polities and to emphasize the significance from a geostrategic perspective of certain key points and the importance of receptive adoption environments for the success of radical new technologies (Hugill and Dickson 1988, 271). The apparent vulnerability of polities to strategic bombing offensives in the mid-1930s lay in the seeming impossibility of defense against an attack that could come from almost any direction and at almost any time and altitude. A haphazard defense seemed likely unless the course, speed, and altitude of the attacking bombers could be detected from the ground and appropriate defensive forces sent up to intercept them. Without standing air patrols or radar this would have been impossible. Although many polities, Nazi Germany in particular, were more advanced in radar technology, it was the British, recognizing that they would soon be forced to fight a defensive battle for London and their industrial cities, who deployed the first effective radar C^3 system, Chain Home.

In the geopolitical conditions of the 1930s Britain was thus simply a more receptive environment for the adoption of radar technology for defensive uses than an aggressively rearming Germany or an isolationist America, the two other countries where telecommunications research was advanced enough to make a quick move to an air defense system based on radar. In Britain in January 1935 the Committee for Imperial Defence established the Committee for the Scientific Survey of Air Defence, more commonly known as the Tizard Committee, after its chair, Henry Tizard. Tizard was convinced that "both aircraft and anti-aircraft guns could be made effective against . . . [air] attack, but only if they were in the right place at the right time" (Bowen 1987, 3). When the Tizard Committee consulted Robert Watson Watt, the superintendent of the Radio Research Station at Slough, about a "death ray" that had been proposed to the committee, he replied that although enough energy could not be broadcast to damage an airplane or its crew, it ought to be possible to devise a system to precisely locate airplanes given their well-reported ability to reflect the shorter radio waves and that in any case such a location system would be necessary even if a death ray were created. In February 1935 this communication reached Hugh Dowding, the air member for research and development at the Air Ministry and later commander in chief of Fighter Command during the Battle of Britain, who recognized its potential and began to fund radar-oriented research (6–8).

Watson Watt had become involved with radio location as a meteorologist in World War I, at which time he had devised a method of locating severe thunderstorms using their radio-frequency emissions. In the mid-1920s he was at the forefront of British radio research of the ionosphere that was vital to the development of long-range radio communications at the time, in particular in the area between the 200 m and the 20 m wavebands. Watson Watt simply proposed a much more sophisticated version of "a well established research procedure in many laboratories around the world" (Bowen 1987, 10). This statement, however, greatly oversimplifies the situation. It was one thing to measure the height of the distant ionosphere, entirely another to locate a fast-moving airplane much nearer the earth's surface. The ionospheric technology, developed in America by G. Breit and M. A. Tuve in 1925, depended on measuring the time taken by a short pulse of radio energy sent up vertically to return to the transmitter. Operating in the 50 m band, the transmitters had a peak power of about 1 kW and a pulse width of 200 μsec. Watson Watt recognized that the wings of a typical bomber with a span of some 25 m would act like a half-wavelength dipole antenna when exposed to radio energy, hence his willingness to begin experiments with low-frequency systems on the long wavelength of 50 m. Other radar designers seem to have considered airplanes as point locations rather than as flying antennae. Watson Watt also recognized the need for a massive increase in the power of the ionospheric transmitter, as well as a decrease in pulse width, because the signal would be traveling a much shorter distance. An increase to a 100 kW output and a decrease to a pulse width of 20 μsec were specified. These were well within the capabilities of existing high-power vacuum tube technology, and by late summer 1935 the first experimental station at Orfordness had achieved 200 kW and 25 μsec (10–13).

This rapid success persuaded the Air Staff that radio location of airplanes was possible, and in December 1935 the Treasury approved an expenditure of £1 million for the construction of a chain of five stations to monitor the approaches to the Thames estuary and London. The radar system moved out of the 50 m band into the 26 m and then the 10–13 m band because of the interference caused by short-wave international radio transmissions in the higher bands. At 10 m, radio waves tend to pass through rather than be reflected by the ionosphere, so the 10 m band was useless for long-range radio. Pushing below 10 m was difficult because of the need for compact vacuum tubes. Large tubes were needed for high power, but compact tubes were needed for the higher frequencies. The biggest problem was that "the transit time for the electrons to move between the cathode and control grid was about one nanosecond [one billionth of a second]. At frequencies of a few MHz, this transit time was virtually insignificant compared to one cycle of the signal frequency. At 100 MHz, however, the time was about ten percent of one cycle and this was very significant . . . the phase lag caused by this time delay resulted in a low input resistance at high frequencies which reduced the gain of the amplifier" (Thrower 1992, 76).

An associated problem was that the first vacuum tubes had been made by manufacturers of electric light bulbs and thus tube design owed much to their practice of using overlarge glass envelopes with little concern for compact electrical circuits. This made for a long electrical path between electrode and terminating pins, which resulted in significant problems in self-inductance and self-capacitance above 30 MHz. Tubes were always the limiting factor in the performance of metric-band radar systems, imposing an effective upward frequency limit of about 600 MHz, about a 50 cm wavelength. German designers were unusually progressive, and part of the reason for the success of German 50 cm radar was excellent tube design. German tube designers also standardized far more and far earlier than designers elsewhere, reducing the number of standard tubes in their communications radio designs to two by World War II and producing 16 million of the more important of these (Bauer 1995, 77).

By the fall of 1938 the basis of the British Chain Home defensive radar system was ready. The stations were built or being built (fig. 7.1). The failures of the 1936 defense exercises revealed lack of ground control as a major weakness of the system, so that the Biggin Hill experiment resulted in 1937 in a system of ground control that allowed those reading the radar to communicate with other radar stations, with Fighter Command, and with the fighters in the air. Although the Battle of Britain was a complex event and many factors contributed to the British victory, Chain Home and the system by which the data it generated were fed to the defending fighters was the single most important factor. Nor should the role of British ground observers, the Royal Observer Corps (ROC), be underestimated, since the British did an excellent job of integrating radar, which tracked raiders approaching the coast, and ground observations, which tracked them once they were inland. In part because of royal patronage, the ROC attracted "a great many retired officers" as well as, in rural areas, members of the "landed gentry" and those who worked for them (Wood 1976, 42). Hugh Dowding recognized that the primitive radar of Chain Home could mislead him in the Battle of Britain if it was not backed up with ground observation; he came to describe radar as his eyes and the ROC as his ears. His use of the word *ears* was almost literal: a good ground observer could identify aircraft by engine sound as well as by sight. These points are generally less well understood by historians of the battle than they should be, although there are exceptions (see Wood and Dempster 1961).

Fig. 7.1. The Chain Home Radar System, the British "Moat Defensive" of 1940 (redrawn from Crowther and Whiddington 1948, 9)

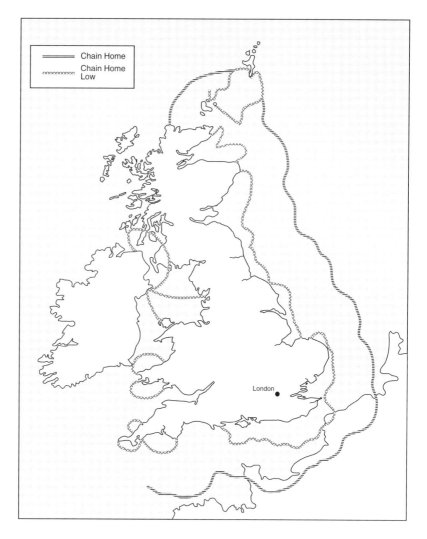

Chain Home

Chain Home Low

London

The British Supermarine Spitfire was one of the most successful fighter planes of its period, and it was a powerful, tangible, even anthropomorphized symbol of a close victory at a time when such symbols were badly needed. It has been appropriately described as the "material symbol of final victory to the British people in their darkest hour" (Green 1957, 25). Yet in combat with Willy Messerschmitt's equally brilliant Bf 109 the outcome was more likely to be determined by the ability of the pilot than by any intrinsic characteristics of the machines. Both fighters have their adherents, and their respective advantages and disadvantages have been argued to death. The Hawker Hurricane, the Spitfire's comrade-in-arms in the Battle of Britain, is certainly worthy of similar lionization. It was easier to build, available in much larger numbers in September 1940, could survive much more battle damage than the Spitfire, and knocked down more German planes. Like their airplanes, the fighter pilots were also easy to lionize. They too were tangible, immediately comprehensible symbols of the struggle. A whole generation of British, German, and American youth had grown up on lurid printed and filmed tales for adolescents of World War I fighter pilots as mythic knights of the sky. It is irrelevant that these tales were almost universally false, that in fighter-to-fighter combat the best pilots killed most of their victims in cold blood and with no warning, not in a duel between equals, and that in fighter-to-bomber combat the crew of the slower, less maneuverable bomber was usually slaughtered. These mythic tales had a particularly pernicious effect in Germany.

It is easy, too, to blame poor tactics for Germany's loss of the Battle of Britain. German fighters should never have been forced to fly close escort when the bomber pilots complained of mounting losses. The Bf 109E's principal advantages over the early Spitfire were speed, rate of climb, and the ability to quickly nose over into a dive. Close escort robbed it of those advantages and forced it to dogfight a superior dogfighter. The inconsistency with which German attacks were pressed home on the radar chain, on forward airfields, and on such critical production facilities as that for the Spitfire have all been properly mentioned as contributions to German defeat.

German technology has also been properly criticized, and British praised. The Daimler-Benz 601 engine, which powered the Bf 109E, was just a little too big and heavy compared with the Merlin. Its variable-drive supercharger made it substantially superior to the Merlin at low altitudes. Below 5,000 ft the Bf 109E was very much the superior fighter. But the Battle of Britain was fought at 15,000 ft and above, where the Merlin came into its own. The larger size of the Daimler-Benz engine made for increased frontal area and therefore a little more drag. The German engine generated about the same horsepower from nearly 34 liters as the British engine did from 27 liters, used more fuel, and had much poorer development potential (Gunston 1989, 46–47, 141). It is all very well to note that the Rolls-Royce was originally designed to perform well on 100-octane when America made it available and that the Mercedes-Benz was optimized for 87-octane fuel and did not do much better on 100-octane. The Merlin was simply a better engine with more potential for development. The Daimler-Benz had been designed to be as light as possible for its size and thus could not stand up to the high internal stresses that the British design, which achieved light weight and low drag by virtue of its smaller size, could handle. Because of its small fuel capacity, the use of the Bf 109 for escort missions over London has been rightly criticized: too many highly trained German pilots died merely because they ran out of fuel in the English Channel on their way back to France. British fighter pilots were fighting for and over their homeland. The civilian population that supported them so strongly saw them fighting and dying in the skies above them, and if they parachuted to safety, they were often back in action in another fighter on the same day.

The failure of Messerschmitt to develop the twin-engine Bf 110 as a strategic long-range escort fighter, or *Zerstörer,* and the more general failure of Göring to implement the designs for a four-engine strategic bomber, which were available by 1936, have also been invoked as reasons for German failure. The German concept of the *Zerstörer* made sense, but the Bf 110 was flawed in overall design and then underdeveloped. It was intended as a destroyer of attacking bombers, hence its name, but it was first used as a long-range escort fighter, a role for which it was ill suited. It lacked maneuverability and was too slow. Yet against Polish and French opposing fighters or British bombers it had done very well indeed, so well that its proponents had become complacent. Against British defensive fighters the Bf 110s were simply outclassed and had to be protected as much as the bombers. The idea of the strategic fighter was proven successful by the American Lockheed P-38 Lightning and the British de Havilland Mosquito, although neither was designed as a strategic fighter.

The main proponent of German strategic bombers, and one of the true visionaries of German strategic airpower, Generalleutnant Walther Wever, died in an air crash in 1936. The four-engine "Ural bombers" Wever had commissioned, with the capacity to reach bomb sites east of the Ural Mountains, died with him. His successor, Albert Kesselring, did not support strategic bombers at the expense of tactical aircraft and canceled the designs that might have been used against Britain in 1940. Göring had often remarked that "the Führer will not ask how big the bombers are, but only how many there are" (Homze 1976, 125). When the Germans once again decided to build strategic bombers in 1942, with such designs as the Heinkel 177 and the Messerschmitt 264, it was too late.

All of these arguments for German failure or British victory, and many more I make no attempt to summarize here, simply pale beside the existence and proper use of the Chain Home radar system and the ROC. Without the early warning and confirmation of enemy aircraft these provided, the British would have lost. The British might well still have won with or without the brilliant Spitfire and its Rolls-Royce Merlin engine, alternatives to both of which existed in the form of the Bristol Mercury–engined Bristol 146 and the Napier Dagger–engined Martin-Baker M.B. 2 (Mason 1992, 264–65); or against the canceled German four-engine Dornier Do 19 and Junkers Ju 89 "Ural bombers" (Homze 1976, 123–25); or even against a better, longer-range German fighter, such as the Focke-Wulf Fw 190, which caused Spitfire pilots so much grief when they were introduced in September 1941, especially when they operated at low to medium altitudes. Chain Home simply put the few but sufficient British fighters in the air in the right place and at the right time. Adolf Galland certainly believed that Chain Home radar cost German fighter pilots the battle: "In the battle we had to rely on our own human eyes. The British fighter pilots could depend on the radar eye, which was far more reliable and had a longer range. When we made contact with the enemy our briefings were already three hours old, the British only as many seconds old—the time it took to assess the latest position by means of radar to the transmission of attacking orders from Fighter Control to the already airborne force" (quoted in Cross 1987, 100).

Finally, how much Hitler really wanted to bring down Britain is debatable. Throughout the 1930s *appeasement* was not the dirty word it is today (Kagan 1995). Many Britons, especially those who belonged to the propertied classes, believed Nazi Germany was a bulwark against Communism. Hitler's geographical sights, at least as expressed in *Mein Kampf*, were set on eastern Europe in any case, and he constantly sought accommodation with the British Empire in the first part of the war. Whereas the invasions of Poland, Norway, Belgium, and France were exceptionally well planned, the invasion of Britain was hardly planned at all and the air Battle of Britain was conducted on a rather ad hoc basis.

THE NIGHT BLITZ, 1940 – 1941

The literature on World War II has tended to focus on British victory in the Battle of Britain to the exclusion of British victory in the night blitz on London that followed the battle from late 1940 to the summer of 1941. There has also been a tendency to assume that the blitz ended because the German effort was turned to the invasion of the Soviet Union, not because the Luftwaffe was defeated. From a technological point of view, however, the British victory in the night blitz was as significant as the British victory in the daytime Battle of Britain. Both were victories won by radar, ground-based in the case of the battle, ground- and air-based in the case of the night blitz. Once the relatively simple technical and far greater organizational problem of ground-based radar had been solved by Chain Home, Tizard and the Committee for Imperial Defence argued that the next problem was night attack, which would be the enemy's logical recourse once day bombing had become untenable. Solving this problem was far more demanding technologically since it required that radar sets be installed in night fighters. In 1936 this problem was turned over by Watson Watt to a small development team under E. G. Bowen, who were able to produce the world's first successful airborne intercept radar just in time for the night blitz in early 1941. Bowen developed guidelines based on the airplanes available. The unit could not weigh more than about 200 lb or use more than 500 W of electric power, the total power available on most British aircraft at the time (Bowen 1987, 32). This meant getting down to wavelengths of 1–2 m with a pulse width of 1 μsec. Bowen eventually settled on 1.5 m.

The switch by the Germans to night bombing in late 1940 also forced a switch to more Douhetian tactics. The unusually clear skies enjoyed by the Germans in the Battle of Britain had allowed precise bombing of strategic targets, and some notable successes were achieved, such as the attack on the Supermarine Spitfire works at Southampton. Night bombing meant attacks on whole cities, the smallest viable targets using then available navigational technology. The Germans had developed sophisticated blind navigation based on the radio-beacon techniques developed for civil airlines. By the summer of 1940 German improvements in RDF technology allowed area bombing of British cities at any time and in any weather. However, the British located the radio beams during German experiments with them in June and thus were able to quickly deploy electronic countermeasures against the first of these systems, Knickebein, and jam it when it was operationally deployed in September. The second German system, X-Verfahren, proved more durable, and attacks began on a variety of strategic targets. Pathfinder lead planes using X-Verfahren lit up a night target with incendiaries, showing the rest of the bombers where to make their drops. Although ECM were developed against X-Verfahren, they were not developed in time to prevent the most successful attack of the night blitz, that on the city of Coventry on 14–15 November 1940, in which the first firestorm was started and civilian morale came near to collapse (Cross 1987, 108). Pilots quickly recognized that X-Verfahren was being interfered with, but after British attacks on Berlin enraged Hitler, German attention was refocused on London, always easy to find by following the River Thames. The last German blind bombing system, Y-Verfahren, entered service in early 1941, but British ECM were deployed against it before operational use began.

Chain Home technology using the 10 m band required vast antennae several hundred feet high and direct connection to the national electricity grid, neither of which was possible in a night fighter. At this point the needs of the RAF coalesced with those of the army and the navy, and Bowen's 1.5 m system was adopted by all. Antiaircraft guns, the responsibility of the army in Britain, needed a precise, easily moved mechanism to predict where and at what altitude a bomber was flying. The navy had early recognized that radar could both warn of the approach of enemy ships and allow more accurate gunfire than could optical sights, with their obvious limitations of range, at night and in bad weather. Ships also needed compact equipment, although not as compact as that required by night fighters. Coastal Command airplanes obviously benefited from air-to-surface vessel radar, which made it possible to locate vessels out at sea. A compact radar set on board ships and airplanes would thus aid the looming maritime war as much as it would the air war. Although Bowen's 1.5 m system could not identify very small targets at sea, such as U-boats and E-boats (fast, light, torpedo boats), it was ideal for locating surface vessels. It was adopted by the navy in 1940 (Howse 1993, 58). Chain Home also had limitations; early defense exercises showed that aircraft flying below 500 ft were not picked up. A 1.5 m supplementary system was needed to catch such low flyers; it became known as Chain Home Low.

The Role of Commercial Television in Britain

The electronic technology used in developing radar was limited by the need for compact, powerful vacuum tubes designed to generate very high frequencies and very short waves. Because long-range radio research had shown very little use for the area below 20 m, television and VHF radio research became prominent. Television and to a much lesser extent VHF radio forced tube manufacturers to develop the much smaller tubes needed to generate and receive such very high frequencies. Television also created a tremendous demand for cathode-ray tubes, which became a vital and integral part of mobile radar systems. In the mid-1930s only the television interests saw the need for the considerably increased bandwidth available in the VHF bands and were willing to accept the line-of-sight broadcast limitations that were a result of VHF transmission. In America, Edwin H. Armstrong's development of frequency-modulated (FM) transmission for radio broadcasts also required a shift to VHF transmission.

*Communications,
Command, and
Control in the War
in the Air*

185

The conventional wisdom in the American radio broadcast industry of the late 1930s was that the extra costs and line-of-sight limitations of FM technology over the existing amplitude-modulated, medium-wave system were not justified. But Armstrong was able, on the promise of his technological breakthrough and his reputation as an innovator, to put together his own FM radio system, the New England Network, demonstrating that the public was indeed willing to pay for the substantial improvement in quality that FM gave. FM operating on VHF bands was also enthusiastically adopted for short-range, two-way mobile radio by American police forces because it was completely free from atmospheric interference.

VHF transmission was also developed by a special group set up for that purpose at the Royal Aircraft Establishment to create line-of-sight ground-to-air communication. This system was developed just in time for the Battle of Britain, in which its contribution to clear communication was vital, although as Callick notes, the RAF showed no interest in switching from AM to FM transmission at the same time and avoiding interference problems (Callick 1995, 153, 155). Callick further notes that he based his development of FM on the work of Hans Rhode at the Standard Telephone Laboratories in Britain. Rhode then returned to Lorenz in Germany, both Lorenz and Standard Telephones being part of IT&T and Lorenz being the major supplier of radio-communications equipment to the German military in World War II. Callick recollects that "Rommel's tanks used an FM communication system" that was almost impossible to jam (155). The excellent high-frequency tubes made by Lorenz for its radio equipment found their way into the Freya and Seetakt radar systems (Bauer, e-mail communication, December 1996).

In America, RCA and Western Electric had tubes capable of very high frequencies on the market by 1933. In the British effort to create AI radar the British General Electric Company's (GEC) American micropup tube was not pursued. By 1940 GEC had a viable but low-power radar system working on a wavelength of 25 cm, but the success of the much more powerful magnetron caused it to be discontinued (Lovell 1991, 30–31, 45).

The link between television and radar technology is clearly brought home by the person most responsible for the development of British airborne radar in the 1.5 m band, E. G. Bowen. Bowen describes the acquisition in 1936 of "a gem beyond price . . . [the] receiver designed by the EMI Company for their projected television service from Alexandra Palace. It operated on a frequency of 45 megacycles per second (Mc/s) [MHz] or a wavelength of 6.7 metres. . . . This receiver formed the basis of our whole airborne radar experimental programme for the next two years" (Bowen 1987, 32–33). Although initial experiments were made in this 6.7 m band, by late 1937 Bowen and his colleagues had moved down to 1.5 m and up to 200 MHz. The 200 MHz signal was then converted for display on the standard television receiver at 45 MHz. Even when centimetric-band radar became available, it was converted to 45 MHz in order to continue use of the commercial television chassis for display. Just as World War I spawned the broadcast radio industry, the end of World War II saw the release of war-surplus radar chassis into radio and then television shops as the basis for many amateur television sets in Britain. Such sets had green screens and, more problematic, a long enduring image so that the radar operator could continue to see a return as the scanner rotated. Television needs a much less enduring image, but sets based on radar chassis were cheap and available. One is on display at the British National Museum of Photography, Film, and Television in Bradford. Those who had serviced radar sets in wartime were quickly able to both convert them to television and set themselves up in the business of repairing television sets when peacetime production resumed.

In fact the link between television and radar is much more crucial than generally has been acknowledged. In the mid-1930s the government-controlled British Broadcasting Corporation (BBC) became embroiled in controversy about whether to institute an electromechanical scanning system of television, the Baird system, named after its developer and promoter, John Logie Baird, or a totally electronic system. Nearly all the pioneering work on electronic television was done by RCA, by an independent American group focused around Philo Farnsworth, and by Electronic and Musical Industries (EMI) in Britain, set up in 1931 through the merger of His Master's Voice Gramophone (HMV) and the Columbia Graphophone Company. EMI was linked to RCA through HMV, and the head of RCA, David Sarnoff, sat on the board of directors of EMI. RCA separated from EMI in 1935, not least because Baird, who was pushing his electromechanical television system, had persuaded the BBC that EMI was not a truly British company. Ironically, Baird had an arrangement with the pioneering German television company Fernseh, which may have hurt Baird when the British had to decide between Baird and EMI. In Germany Telefunken obtained basic patents on television that it shared with RCA and EMI to become part of the same patent pool (Abramson 1987, 164, 203, 208, 223). RCA, EMI, and Telefunken all depended on Vladimir Zworykin's iconoscope, developed at RCA, to convert light falling on an electronic grid into electrons.

Part of EMI's success may also have been that it linked up with Marconi to become the Marconi-EMI Television Company in May 1934. Marconi had, of course, laid the foundation for many of the electronic techniques needed in high-definition television: shortwave and ultra short wave transmitters, modulators, and focused antennas (Burns 1986, 316). Although Marconi originally preferred a connection with Baird's company for television purposes, the chair of EMI, Isaac Shoenberg, formerly the joint general manager of Marconi's, was able to persuade his old employer that EMI was a serious proposition. Baird later wrote that "if we had joined Marconi we should have been with this combine, not against. . . . We should have made terms" (quoted in Burns 1986, 318).

Although the electronic system worked, at first its definition was not greatly superior to that of the mechanical system. The two systems chosen for experimental transmissions by the Television Advisory Committee set up to guide the BBC's development of television in the mid-1930s were the 180-line Baird electromechanical system and the 240-line EMI electronic system. The ease with which Baird's electromechanical system could transmit movies was a marked advantage, and it was improved to

240 lines by late 1936, when experimental broadcasts began using electromechanical and electronic systems in alternate weeks. EMI's electronic system offered a mobile camera and the possibility of transmitting live such important outdoor events as horse races and cricket matches. Two things brought the BBC down on the side of the electronic system: one was the fact that EMI was able to virtually double its initial 240-line definition to 405 lines; the other may have been direct interference by the Cabinet. By 1936 Watson Watt, and thus almost certainly the Tizard Committee, the Committee for Imperial Defence, and the Cabinet, knew that Bowen's attempt to develop AI radar would be greatly aided by 45 MHz television. Radar was an item of key interest to the Cabinet, then on the verge of committing vast sums to Chain Home. Tizard was quite clear about the need to develop AI radar as the step after Chain Home, and for him not to have communicated electronic television's centrality to the effort seems unlikely. The current curator of the television section of the British National Museum of Photography, Film, and Television, John Trenouth, certainly believes it possible that the Cabinet intervened, although documentary proof is probably still classified (Trenouth, interview by author, Bradford, 3 March 1995). From 1936 on, British Air Force officers in civilian dress visited London's electrical factories to recruit electrical technicians for the coming electronics war. Jack Nissen, a 1936 recruit who ended up at Bawdsey Manor helping to develop Chain Home, was familiar with early work on television because his father had given him a Nipkow disc, a vital component of electromechanical television, as a Bar Mitzvah present. By 1938, "when barely nineteen," he was "teaching television theory to men twice my age" (Nissen and Cockerill 1987, 3, 16).

Television was also the point of entry to radar for G. H. Dummer, who would develop the plan position indicator (PPI), "one of the most important advances in military and civil radar technology" (Callick 1990, 118). Together with the magnetron, the PPI gave the British their crucial lead in radar technology by late 1940. Dummer worked for the British television company Cossor before World War II, where he was concerned primarily with problems of timebase. He entered military service by responding to an advertisement in an electronics journal for people who knew about cathode-ray tubes. In Dummer's opinion, only someone with his sort of experience could have come up with the idea of the PPI because of PPI technology's inherent dependence upon a thorough understanding of timebase problems (Dummer, interview by author, Bournemouth, December 1995). The first PPIs of 1940 used Cossor 12-inch television display tubes (Callick 1990, 111).

The events surrounding the onset of World War II in September 1939 certainly suggest that the British government took television very seriously. Television broadcasts from the large station at London's Alexandria Palace were stopped mid-cartoon because German bombers might use the signals to radiolocate central London. Yet other, equally powerful radio emitters were not closed down. The next day, Alexandria Palace personnel found themselves in uniform and dispersed throughout the radar industry. Bernard Lovell notes that the GEC Research Laboratory was a fertile source of radar personnel and that "soon after the outbreak of war the team on television at Wembley was at the Government's request diverted to work on airborne radar" (1991, 45n).

Other sources corroborate this conclusion. Radar technicians were needed by the army and navy as well as by the air force. The 1.5 m system became the basis for excellent gun-laying radar for ships and antiaircraft guns. Before the war Dick Gargan's shop in Glasgow sold and serviced radio equipment. Recognizing that television would be a whole new market, he took courses on television at local technical schools. Given that the London station had a radius of perhaps 30 mi and that Glasgow is more than 400 mi away, it is remarkable that such courses were offered so early. Gargan notes,

> When we had completed the obligatory "square-bashing" . . . we began to see why the Royal Army Ordnance Corps were so keen for Radio Engineers to join them. After endless quizzing re our knowledge of basic principles and experience carried out at several Technical Centers with a view to assessing our abilities, we were graded by the officer, who appeared to know less than we did about modern techniques! Not surprising, since the reference work he referred to was the Admiralty Handbook, which from what I could see appeared to be based on W.W. I equipment! . . . We were eventually subject to a six months course at the Military College which was a three year course telescoped down because of the urgent need for field technicians to keep the [antiaircraft gun-laying] equipment in action. This was so urgent we quite often found ourselves having to leave the classroom to repair equipment we had never seen before. (Gargan, letter to author, March 1995)

Finally, the mere existence of a large number of commercial engineers who could be rapidly diverted from television and radio interest to war work was crucial in all areas of the electronics war. Callick notes, for example, that most of the personnel who joined the VHF development group that produced VHF radio communications for RAF fighters in time for the Battle of Britain "brought with them the latest techniques in the design of radio, TV and communications equipment and several years experience in industry or other relevant activity" (1995, 159).

The Social Organization of Research in British Television

Another reason for the success of British radar may well have been the number of Jews involved in the British electronics industry. Many of Marconi's vacuum tube factories were set up by Serge M. Aisenstein, a Russian Jew and a Menshevik revolutionary who built communications equipment for the French military in World War I. Aisenstein and his British partner in what became the English Electric Valve Company, a Jew by the name of Bill Young, were persuaded by Aisenstein's fellow Russian Jew Zworykin, whom Aisenstein had known well in Russia, to manufacture in Britain the iconoscope Zworykin had developed at RCA (Abramson 1995, 31). David Sarnoff, the head of RCA, was also a Russian Jew. After some hesitation because a £1 million investment was required,

Aisenstein accepted Zworykin's offer (Callick, interview by author, Bournemouth, December 1995). This decision and the subsequent improvements to Zworykin's technology made at EMI helped place Britain at the forefront of the development of electronic television. Isaac Shoenberg, the head of EMI, was also a Russian Jew, and Jewish employees were commonplace at EMI. Perhaps the most significant of these was Alan Blumlein, who worked on the first television cameras built by EMI using English Electric's iconoscopes, developed 405-line high-definition television for EMI and went on to head the H2S (for "Home Sweet Home") ground-imaging-radar program. Regrettably, he died in the tragic crash of a Halifax bomber carrying the prototype H2S and most of its development team in May 1942 (Lovell 1991, 126–30).

1.5 m AI Radar

Although 1.5 m radar worked in AI applications, its range was limited by height and ground returns. The AI radar picked up only data in a sphere around the night fighter that was equal in diameter to the fighter's altitude. Beyond that distance it merely read the ground. Ground radar and a competent system of ground-controlled interception (GCI) could vector a night fighter to within about 3 mi of an intruder, so the minimum range for AI sets was defined as 3 mi (Putley 1988). The height at which most attacking bombers could be expected to fly was 15,000 ft (Bowen 1987, 67–68). AI in a fighter at 15,000 ft would have a 15,000 ft range and could then guide the fighter to within 1,000 ft so that a positive visual identification could be made before the attack. The latter was needed because no electronic Identification, Friend or Foe (IFF) equipment was available in 1940 and there were too many cases of friendly airplanes' being shot down mistakenly in the night skies. Despite its limitations, the 1.5 m system was put into production in late 1939. It took two years of development before it could be called a success.

The receiver proved a serious bottleneck. As late as mid-1939 the only working receiver in Bowen's experimental unit was the modified experimental EMI television chassis from 1936. Bowen notes that he heard "quite by chance . . . [that] the Pye Radio Company, still hoping there would be a television industry in Britain, had set up a production line for 45 Mc/s . . . chassis and had actually made a trial run" (1987, 77). The Pye receivers were much better suited to airborne radar than the 1936 chassis because they used a very sophisticated new vacuum tube made initially by Philips in Holland. According to Bowen, this tube, the EF 50, "was destined to play almost as great a part in the radar war as the magnetron itself" (77). Bowen's comments about television seem a little disingenuous: television was a massive growth industry in the late 1930s, and the EF 50 was hardly a secret. Philips would publish full technical details of the EF 50 in the June 1940 issue of *Philips Technical Review* (Strutt and van der Ziel 1940). As the Germans invaded Holland, the British took some 25,000 tubes from Philips to ensure that 1.5 m radar could be quickly produced (Bowen 1987, 77). At the same time, "members of the Philips family escaped to Britain with the necessary industrial diamonds and dies to produce drawn tungsten wire" of the type needed to continue production of the EF 50 (Trenouth, personal communication, December 1996). The British company Mullard put the EF 50 and the EE 50 into production in Britain in 1939 for television use, although the bases were different than those of the Philips tubes (*WW* 45 [31 August 1939]: 202; Callick 1990, 144–45). The EF 50 as well as the EFF 50 and the EFF 51 were listed as in service on a German naval stock list of the period (Bauer, e-mail communication). In any case, Pye's contribution to wartime radar included not only the receiver chassis but also a cathode-ray tube suitable for airborne radar (Bowen 1987, 78).

Although the C^3 system for Chain Home developed in the Biggin Hill experiment of 1937 was perfectly adequate for massed daylight raids (Bowen 1987, 26), it had problems. Chain Home Low working on 1.5 m began to solve the problem of individual, daylight, low-level raiders and promised to enable detection of night raiders. But the early Chain Home Low stations could not be rotated, nor could they find the height of their targets until late 1940. Nor was Chain Home Low "sufficiently accurate to bring a fighter within visual range of its target at night and . . . the operation of the complex network was too slow. Finally, the original chain only looked out to sea, so could not deal with interceptions inland. Studies of interception attempts by nightfighters fitted with the early AI sets in November 1939 showed the need for a new type of radar along the lines of the 'radio lighthouse' proposed by Bowen as long ago as 1935 and for the controller to work directly from the radar" (Putley 1988, 162).

Rotatable Chain Home Low coupled with the PPI developed by Dummer, a cathode-ray-tube display on which the fighter controller could locate both pursuer and pursued, the former through its IFF equipment, solved the C^3 problem in time for the 1941 night blitz. Fighter controllers were limited to one interception at a time, which proved adequate for the relatively small night raids of which the Luftwaffe was capable. Dummer also solved the problem of training large numbers of aircrew to use AI radar by devising an extremely sophisticated ground simulation system at TRE. This meant that valuable flight time did not have to be wasted on night training flights, and the inevitable accidents on such flights could be prevented (Dummer 1995).

AI and GCI using a PPI did not solve all the problems. The sets had to be mounted in a suitable airplane. After his first experience with 1.5 m AI radar in 1939, Dowding recognized that long-endurance, multiseat, twin-engine airplanes were a virtual necessity for night fighting (Bowen 1987, 71–72). (In 1939 the only such plane designed as a fighter to have entered service anywhere in the world was the Messerschmitt Bf 110. Although it was a failure as a bomber escort in the Battle of Britain, the Bf 110 distinguished itself as a night fighter in the air war over Germany.) Dowding assigned the Bristol Blenheim I to the task because, with a top speed of 285 mph, it had been the fastest medium bomber in the world in 1936 and thus ought to be able to catch the German bombers it was likely to encounter, the Dornier Do 17 and the Heinkel He 111, with top speeds of 259 mph and 258 mph, respectively (Barnes 1988, 282; Green

1959, 51, 16). Based on the same reasoning, the Germans modified the Dornier Do 17 (as the Do 215) for night fighting. But such factors as top speeds and rates of climb are elusive categories in airplanes. Carefully built and maintained examples could easily be 20–30 mph faster than planes off a production line operating in service conditions. The Blenheim I's apparent speed advantage evaporated in such circumstances because of the drag created by the external radar antennae. As radar-equipped night fighting began over Britain in 1940, Blenheims scored a few early victories and then went into decline. They could catch opponents who were flying slowly with their engines on cruise settings, but such dawdling soon stopped. Even when the Blenheim could close in for the kill, its single 9 mm machine gun armament usually could not do the job before the German pilot used full power to escape. One commanding officer of a night-fighter squadron used his own money to equip his squadron's planes with four additional 9 mm guns (Bowen 1987, 118). The night fighter had to be able to destroy its target with its first burst of fire, hence the need for very heavy armament. The plane that won the night blitz for Britain was thus the Bristol Beaufighter, a derivative of the Blenheim with nearly twice the engine power, a lightweight fuselage, and an armament of four fast-firing 20 mm cannon and six 9 mm machine guns. The prototype was capable of 330 mph, but the 309 mph of production airplanes was more than adequate for night fighting (Barnes 1988, 307, 293).

Even with the Beaufighters night kills were slow in coming because of ground-radar problems. In January 1941 much improved GCI allowed the ground controller to follow both night fighter and quarry on the same PPI. This was as important a technological breakthrough as the magnetron. The ground controller could distinguish between pursuer and pursued because British night fighters were being fitted with IFF equipment. The PPI was basically a television picture tube modified to display ground-radar impulses and is thus another example of the importance of television in the development of radar. Night kills climbed steadily: 22 in March 1941, 52 kills and 88 probable kills in April, and 102 kills and 172 probable kills in May (Bowen 1987, 130). Eleven of the 22 kills in March were by Beaufighters using radar, a rate the Luftwaffe could afford. But by May losses were intolerable: 16 night bombers fell to London's defenders on 10 May (Cross 1987, 110), and in the even fiercer fighting on the night of 19 May, 24 enemy bombers fell to radar-equipped Beaufighters and 2 to antiaircraft guns (Barnes 1988, 296). Attrition rates greater than 10 percent were too great for the Luftwaffe (or any other air force) to afford, even before the transfer of most of the bombers to the eastern front when the invasion of the Soviet Union began on 22 June.

Still being developed at the end of the night blitz was centimetric radar using the magnetron valve developed by two British physicists at the University of Birmingham, H. A. Boot and John Randall. It was the great breakthrough that made possible powerful radar systems operating in the 10 cm waveband, hence the term *centimetric radar*. In and of itself the magnetron was not sufficient, however. Boot and Randall's magnetron was inherently unstable and had to be substantially modified to make it part of a workable radar system; microwave radar needed a whole family of technologies to make it an operational success (Callick 1990, 55–58).

In his search for the ideal AI radar, with its antenna mounted entirely within the aircraft, Bowen set 10 cm as a goal as early as 1939 on the basis of desired beam characteristics and available airframes: "I did a back-of-an-envelope calculation of the best wavelength for the purpose . . . by far the most limiting factor in airborne radar at 200 Mc/s was the enormous signal which came from the ground directly below the aircraft. . . . The only way of improving the system would be to project a narrow beam forward to get rid of the ground returns. I estimated that a beam width of 10 degrees was required to achieve this. Given an aperture of 30 inches—the maximum available in the nose of a [twin-engine] fighter—this called for an operating wavelength of 10 centimetres" (1987, 145). The first magnetron had a greater power output than the first, experimental ground stations of Chain Home, remarkable progress in five years.

THE GERMAN CRUISE MISSILE ATTACKS, JUNE–SEPTEMBER 1944

The Battle of Britain and the night blitz proved that a C^3 system based on radar was the key to victory. The Germans never fully returned to a manned offensive against Britain. Attempts to do so in the "little blitz" of early 1944 were too expensive, so they switched to missile attacks. The use of V-1 and V-2 weapons demonstrated again the vital defensive role of radar. In the case of the V-2 ballistic missile, nothing could be done once it was launched, but because the ballistic track of the V-2 was predictable, Chain Home could give the civilian population in the target area a warning of several minutes so that they could reach their shelters.

The targets of unguided, pulsejet-propelled V-1 cruise missiles could not be predicted, so they had to be shot down. The V-1 was cheap, deployed in large numbers, and hard to intercept. Most piston-engine fighters were too slow, and those that were fast enough had only a very small speed advantage. Even the first British jet fighter, the Gloster Meteor, was only just fast enough. Large numbers of pilots were killed by debris when they successfully exploded the missiles by gunfire, which led to the technique of flipping the missiles over by the wing tip. The real solution was the radar-directed antiaircraft gun and the proximity-fuse shell, which depended on the American SCR 584 gun-laying radar "for which [John] Cockroft [a member of the Tizard Mission] wrote an outline specification" (Bowen 1988, 306). The SCR 584 was designed to control antiaircraft guns, which largely accounts for its superiority over earlier British and German equipment, which had been merely modified for such service. Since total accuracy in height measurement was rare, the development by the British of a tiny radar to fit inside an antiaircraft shell—the proximity fuse—had a major impact on the defensive war. Bowen notes that close to half of the V-1 missiles to cross the English Channel were shot down with SCR 584 radar and proximity-fuse shells (Bowen 1988, 306).

Between 12 June and 1 September 1944, when the Allied troops over-ran their Pas de Calais launch sites, a total of 8,564 V-1s were launched against London. Of these, 1,006 crashed shortly after launch and 2,340 reached their target, just over 30 percent of the successful launches. After September 1,200 V-1s were air launched from Heinkel He 111 bombers. Of these, 66 hit London, 1 hit Manchester, and 168 fell elsewhere in Britain, less than 20 percent of launches (Pocock 1967, 49).

THE AIR WAR OVER GERMANY, 1943–1945

The air war over Germany saw a replay of the pattern established over Britain, although with somewhat different results: for the most part, the Germans lost. When America entered the war, British experience was used to guide American technology and doctrine. This was not always for the best, since American needs for radar were often quite different from those of the British, and the resemblance between the American and the German systems of developing new technology was too close for comfort.

British and American bombers delivered about the same total bombloads on Nazi Germany, with the Americans slightly in the lead. More than half of the American bombs were delivered with radar guidance in overcast conditions to targets no smaller than a city: whatever they preached, Americans practiced area bombing almost as much as the British. In one important sense the two doctrines and their resultant technologies complemented each other, since the German population underwent constant interruption day and night.

The first German attempts at aerial defense were as crude as those being considered by Britain's Committee for Imperial Defence before Chain Home. As late as 1937 "it was believed that all that was necessary were a few flak guns, a string of sound locators and the eyes of the Observer Corps." The prevailing attitude was that "the Luftwaffe attacks but does not defend" (Pritchard 1989, 204). Only after the British night raids began was General Josef Kammhuber permitted to form the first group of night fighters. By September 1940 Freya radar units had been installed along the south of the Zuider Zee to locate British bombers as they crossed the Channel, and by October the first Würzburg radar units had been installed to replace sound locators in the illuminated night-fighting zones east of the coast. Even though the Freya and Würzburg units were reliable, they were not deployed in an effective C^3 system. Their first use was merely to direct flak (70–71). It was easy for British planes to evade the system, even when Lichtenstein AI radar sets were installed in German night fighters. The 120 British planes lost to radar-equipped night fighters by early 1942 were clearly not enough. The area covered by radar was too narrow, and each night fighter had to fight within a narrowly defined control "box" of airspace and then hand an intruder on to the next box (208). By May 1943 Kammhuber had designed a much more sophisticated system, Himmelbett. Himmelbett and the C^3 system that preceded it were jointly referred to by the British as the Kammhuber line. Himmelbett gave good results: in 1943, 76 percent of night interceptions resulted in kills, up from 65 percent in 1942 (Aders 1992, 82). But Himmelbett had drawbacks. One hundred forty ground personnel were required to vector one fighter onto one bomber (59). It was susceptible to such simple ECM as appropriate-length aluminum strips, called Window. And it was easily swamped by a large numbers of targets. The 1,000 bomber raids that almost annihilated Hamburg in a firestorm of unprecedented dimensions in July 1943 marked the first use of Window. Würzburg read each strip as a separate airplane. With Himmelbett's utility curtailed, the Germans reverted to an extremely crude form of night interception, Wilde Sau, which used the light of the burning cities to illuminate British bombers for fighters cruising overhead. Unfortunately, few of the day-fighter pilots flying Wilde Sau missions were good at night flying and especially landing. There were almost as many German casualties as British.

German scientists, more technologically advanced in the development of radar and working from the mid-1930s with wavelengths below 1 m, deluded themselves into believing that the British had no radar before the Battle of Britain. In part their problem was that they were so advanced: they believed that their own research had set the upper limit of scanning radar at around 3 m (Freya) and the lower limit, set by vacuum tube capabilities, at about 50 cm (Würzburg). Thus, when the Germans sent the *Graf Zeppelin* to carry out electronic surveillance over Britain in 1939, they were looking for radar signals with a pulse repetition frequency (PRF) of at least 1,000 Hz, the lowest frequency at which Freya operated. "The British system . . . used a much longer wavelength, and a much lower [PRF] (50 Hz)" (Nissen and Cockerill 1987, 36). The German engineers certainly picked up 50 Hz signals, but they assumed that they were the background noise of the British power grid, delivering alternating electric current at precisely 50 Hz. The duralumin structure of the Zeppelins "acted as aerial systems that absorbed, like vast sponges, giant quantities of energy from the CH transmission that saturated the on-board receivers" (37). This failure to identify Chain Home as a crude but effective radar system was a fatal error, since the Germans thereby assumed that their bombers would always get through, and when they found otherwise it was too late to develop effective ECM.

German radar certainly was technically superior to anything the British had before Chain Home Low, but as part of an effective system of defense the British system of 1939 was superior to German efforts up to 1943. When it finally became clear that Chain Home was a crude low-frequency radar, and attacks upon its huge stations by the Luftwaffe began, the British were saved by the fact that they had developed a system of overlapping stations that could cover gaps while repairs were being made. When the Germans realized that Fighter Command was not "blinded" by attacks on Chain Home, they failed to press home such attacks on a regular basis. Göring, who believed that "wars are won by men, not by equipment," ordered that the attacks on Chain Home be discontinued in favor of attacks on airfields (Pritchard 1989, 120). The British had been forced to favor defensive doctrines, radar in the late 1930s worked best in a defensive battle, and the British had had to develop an effective C³ system of defense, the operative word being *system*.

The Germans were not put on the air defensive until late 1941, and even then it was only in a minor way, but at that point they too began to recognize the crucial role to be played by radar in such a system. However, it took them until mid-1943, by which time the success of the British strategic offensive had become all too evident, to work out a C³ system despite their initial superiority in radar itself. The Luftwaffe paid a very heavy price for an offensive doctrine and technological hubris when it was beaten back in the Battle of Britain, but once the air war over Germany was begun, it should have enjoyed the advantages that accrued to the defense. The key question that must be asked about the air war over Germany is thus not so much why the Allies won but why the Germans lost. Perhaps the main reason was that they were not at first prepared to fight such a defensive battle and had to make the critical adjustment to such a fight while engaged in a two-front war. The British had the luxury of several years of peace, 1936 to early 1940 if one includes the period of military inactivity known as the "phony war," between September 1939 and the Battle of Britain, in which to develop their radar hardware and software.

One reason given for Germany's loss is the quantitative difference between the German bombing of Britain using tactical bombers and the Allied bombing of Germany using strategic bombers. "For every ton of bombs dropped by the *Luftwaffe* on Britain, Germany . . . received 315 tons" (Cross 1987, 160). In all, the British and the Americans dropped 1.4 million tn of bombs on Germany proper (Guerlac 1987, 757). Perhaps Germany's failure can best be understood by looking at it from the German point of view as stated by Aders (1992). Aders blames Germany's poor technology, its misuse of human resources, and the poor defensive doctrine that brought night-fighter groups into being only after it had become clear that ground-based antiaircraft guns (flak) could not cope with the night intruders. He further blames the Germans' failure to develop high-performance night fighters until it was too late. The two main German night fighters were the Messerschmitt Bf 110 and the Junkers Ju 88. The first was a modified *Zerstörer* design dating back to 1934. The latter was a *Schnellbomber* design dating back to 1936, one of the finest all-purpose aircraft ever produced. By late 1940 the Bf 110 was outperformed by the British four-engine bombers then entering service, which, says Aders, "pulled away from the Bf 110 at higher altitudes and the fighters' climbing speed was not enough to bring them within range of the bombers when their altitude was measured by *Würzburg-Riesen* [radar] 80 km in front of the night-fighter stand-by zone" (35). When the Bf 110 was equipped with Lichtenstein AI radar, its speed was further reduced by the drag caused by the complex, nose-mounted antennae. Attempts to develop internal antenna systems to reduce drag not only failed but took almost a year, from early 1940 to February 1941, time that could have been spent more fruitfully on something else (Pritchard 1989, 136). Although the Bf 110 was given more powerful engines in early 1943, it took a year to solve the problem of fires in the new engines (Aders 1992, 68–69). The

Ju 88 was also too slow when equipped with external-antenna AI radar, even with the improved engines fitted by 1943. Only the British bombers' consistently cruising below their maximum cruising speed to conserve fuel and reduce engine wear allowed either fighter to be effective at all (69). The numerous attempts to produce better night fighters on the basis of later *Zerstörer* designs than the Bf 110 all foundered. It was always a case of too little too late: lack of focus, inadequate development, or political interference with what should have been technical decisions, the last named producing the saddest errors. Kurt Tank's potentially brilliant design for Focke-Wulf, the Ta 154, was built to a specification calling for a wooden airplane to answer de Havilland's Mosquito, raids by which were a constant nuisance to the Germans. Despite clear statements by the Luftwaffe's research branch that Germany's woodworking industry was not up to the job, the plane always referred to in propaganda material as the German Moskito went ahead. Problems with the glue used on production aircraft led to wings that disintegrated in flight (Aders 1992, 192; Green 1970, 241–44). Aders lists only a single Moskito as operational on 30 September 1944, compared with 574 Bf 110s and 740 Ju 88s (Aders 1992, 274). The Heinkel He 219 Uhu (Night Owl) fared little better, with only 32 operational in late 1944. Only General Kammhuber's farsighted insistence in 1941 that the Bf 110 would be unable to combat the new British four-engine bombers resulted in the He 219's being saved from oblivion in the first place. When Kammhuber fell from power in 1943, his successor, Field Marshal Erhard Milch, did his best to kill the He 219. Although Milch's resignation in July 1944 allowed the He 219 to return to favor, it was too late: total production for 1944 and 1945 amounted to only 268 operational aircraft (Aders 1992, 187–91).

The Germans proved almost as adept at ECM as the British, and the radar war evolved into one of electronic move and countermove. German night fighters were able to home in, successively, on British IFF equipment, the tail-warning Monica 1.5 m radar systems used by British bombers to locate pursuing night fighters, and the 10 cm emissions of the British ground-imaging radar, H2S (Pritchard 1989, 107, 214). At the end of the war General "Beppo" Schmidt stated that Window had been the biggest British success in the electronic war and H2S the biggest mistake because it eventually allowed German night fighters to use a passive homing system, Naxos, to easily locate the British bombers by their H2S emissions (216). He might have added that the British use of Monica was at least as devastating, since by the end of 1943 most German night fighters carried Flensburg receivers to home in on Monica (Gunston 1976, 104, 124–25).

The Role of Commercial Television in Germany

German ground radar did relatively well technically, although it was not quickly made part of a comprehensive operating system. The real development of radar in Germany began in 1934. Despite successful experiments on radar imaging of ships by the Pintsch Company with a 13.5 cm system as early as May, it was Dr. R. Kühnhold who, in close cooperation with the Nachrichtenmittelversuchsanstalt (NVA, the German Naval Research Establishment), created the first German radar company, the Gesellschaft für Elektroakustische und Mechanische Apparate (GEMA), that same year. GEMA obtained radar reflections from airplanes as early as 24 October 1934 (Bauer, e-mail communication).

In 1936 the chief engineer at Telefunken Wilhelm Runge, began experiments at wavelengths around 50 cm. As in Britain, the merger of high-definition-television interests with those of radar was crucial, albeit much later. German television of the early 1930s had been conceived as a propaganda tool rather than an entertainment device. A limited number of cinemas were equipped with 180-line projector receivers so that Nazi Party propaganda could be disseminated easily, and "cinema television was used throughout the war for troop entertainment" (BIOS 1950, 35). The German Post Office claimed to have opened "a 'regular' public service from Berlin" in March 1935, although the system could transmit only from film images and went off the air after a disastrous fire in August 1935 (Abramson 1987, 217–18). In 1936 the Berlin correspondent for *Wireless World* reported that there were no more than 100 television receivers in Berlin and that "in spite of announcements to the contrary, no television receivers are sold to the general public at the moment" (39 [10 July 1936]: 29). Later that year *Wireless World* described the televising of the opening ceremonies of the Berlin Olympic Games on the 180-line system as "so poor . . . that even the most sympathetic members of the audience turned away" (39 [21 August 1936]: 191). A later report claimed that "the quality of the 'direct' television transmissions improved almost from day to day" and included photographs of the television screen—"photographed in the writer's laboratory. They are reproduced here without retouching"—as evidence (39 [11 September 1936]: 280). The still images of the scoreboards were good, but the image of swimmers about to start a race was poor. The live image does not bear comparison to photographs of the screen images transmitted on the new British 405-line service early the next year (40 [5 March 1937]: 234). *Wireless World* noted, however, at the time of the switch to the 405-line service, that "during the last few weeks particularly, the programmes from Alexandra Palace have improved so much that we feel it is unfortunate that there is no record of what they were like at the beginning of the service, in order that direct comparison could be made" (40 [19 February 1937]: 173).

An "untouched posed portrait taken in New York of the reception of the N.B.C. 441 line, 60 frame transmission" published later that year was of even better quality (40 [26 March 1937]: 310). The German television industry showed its version of the 441-line system for the first time at the Berlin Radio Exhibition in 1937. It was reported, however, that "the German government was in no hurry to start an extensive program service and that it was probable that no developments would take place on the commercial side for the next two years" (Abramson 1987, 239). *Wireless World* reported in 1939 that "only one television receiver will be seen at this year's Berlin Radio Exhibition which opens on July 28th. It is to be called the television standard receiver and has been designed by the five firms concerned in German television developments who will manufacture some 5,000 by the end of the year" (45 [20 July 1939]: 63). However, at the outset of World War II, "civilian television development in Germany was suspended just as work had progressed so far that television could have been generally introduced" (Bruch 1959, 291).

By 1945 two cable systems had been completed, one for 441-line transmission and one for the old 180-line system. Berlin and Munich were both centers for the 441-line system (see fig. 3.9). The primary purpose of the cable, however, was high-grade telephone transmission. The cable approach to television mirrored the German preference for transmitting three channels of radio entertainment through standard telephone lines to individual subscribers (BIOS 1950, 10–11, 34, 35).

As in Britain the link between television and radar was important. Telefunken may have done better than GEMA in the long run simply because it started to experiment with electronic television in the 1920s (Runge 1959, 70). According to W. T. Runge, "In the summer of 1940 the Television Development Group [of Telefunken] under Dr Urtel came to join us, and this was a very valuable reinforcement. Urtel got busy straight away by forming a team for developing centimetric measuring techniques. One of the team, a Dr Kettel, turned out to be an excellent researcher into pulse problems, circuitry, precise pulse-time measurements and so on. And of course, these television people understood much more about cathode-ray tubes than we did" (quoted in Pritchard 1989, 66).

Despite Runge's comments, the difference between the commercial television industry in Britain and Germany is clear. The British were using high-definition television commercially several years earlier than the Germans and had many more broadcast radio receivers to boot. Thus by the late 1930s there was a cadre of well-trained television and radio engineers in Britain but not in Germany. This cadre was easily identified and conscripted into the radar war through night courses in radio and television repair and through the RAF's practice after 1936 of visiting and recruiting from amateur radio and television societies. The difference was exacerbated by the different social forces at work in the two countries, with British Jews especially eager to find a way to avenge the Nazi persecution of their German coreligionists. The German failure to fully mobilize human resources was thus particularly problematic in the area of high-technology electronics. As already indicated, the Germans made no effort to locate electronics specialists in the general military draft; they neither offered them protected job status nor recruited them for specific electronics tasks. Thus, by late 1943 the Luftwaffe was reduced to advertising in electronics magazines for such people, even asking parents whether their drafted children had electronics skills. Presumably more than one parent was glad to save a child from the Russian front.

Origins of German Ground Radar

German superiority in vacuum tube design and production allowed German designers to move directly to what would be the limits of vacuum tube–powered radar technology. It would take the magnetron to go

*Communications,
Command, and
Control in the War
in the Air*

199

above 600 MHz. Early, unsuccessful experimentation with wavelengths as short as 13.5 cm may have resulted in later German unwillingness to return to them (Kummritz 1988, 207). Such considerations pushed the excellent German naval gun-laying radar, Seetakt, up to 82 cm and the air surveillance radar, Freya, to 2.4 m. Freya was the best and most successful air-surveillance radar of the war, spawning two, much more powerful descendants, Wassermann and Mammut. Freya was responsible for the first really effective radar-controlled interception of the war, over the Heligoland Bight during a British attack on the German navy on 18 December 1939. Only 10 of the 24 Wellington bombers dispatched made it home, and the Germans lost only 2 fighters (Levine 1992, 22). At the time Freya was not, however, part of a system comparable to Chain Home because "there was . . . no central command post at which all incoming flight reports could be evaluated and orders issued to the fighters and *Flak. Freya* data went to the Navy and the *Luftwaffe;* in the latter case the reports were received partly by the *Luftlotten* and partly by the *Luftgaukommandos.* . . . This division of responsibilities was never completely overcome" (Aders 1992, 22). Moreover, the interception of the Wellingtons at Heligoland Bight was improvised locally through the cooperation between a Luftwaffe fighter leader and a German naval radar gunnery officer; it was a tactically brilliant improvisation with no base in strategy (John Guilmartin, personal communication, August 1997). After 1943 information from Freya and Würzburg was combined by the Reichsluftverteidigung (Reich Air Defense) with Freya serving as the early-warning radar and Würzburg serving as a radar "spotlight" (Bauer, e-mail communication). This new system served Germany well, although the lack of a PPI made the system cumbersome for night interception. As many as 1,500 people worked to integrate the radar information gathered (Bauer n.d., 125; Bauer, e-mail communication).

Würzburg remained truer to Runge's original ideas for 50 cm radar at Telefunken. It operated at 53 cm and was sold in large numbers to the antiaircraft gunnery arm of the Luftwaffe, the Flak, to direct searchlights and guns. Its principal problem was lack of range. When a Würzburg unit was captured by the British in the Bruneval commando raid in February 1942, it became the basis for British ECM. Like Freya, Würzburg spawned a series of successors, most of them concerned with increasing range, most notably the greatly enlarged Würzburg, Würzburg-Riese (Kummritz 1988). Würzburg in particular was a technical tour de force. Had the data Würzburg and Freya were capable of producing been properly integrated into a C^3 system of air defense of the sort developed in Britain in the Biggin Hill experiment of 1937, the course of the air war would have been very different. The great German failure was one of software, not hardware.

The British were almost as slow to realize that German radar existed as the Germans were to recognize the existence of British radar because German radar operated at much higher frequencies than did British radar. Although a clear statement of German radar capabilities, the so-called Oslo Report, had been received by British intelligence in November 1939, it had not been believed. The report described a coastal early-warning system with a range of 75 mi, that is, Freya; and a "second type of aircraft detection device working on a frequency of about 600 mega-cycles and using a parabolic . . . reflector," Würzburg (Price 1978, 72). By July 1940 the term *Freya-Meldung* (Freya warning) was cropping up in German reports. Price notes that the name Freya was suggestive to R. V. Jones, the chief scientist at air intelligence, whose first responsibility was to interpret the Oslo Report. Freya was the Nordic goddess of beauty, love, and fertility, whose most prized possession, the necklace Brisinga-men, was guarded by Heimdal, watchman for the gods, who could see 100 mi in all directions, day and night. Jones noted that the code name Heimdal might have been too obvious (73).

As the British prepared to go on the offensive in 1941 they diligently began to search for defensive ground radar. They found it in three forms: the naval gun-laying radar, Seetakt, which turned up on the scuttled *Graf Spee* in December 1939; Freya; and Würzburg. Once Seetakt was identi-fied, it became clear that it was being used to sink ships in the Channel with coastal guns. For some time the Air Ministry worried that the accu-rate, long-range Seetakt would be turned to air-defense uses, but the Luft-waffe and the Kriegsmarine were never on good terms, and the Luftwaffe preferred to modify Freya instead. Freya and Seetakt, both produced by GEMA, were actually very similar, although they operated on different frequencies. As bomber casualties mounted over Europe, the British cap-tured a Würzburg in the Bruneval raid and went on to investigate what they feared was a Freya modified to be dangerously precise. Jack Nissen gained useful firsthand knowledge of Freya in the Dieppe and Pourville raid in August 1942. Once it became clear that the commandos could not capture the Freya or any part of it, Nissen cut the telephone wires that connected it to the Luftwaffe. This forced the Germans to control their fighters by radio, which was being closely monitored by the British Y Ser-vice. This made Freya's limitations plain and allowed ECM to be intro-duced in the form of the Mandrel jammer (Nissen and Cockerill 1987, 192–93). It was also possible to deploy what the British called "spoofs" against German radar. The British would send out a small force equipped with ECM to make it seem like a much larger force than it really was. This would persuade the Germans to concentrate their defense against an at-tack in one direction, when the real raid was elsewhere. Freya was usual-ly "spoofed" by a device code-named Moonshine, which was used only in daylight (Price 1978, 88).

The German obsession with 50 cm systems makes sense in the light of the Nazi projection of a two-front war against Poland and France. Würzburg was compact and highly mobile, easily towed behind a small truck and powered by a mobile diesel generator. Britain's Chain Home was a classic "moat defensive" whose clearly stated aim was to turn Britain into an island again (Crowther and Whiddington 1948, 10). Germany needed radars suited to a war of rapid expansion and mobility. Britain's "moat defensive" geostrategy had clear implications for the sort of tech-nical solutions Britain chose to meet the threat of a strategic air offen-sive.

German AI Radar

Despite the success of German ground radar, German AI radar fared little better than German airplanes. It developed from "an airborne radar for dive bombers to enable them to pull out of a dive automatically" that Runge was working on before the war (Kummritz 1988, 217). Telefunken's FuG 202 Lichtenstein BC, operating on 61 cm, had its first operational trials in August 1941. It was hard to interpret because it used three cathode-ray tubes, to display range, azimuth, and elevation, and it was disliked by pilots because of its cumbersome external antenna array. After a year was spent trying to devise an internal antenna to reduce drag, the set entered service with its original array. Some pilots simply sawed the external antennas off, and not until the end of 1942 were such antennas really accepted (219). Runge recounts an even more ridiculous problem: at the request of General Martini, the head of the Luftwaffe signal corps, Runge had an experimental AI radar fitted in a Bf 110 in 1940 and sent for testing to a squadron "made up of officers mainly from the highest ranks of German society . . . [who] considered this new-fangled technical stuff to be unsportsmanlike" (quoted in Pritchard 1989, 68). About six months later, when Runge asked Martini how it had worked, he was told that it had been assigned to a pilot from a working-class background. Once he had learned how to work the system, he had made three kills one night and two the next. He was then grounded so that he would not have to be decorated before higher-status members of the same squadron. Both the lack of technical education of night-fighter pilots and an obsession with preserving social status come from the same source. A war based on science and the new technology it made possible discredited the society of German fighter pilots, headed by men such as Göring, who thought of themselves as descendants of the Teutonic knights and saw air combat as a noble calling in which bravery and skill counted far more than technology (Bauer, e-mail communication). A working-class pilot who did well because he understood the new technology was no more likely to be tolerated in such a society than was AI radar. Presumably, however, this reluctance of the elite to embrace technology ended after 1941, since the top-scoring night-fighter pilot was Major Heinrich *Prinz* zu Sayn-Wittgenstein.

The successor to Lichtenstein BC (FuG 202) was Lichtenstein SN-2 (FuG 220), introduced in the fall of 1943, after the Hamburg raids. SN-2, which represented a return to lower-frequency operation to avoid Window jamming, used an ingenious electrically extended dipole to further resist ECM (Bauer, e-mail communication). Since reconnaissance photographs could measure the length of the dipoles on the night fighter's antennas, it was possible to calculate the frequency at which the radar operated. This did not work for SN-2, and the British were unable to jam it until they were presented with one by a German night-fighter pilot seeking asylum. It was also much easier to operate because it had only two display screens, compared with the three screens used by the FuG 220.

All Lichtensteins suffered immense quality-control problems. As late as August 1943 about 80 percent of the units delivered were defective

(Aders 1992, 247, 77). Exton (1995) notes that the British company Ekco had substantial quality-control problems when it created a shadow factory deep in the British countryside to build electronic equipment using largely unskilled labor with no background in electronics. The problems were exacerbated by tight security, so that it was not immediately clear where on the production line problems were originating. In one case workers in one part of the factory were openly repainting color-coded resistors to be the "right" color for use on circuit boards because they did not understand that the combination of colors signified the exact resistance of that component. No one could understand why so many units failed until someone with sufficiently high security clearance backtracked to the storage area and found the paint cans. Fortunately, such problems were rare in Britain, but they may not have been rare in Germany.

Other problems are also worth mentioning. After 1941 German industry suffered increasingly from raw-material shortages, and production capacity was often pushed to the limit. National Socialism had its internal detractors, and product quality declined because of sabotage as the use of slave labor increased (Bauer, interview by author, Bournemouth, December 1995). The first British AI radar began operational trials in 1937. It used only two cathode-ray tubes from the start, entered service in Mark IV form in time for the night blitz of early 1941, ran into no problems caused by the social ranking of pilots, and seems to have enjoyed good quality control (Hanbury Brown 1991, 26–28). Finally, the Germans so misused human resources by their excessively complex operational system for night defense that each interception required 140 signal corps personnel.

The Social Organization of Research on German Radar

Aders' main criticism is reserved for the Germans' loss of the technical leadership in radar that they had clearly held in 1939 and their failure to catch up in "the high-frequency war" (Aders 1992, 218–21). In 1943, when the British first deployed Window ECM at Hamburg, the Germans' response was to move their AI radar to the longer wavelengths. FuG 220, Lichtenstein SN-2, operated on 3.7–4.1 m. However sensible this may have been, it was also a technological move backward since SN-2 could not vector the night fighter as close to the bomber. Fighters thus often carried the old Lichtenstein BC as well to enable them to make their final run-in. This made life even harder for the radar operators in the already cramped cockpits of most night fighters.

This hardware problem arose from two software problems. The first was the lack of both a comprehensive research effort and a close relationship between serving officers and scientists (Pritchard 1989, 35). Aders is particularly clear about the lack of a comprehensive research effort, noting that the Germans had "only a tenth as much research capacity" and that "whatever research was going on was dispersed among 100 small laboratories which for reasons of secrecy were not allowed to exchange their practical knowledge and experience" (74). This situation, as Kern (1994) makes clear, was a result of the Nazi attitude to research: the Nazis tended to fund ideologically driven work on folklore, myths, and racial origins before they funded physics, electronics, and other technologies. The output of the technical universities was consequently halved. Wholesale conscription occurred without regard to job skills needed by industry and, apparently, without adequately recording who possessed those skills. Finally, the Nazis effectively drove Jewish scientists into the arms of their future opponents. Such policies persuaded many engineers to seek haven in industrial laboratories, where they were less subject to ideological control (Overy 1980, 239). It is impossible to describe this "approach" to research as other than pure, self-destructive lunacy. The situation was exacerbated by the wholly offensive doctrine subscribed to by Hitler and the Nazi leadership. This doctrine resulted in an order first mentioned in February 1940 during a meeting at Karinhall, the home of Göring, forbidding research that would not come to fruition in a short time (Bauer, e-mail communication).

The second problem was the squabble lasting nearly a year over which wavebands should be used for radar (Pritchard 1989, 87–88). Such squabbles seem to have been frequent in the Nazi state, which recent work has described as comprising at least four major power blocs—the Party, the state bureaucracy, industry, and the army—that variously cooperated and competed. In this interpretation, National Socialism began as an antimodernist revolt by those who feared or were subject to marginalization (Mehrtens 1994). In mathematics and physics, for example, there were vicious arguments about what constituted an "Aryan" science, and those desirous of rising to power in the universities could always count on Party help if they accused their opponents of being *Jude im Geiste* (Jewish in spirit). This had real implications in physics, where Werner Heisenberg's appointment to the most important chair in theoretical physics in the country was blocked by just such a tactic (Trischler 1994, 81). Such arguments were at a peak in the early years of Nazism. After 1936 or so the intensified preparation for war strengthened the technocrats, and after the reverses of 1943 "even party agencies that had stubbornly insisted on the value of a romanticist 'German' science dropped this notion in favour of technological and scientific rationality" (Mehrtens 1994, 295).

What happened in the radar community is less clear, perhaps because radar research was embedded within industry, where the Nazi Party was less powerful, perhaps because most of the development occurred after 1936 and the relative triumph of technocratic reasoning. Nevertheless, what originated as an academic argument reached the point of tragedy or low comedy, depending upon one's viewpoint, when those supporting high-frequency systems began to argue that they were more ideologically pure than low-frequency ones. Runge's group stayed with the lower frequencies, but the post of "special commissioner for high frequency technique" was created in late 1941. This post was occupied first by Hans Plendl, then by Abraham Esau, whose "rapid rise during the Third Reich was due to his party membership" (Kern 1994, 180). Although Esau was a capable engineer who had worked for Telefunken before becoming a professor at the University of Jena, the very structure of the system polarized research in a very unhealthy way (Bauer, e-mail communication; Pritchard 1989, 92). In mitigation of Runge's resistance to higher frequencies Bauer comments that Clausewitz argued that generals who switch from their own tactics to those of their opponent run the risk of not understanding the basis for their opponent's tactics (Bauer n.d., app. B; Bauer, interview), in effect to beware changing horses in midstream. The discovery of British centimetric H2S radar in a Stirling bomber downed near Rotterdam in 1943 ended the paralysis but did not remove the political divide in the radar research establishment. Göring was so shocked at how far ahead the British were that he said, "I did hope that if we were behind we could at least be in the same race" (quoted in Pritchard 1989, 88). In the event, German magnetrons were not widely available before the end of the war, and the Berlin radar system entered service in only one squadron in March 1945, accounting for only 10 victories (Pritchard 1989, 162). Bauer notes that the Berlin system was also used by the Kriegsmarine and that approximately 100 sets were in service at war's end (e-mail communication).

THE FAILURE OF NAZI NIGHT AIR DEFENSE

The crucial German failure was that Germany took too long to conceive and develop an effective C^3 system of night defense against a committed opponent because Nazi air doctrine favored the tactical offense to the virtual exclusion of all else and did not shift quickly enough to defensive doctrines when they clearly were needed. The one person in authority who fully understood the need for defensive doctrine in the night war and who did the most to develop German night defenses, General Kammhuber, fell out of favor for three main reasons. First, the Germans used ground-controlled night fighting without an IFF system or an adequate AI radar. This produced a cumbersome system of tightly defined "boxes" through which British bombers quickly passed (fig. 7.2). If one imagines the four corners of these "boxes" extending skyward, the German name for the system, Himmelbett, translated as "four-poster bed" or "bed in the heavens," makes sense. It was approximately the technical and organizational level that the British had reached in late 1939. Adding Lichtenstein AI brought it up to the British level of mid-1940, since the Germans persisted in an extremely complex system of display using the Seeburg plotting table. Seeburg had none of the technical elegance or ease of use of the British PPI of late 1940.

Once the British understood the German system through the Y Service, they reasoned correctly that huge numbers of bombers concentrated in a narrow stream would overwhelm the defenses (Clayton 1993, 101). Three "thousand bomber" raids in late May and early June, on Cologne, Essen, and Bremen, proved them right. In the Cologne raid only eight night-fighting zones were touched and only 25 night fighters were able to reach the bombers (Aders 1992, 56). Second, Kammhuber had believed entirely accurate reports of the buildup of American bombers in Britain in early 1943 and had persuaded Göring to support

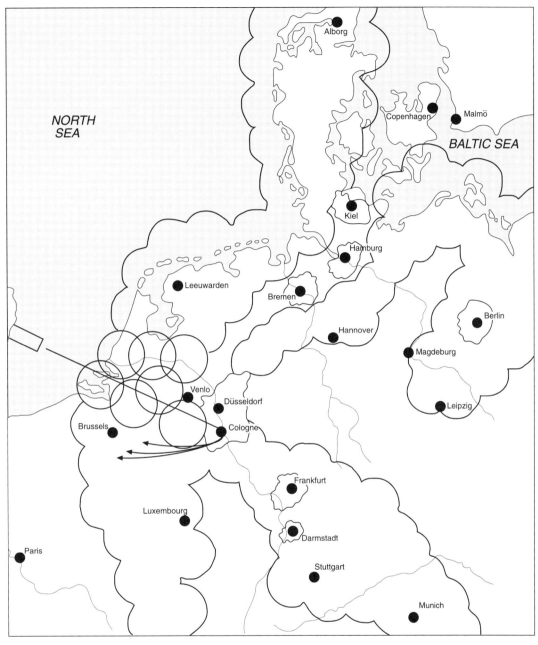

Fig. 7.2. The German Himmelbett Night-Fighting System, Summer 1943 (redrawn from Aders 1992, 86)

a massive increase in spending on Himmelbett. Hitler rejected this increase on the grounds that the reports of the American buildup were obviously impossible exaggerations. The estimates of the coming American strategic offensive were considerably reduced, and Kammhuber fell out of favor with Göring because Göring lost face with the Führer (61). The third force contributing to Kammhuber's downfall was the Hamburg raid of late July 1943. At least 40,000 died in a firestorm of unprecedented proportions, the worst of the European war (Levine 1992, 62). A major reason why the British suffered negligible losses was that they deployed Window for the first time to interfere with Würzburg and Lichtenstein. Several changes were needed to restore German night fighters to the effectiveness achieved by mid-1943, most notably more flexible ground radar control. New AI radar was developed that was harder for the British to jam. The planes themselves needed more power to carry more electrical equipment into battle, and the Bf 110 was thus reequipped with more powerful engines, although it took time to make these work properly. Finally, the Germans adopted a new armament, twin cannon firing upward at a 30° angle, which they called *schrage musik* (slanting music), after their name for jazz. Armed with these a German night fighter would maneuver below a British bomber, unseen by the crew since British bombers carried no ventral gun turret, and aim at the wing root. A few shells were usually enough to open up the critical fuel and hydraulic control lines and send the victim down in flames or out of control. For several months the British literally had no idea that German night fighters were causing such losses. By the time of the Nuremberg raid such changes had made the Germans formidable opponents again.

But the Germans suffered from increasing technical problems as the war wore on. German night fighters equipped with Lichtenstein were just as easily identified by their radar emissions as British bombers were by Monica or H2S. In the last months of the war the British simply did what they should have done much earlier: along with the bomber stream they sent Mosquito night fighters with their AI system switched on, since centimetric AI and H2S had identical emissions. This made them look like British bombers to the German night fighters. When the Mosquitoes came upon the inevitable Lichtenstein emissions on their own passive receivers—Boozer for the early Lichtensteins, Serrate IV for the later ones—the hunters suddenly became the hunted, and no German night fighter could cope with the Mosquito. Unfortunately, centimetric AI was not cleared for use in Europe until the disastrous Nuremberg raid forced Bomber Command to recognize the need for night escort fighters (Chisholm 1953, 182, 184). Until then the new centimetric radar was restricted to home defense, in part on the grounds that it could not fall into enemy hands that way. This was perhaps the biggest strategic error made by Britain in the air war, and it was certainly responsible for the unnecessary deaths of many Bomber Command aircrews flying unescorted into hostile skies. British bombers had been flying over Europe with centimetric H2S since early 1943, and the Germans had captured one almost immediately. The British were well aware that the magnetron at the heart of the centimetric system was indestructible and that once the Germans had one, it was only a matter of time before they too had centimetric radar. However, the Germans seem to have initially regarded H2S as merely a ground-imaging radar for bombers.

In the last few months of the war Mosquito night fighters equipped with centimetric radar mounted standing patrols over German airfields. Any German night fighters that got through the swarm of patrolling Mosquitoes over their airfields had to risk closing on a Mosquito masquerading as a bomber and then landing through another swarm of Mosquitoes. Strangely, one otherwise fine recent history of the night battle for German skies (by a survivor of Bomber Command) fails to mention the depredations of the Mosquito night fighters against their German counterparts (Hinchliffe 1996). One of the few German accounts of the intense night fighting of the last few months of the war is that of Wilhelm Johnen, an *expert* (ace) with 34 victories, who describes the Mosquito as "the night fighters' greatest plague" and its radar system as "technically so perfect that at a distance of five miles they could pick the German night fighters out of the bomber stream like currants out of a cake" (1994, 188). Johnen's account of his last mission is a chilling reminder of how the Germans came to lose 115 out of their 283 night fighter *experten*, 61 in 1944 and 19 in 1945: like many pilots, Johnen survived because he did not fly after 17 March 1945 because his unit was out of fuel. With a combined total of at least 5,723 night victories, these 283 *experten* produced an average of 20 kills each, more than 80 percent of the 7,192 victories by radar-equipped night-fighter units flying both by day and by night (calculated from Aders 1992, 232–39). Of these, 5,833 were night victories over Germany and its western approaches.

British night-fighter pilots over Germany usually fought other fighters, not bombers, so their scores were lower. The main bomber support came from 100 Group, which shot down some 264 German airplanes over Germany between 24 December 1943 and 25 April 1945, 222 of them night fighters. Of the night fighters 98 were Bf 110s, 84 were Junkers 88s, and 23 were Junkers 188s, which were fairly easily confused with Ju 88s at night. In addition, 100 Group was credited with 4 Dornier 217s, 8 Heinkel 219s, and 5 Messerschmitt 410s. The remainder of 100 Group's victories were against day fighters (22), bombers (8), transports (4), and unidentified airplanes (8) (Bowman and Cushing 1996, app. 2). All but the very first of these victories occurred when flying Mosquitoes. Most (217) were won after centimetric radar was cleared for use over Germany in June 1944.

Nor were the Germans able to win the production war. Exact figures are hard to come by, but Aders puts total production of German night fighters between 1941 and 1945 at 8,174 (1992, 273). He puts Reich Air Defence night-fighter kills for the same period at 5,791. Thus, each night fighter built accounted for 0.7 of an enemy airplane, a depressing statistic in a war of attrition (239). Johnen summarizes the late war position well from the point of view of the pilot.

> We were fighting against enormous odds. Against a formation of 600 to 800 four-engined bombers and 150 to 200 long-range night fighters we could put into the air 60 to 80 night fighters which rarely managed to penetrate the stream. It was incredibly difficult to get a bomber in our sights for the Mosquitoes sought us out and sped like rockets to the aid of the bomber. Not only had we the enemy in front of us but also in our backs. The losses rose so appallingly that science eventually had to come to our aid. It came at last in the form of the *Naxos* apparatus. This radar [it was actually a passive receiver] with antennae mounted on the tail of the aircraft warned the night fighter by acoustic signals in his headphones and a flicker in the cathode ray tubes of the presence of an enemy fighter to stern. (1994, 189)

According to Aders, "German night fighters would have been restricted by the nature of the system to a maximum kill ratio of six percent" over western Germany (1992, 83). The actual average British losses of 5.6 percent were thus evidence of the efficiency of the system of 1943 as well as its inadequacy against an opponent willing to accept losses of up to 10 percent. Only on longer missions, when the bombers had to penetrate more night-fighting zones, was the ratio likely to reach 9 percent. "Only once during the entire war did the Luftwaffe manage to achieve a real defensive success against a major . . . night raid" (Aders 1992, 145): when nearly 12 percent of 795 British bombers went down over Nuremberg on the night of 30–31 March 1944 (Levine 1992, 68). But weather conditions conspired to make the bombers easy to see, "a half-moon silhouetting aircraft against an 8/10ths layer of altostratus at 14,000 ft and a strong, unforecast jet stream blowing from the northwest across Germany. In the extremely cold conditions each bomber left a tell-tale condensation trail in its wake, glowing phosphorescent in the moonlight" (Cross 1987, 150).

At Nuremberg the German fighters needed almost no radar to achieve excellent results. Levine is negative about British losses in this period, claiming that they were the equivalent of defeat and that Bomber Command was only saved by switching from long-range missions to Berlin to the support of the invasion of Europe. Certainly Bomber Command morale plummeted after Nuremberg and the Battle of Berlin. But to equate this to defeat is shortsighted. Nuremberg was a fluke, and the British bombing of Berlin would have stopped as the nights grew shorter: Berlin was so far that for a bomber requiring the cover of darkness it was a winter-only target. Levine's bibliography does not include Aders' book, available in English translation since 1979, and he takes no account of the vital fact that the British clearly were winning the war of attrition against the German night *experten*. Nor does he account for the remarkable drop in night-bomber losses after D-day, when Mosquitoes equipped with centimetric AI began to escort the bombers over Germany. As with the much more stunning American losses in unescorted bombing in 1943, the key to reducing bomber losses was escort by long-range strategic fighters.

Aders lists as confirmed only 39 out of 50 estimated kills of Mosquitoes by German night fighters (1992, 245). Johnen's account of his last combat mission, the night of 16–17 March 1945, describes a wild death ride through Mosquito-infested skies during the raid on Würzburg. Even with Naxos to warn him that he had become the target of a Mosquito's radar, he was lucky to survive. He recounts a minimum of six interceptions by radar, although at two points in the narrative he merely notes that the Naxos signal was constant and that he took constant avoiding action. He had to take off through patrolling Mosquitoes, fight them constantly to get into the bomber stream, where he downed a Lancaster bomber, his last kill of the war, fight his way out of the bomber stream, and then fight off a standing patrol as he landed his damaged Bf 110. No sooner had he landed than the Mosquitoes destroyed his plane along with two others on the ground (1994, 193–98).

It is certainly appropriate to consider, as most air war buffs do, what might have happened had the Germans been able to introduce jet fighters equipped with centimetric AI radar. Even with Lichtenstein the Messerschmitt 262 jet fighter could master a Mosquito, and 9 of the 39 confirmed night kills of Mosquitoes were by Me 262s (Aders 1992, 245). An Me 262 with Berlin centimetric radar would have been a formidable opponent indeed. But the Me 262, like all new technologies, proved hard to develop, suffering severe problems with its revolutionary engines. It was never easy to fly, and too many pilots were killed as they made the transition from conventional airplanes. Finally, the massive strategic bombing effort simply reduced fuel production to the point where most planes were grounded.

In the creation of the Kammhuber line of radar defense, the Himmelbett system of ground control, and their successors the Germans merely replicated British systems of radar-oriented defence, usually several years later. The same was true of night fighting. The Germans turned first to such fast twin-engine medium bombers as the Dornier Do 17 as a source of night fighters, then to the Messerschmitt Bf 110, which could catch and kill a British four-engine bomber as long as it did not fly at the maximum cruising speed or take the sickening evasive action the British called "corkscrewing." Once the British started to send Beaufighters equipped with 1.5 m AI Mk IV radar along with the bomber stream to intercept German night fighters, the Bf 110 could usually hold its own against them if it was not surprised, and the Beaufighter's range was limited (Chisholm 1953, 135). But when the British began to send large numbers of Mosquitoes, with 1.5 m AI Mk IV before June 1944 and then with centimetric radar, both against Bf 110s in the air and as standing patrols over their airfields by night, the Bf 110 was doomed, as such *experten* as Johnen quickly recognized.

The one night fighter that Green claims might have held its own against the Mosquito at the height of the British night offensive was the Heinkel He 219 Uhu (1970, 349). However, Aders says that it was not well regarded by experienced night-fighter pilots, who found its performance "not significantly better" than the Bf 110's. It suffered from protracted development, too many changes in specification, and the High Command's refusal to recognize its need. The 268 He 219s built accounted for only 10 Mosquitoes (Aders 1992, 190, 245). The He 219 was ignored because of a combination of factors: German offensive doctrine was slow to give way to a defensive one; Heinkel himself fell out of political favor with the Nazis; and because the Heinkel Company's reputation was built on its bombers, it was not looked upon as a supplier of fighters.

On this last point, the British case is interesting. In 1940 Bristol, which made the first two British night fighters, the Blenheim and the Beaufighter, was producing mostly bombers. Certainly Bristol was never regarded as specializing in fighter planes. In the 1920s and 1930s de Havilland had successfully concentrated on civil aviation, although it had provided the RAF with a large number of trainers. Between the end of World War I and 1939 it tendered for two fighters, only one of which, the D.H.77 to specification F.20/27, was even built in prototype form (Jackson 1987, 289, 514). Even the Mosquito was initially intended not as a fighter but as a high-speed bomber using technology developed for the

Albatross airliner and carrying no defensive armament. That it could be developed into by far the best night fighter of the war was fortuitous. Only 2,119 night-fighter variants were built, representing just over 27 percent of total production (421–22). The British were both more flexible in doctrine than the Germans and better at procurement of the technologies needed to pursue those doctrines. Much of that procurement came from the British ability to draw on American production capacity and expertise.

In the most overarching attempt thus far to examine the progress of World War II in the air, although one sorely lacking in a full study of the role of radar, Overy arrives at some useful conclusions about the role of science in the British and American war efforts compared with its role in the German effort. Air war required "movement forward on a broad front of research," and this could only be guaranteed by a "close alliance of the scientific and military elites." Even though the British scientific elite was "smaller and less well-funded than its German equivalent it was recruited much more energetically for war." In Germany, "resistance to National Socialism . . . led to an emigration of scientific personnel in large numbers and to an internal struggle over Nazi intrusion into appointments and projected research" (Overy 1980, 237–39). Despite the Germanic character of American universities, there was little political intrusion into research decisions and generally the same level of cooperation between the scientific and military elites as in Britain. Much of this can be put down to the fundamental differences between the liberal capitalist economies of the West and the command economies that prevailed in the Axis powers and the Soviet Union:

> In Britain and the United States the political structure itself, coupled with the high degree of social modernization characteristic of such a structure, dictated that the war, while fought by soldiers, would be planned, supplied, researched for and profited from by civilians. Hence the stress in the western powers on economic effort, on massive production as the key to victory; hence the concentration on scientific mobilization; hence, too, the degree to which the war effort of both countries relied from the start on the co-operation and participation of bureaucrats, managers, engineers or, for that matter, women. (Overy 1980, 270)

THE AMERICAN CONTRIBUTION

The American perspective is worth noting since it came to dominate the British view of the strategic air war (McFarland and Newton 1991). The Americans entered the strategic air war convinced that they would bring Germany to its knees by daylight precision bombing of German infrastructure. This view was consistent with American air doctrine as developed in the 1920s, but it collapsed almost immediately in the light of four problems. First, because of weather conditions over Germany the use of the much vaunted Norden optical bombsight was rarely possible. Second, American bombers were opposed by Luftwaffe radar controlled fighters. Third, the performance of the bombers was never high enough to evade the fighters. Fourth America initially lacked strategic escort fighters. If, like both the Germans and the British, one regards loss rates of 10 percent per mission as insupportable, then the American strategic offensive was on the verge of defeat by October 1943, after the Ploesti, Regensburg, and two Schweinfurt raids. Losses over Europe alone (not counting aborts and write-offs) amounted to more than 20 percent: 54 out of 177 at Ploesti; 24 out of 146 at Regensburg; 36 out of 230 and 60 out of 291 in the two Schweinfurt raids. More than 1,500 skilled aircrew died. Morale was so low by the second Schweinfurt raid that 92 out of the 383 B-17s that took off from Britain aborted (Levine 1992, ch. 6).

Some comfort was gained by the use of borrowed H2S radar equipment in lieu of American-made H2X to see through overcast in the Emden raid, and the German defense was absent because of the bad weather, but H2S was too inaccurate to be other than an area bombing weapon. Nothing could be done about the poor performance of the bombers, since every increase in engine performance was wiped out by the addition of more defensive guns, fuel, and other equipment. The B-17 was properly lauded because it flew high and fast and was able to absorb a huge amount of battle damage. It was certainly much safer than the other strategic bomber used in the European theater, the B-24. Yet, outside of the 1943 disasters, the crews of American bombers had about the same survival rate on a 25-mission tour as the crews of British night bombers, about one-third of whom survived a tour of duty of 30 missions. The appalling attrition of flight crew in strategic air offensives can never be truly quantified. Randall Jarrell said it best in his poem "The Death of the Ball Turret Gunner":

> From my mother's sleep I fell into the State
> And I hunched in its belly till my wet fur froze.
> Six miles from earth, loosed from its dream of life,
> I woke to black flak and the nightmare fighters.
> When I died they washed me out of the turret with a hose.

(Williams 1945, 157)

The strategic escort fighter was obviously the key to the problem by mid-1943, and the Lockheed P-38 Lightning, intended as a high-altitude, long-endurance fighter from its origin in 1937, was becoming available. As it turned out, the P-38 was the wrong airplane. However good its range—and it was never good enough over Berlin—its Allison engines lacked performance and German fighter pilots regarded it as "easy meat" (Green 1957, 71–75; Levine 1992, 111–12). The Republic P-47 was also touted as a strategic fighter, but its huge engine gulped fuel (Levine 1992, 113). It was unable to penetrate deep into Germany, and its best use was as a ground attack fighter. In the absence of an American Mosquito, the emergence of the strategic fighter par excellence was a surprise to almost everyone. North American Aviation's P-51 Mustang had been designed in a hurry for the British Purchasing Mission in 1940. With its original Allison engine it was adequate; however, it was quickly recognized as a brilliant, low-drag, high-performance airframe in search of an engine. Fitted with the fuel-efficient Merlin in 1942, it became the best all-round fighter of the war, able with drop tanks to escort bombers to Berlin and back and take on almost any German fighter, even the later Focke-Wulf 190s, on equal or better terms (Green 1957, 98). Because of ineptness on the part of the American air force, it was, however, 1944 before any number of P-51s became available for such duty (Levine 1992, 112–14). Much of this ineptness was induced by the proponents of the self-defending bomber, who were not proven false until the summer of 1943.

But the Americans quickly came to recognize, as the Germans had done in 1940, that the defending air force had to be smashed before the strategic offensive could be successful or a ground invasion could occur. In this light the endless sorties of escorted American bombers over Germany were part of a battle of attrition against the Luftwaffe. In late February 1944 the Americans launched Big Week, a smashing offensive against the Luftwaffe with 1,180 bombers and 676 long-range escort fighters, the former as much to entice the German fighters into the air as to bomb German aircraft plants (McFarland and Newton 1991, 172). By March the Americans had come to see the bomber's main role to be to entice the defending fighters up so that they could be cut down by the escort fighters; in this way, said "Beppo" Schmidt, the general responsible for the defense of Berlin, the Americans "captured air supremacy over almost the entire Reich" (quoted in ibid., 219). While it is accurate to say that in the strategic sense the air Battle of Berlin, unlike that for the Ruhr, or Hamburg, or numerous other targets, was never won by the aerial destruction of the city or key elements of its infrastructure, it was a significant victory because it reduced the Luftwaffe to near helplessness.

American night-fighter efforts deserve a brief mention. The Northrop P-61 Black Widow was the only night fighter designed as such to make it into the war. It was a huge, powerful plane with very heavy armament, but it never had a real chance to prove itself because it was never deployed in serious action over Germany. This may well have been fortunate. The flight commander of the RAF's night experimental unit, Jeremy Howard-Williams, described it as a "brute" and was highly critical of its dependence on electrical systems (1992, 155). Since he and other RAF night-fighter pilots had had unexplainable trouble with the electric actuation of the cannon on their Mosquitoes, he had some grounds for his criticism.

THE OFFENSIVE USE OF RADAR IN THE AIR BATTLE FOR GERMANY, 1943–1945

Three principal C^3 technologies were deployed by British and American bombers in the strategic bombing of Germany: Gee, Oboe, and H2S. Gee and Oboe were navigational systems using transmitters in Britain. H2S was ground imaging radar carried in the bombers themselves. Gee was a sophisticated and updated version of the German blind-navigation technologies of the night blitz period that originated in the adaptation of American blind-landing radio technology by the German Lorenz Company. Gee used three stations to lay a grid of beams across Europe.

Bombers could determine their location by reference to a Gee map, and so as not to warn the Germans, no beams were laid over the target. At a distance of 400 mi from home a bomber could determine its position to within 6 mi. As the British developed Gee, they reasoned that their adversaries would be as astute as they had been in electronic intelligence and thus prepared an elaborate but nonexistent system called J, which they allowed the Germans to discover. After Gee suddenly entered service in March 1942, it was six months before the Germans could develop their own ECM to deal with it (Price 1978, 98–104). By late 1942, however, Himmelbett- and Lichtenstein-equipped night fighters were causing trouble for the British. New ECM and new, more accurate navigational aids that were harder to jam were needed.

The best of the navigational aids was Oboe, developed by F. E. Jones and A. H. Reeves. Oboe used two radar transmitters to position airplanes over western Germany. Mark I Oboe used Chain Home transmitters, and Mark II used the 10 cm band. In the latter Reeves' PCM technique was used to convey instructions to aircrew. PCM was a technique of digitizing information so that it could not be easily decoded without extremely high speed modern computers and thus was unreadable by an opponent. Since the accuracy of radar does not fall off with distance, Oboe-directed pathfinder airplanes would mark targets with lurid incendiary bombs. They could place these with an average radial error of 135 m from a height of 9,000 m (Jones 1988, 320). The lower-flying heavy bombers then aimed at the colored fires. Until late in the war Oboe could handle only one airplane at a time, and a reasonable time interval was needed between marker airplanes. Because of limitations imposed by the curvature of the earth and the altitude at which the pathfinder airplanes could fly, and as long as transmitters had to be sited in the British Isles, Oboe's range was just beyond the industrial cities of the River Ruhr. British bombers flew 9,626 Oboe-controlled sorties, American bombers a further 1,663 (321).

When Mark I Oboe operating on 200 MHz was jammed by the Germans in late 1943, Mark II Oboe was deployed. From 1943 on, the massive destruction of the Ruhr cities through their usual pall of industrial pollution and cloud cover was made possible largely by Oboe. Only after the invasion did forward Oboe sites allow deeper penetration raids, and these were never properly carried out because of the bizarre opposition to Oboe among some British officers and scientists (Jones 1988, 317, 327). In this instance British behavior was almost as inexplicably foolish as that of the Germans. The Bomber Command hierarchy was well aware of the importance of Oboe and of the need to conceal it from the Germans; as one high-ranking officer commented, "The longer the enemy thought they were depending on H2S the better" (Maynard 1996, 111). Group Captain Don Bennett, the head of the Pathfinder Force, wrote a memorandum to Arthur Harris, in charge of Bomber Command, in which he accused Harris of seeing repeater Oboe as a panacea—Harris denied the charge in the margin of this particular memorandum. Bennett was clear that he preferred to see Mark II Oboe fully developed and then put into service as quickly as possible (135).

Repeater Oboe was possible as early as late 1943, an airplane within line-of-sight range of British transmitters circling and passing the signal on to the pathfinder aircraft. Maynard notes that "for reasons that remain totally obscure, the Bomber Command reaction was lukewarm!" (1996, 116). One possible reason was that Oboe Mk I repeater, operating on 200 MHz, would not allow more than one pathfinder airplane to mark the target within 15 min, the length of time considered safe for a raid by late 1943. It was also an obvious candidate for jamming. Bomber Command gave its highest priority to "improved H2S 3 cm. equipment" for long-range operations on the principle that operation at higher frequencies than the 10 cm H2S would result in better ground imaging; second priority was given to "Oboe Mk II non-repeater for short range"; and third priority was given to "Oboe Mk II repeater, but not at the expense of Oboe Mk II non-repeater" (Maynard 1996, 140–41).

Until forward Oboe was deployed, reasonably accurate area bombing of targets past the Ruhr at night or during bad weather could be achieved only by centimetric radar systems carried on board the attacking airplanes. These systems evolved from the use of the magnetron and the development of centimetric airborne AI and ASV radar systems. The versions carried by bombers and used to identify ground targets were called H2S by the British and H2X by the Americans. The 10 cm radars worked well against port or riverine cities because they distinguished the interface between water and land very well. They failed dismally against Berlin, set in a sea of swamp, and against an opponent who knew by then which system was in use and used gigantic movable steel sheets to provide false targets. The theory that higher-frequency systems operating at 3 cm would improve this definition was not borne out in practice, nor were many such sets ready in time for the war. Being an active system, H2S also sent out pulses of electrons that Naxos-equipped German night fighters could follow directly to their targets without using their own active radar systems and attracting the attention of British night fighters. But Naxos also had its detractors. Saward notes that interrogation of a captured German night-fighter pilot indicated that Naxos provided "no measurement of range or the accurate bearing of the detected H2S aircraft" and that in any case "there was no evidence . . . that losses were greater amongst H2S aircraft than amongst aircraft not equipped with H2S" (1959, 241). Aders' definitive account indicates that Naxos gave "azimuth measurement only" (1992, 252). Certainly by the last months of the war Johnen regarded Naxos as primarily a defensive technology against the dreaded Mosquitoes (see Johnen 1994, 189).

The strategic offensives favored by both the British and the Americans required an unrelenting aerial assault on German cities and infrastructure. This did not really begin until 1943. Radar offered accuracy, even some possibility of precision, in overcast conditions and at night. Most of the 858,700 tn of bombs delivered by British airplanes in 1943, 1944, and the first four months of 1945 were guided by radar (Saward 1959, 252), at least in the sense that the pathfinder airplanes that marked the target with incendiaries always did so with the aid of Gee, Oboe, or H2S (Guerlac 1987, 773–77). Given America's preference before the war for using its very accurate Norden optical bombsight, it is somewhat surprising to find that radar accounted for 61 percent of all American bombs delivered and that H2X alone accounted for nearly 50 percent (799). More than three-quarters of the bombs dropped on Nazi Germany were thus aimed by radar; of these, probably two-thirds used radar guidance of the pathfinders and perhaps one-third used ground-imaging radar. Although H2S and H2X may have caused problems for the planes that carried them, there were never enough Naxos-equipped German fighters, the German air-defense system was not well coordinated, and much of the tonnage of bombs dropped would never have been dropped on target in bad weather or at night without radar guidance. Moreover, German night fighters were not used, as they should have been, as all-weather fighters, so that in bad weather German day fighters were rarely or never seen, which meant much lower losses for the Americans on such missions (781–82). It is certainly reasonable to claim that without radar guidance of bombing, neither Britain nor America could have achieved its strategic aims. At the least the war would have lasted longer; at the worst, the Germans might have caught up with Allied centimetric radar technology and jumped ahead into the jet era with the full deployment of the Messerschmitt Me 262. Such a protraction of the war in Europe would almost certainly have meant that the first nuclear weapons would have exploded over such cities as Berlin, as was originally the intent, rather than over Hiroshima and Nagasaki.

The development of centimetric radar allowed the British to develop offensive radars as well as much better defensive radar systems than the Germans by 1943. Since Britain was by then on the offensive, the defensive role of British centimetric radar generally has gone unremarked. However, British night fighters equipped with centimetric radar essentially denied the Germans any chance of continuing a night bombing offensive over Britain after 1943. Centimetric radar–controlled antiaircraft guns also allowed the British to intercept and destroy the pulsejet-propelled V-1 cruise missile, which was launched against Britain in large numbers from 1944 on. German radar and gunnery never even came close to this capability. British Mosquitoes, which flew well above the reach of the German Flak and most German night fighters, were capable of operating deep inside Nazi Germany by 1945.

Among the main belligerents in World War II only Britain recognized that Mahanian geopolitics had to be reworked to suit air warfare and that radar rendered at least some of Mackinder's ideas no longer valid. It was too easy for both America and Japan to maintain a belief that the surrounding seas and their battle fleets constituted an adequate "moat defensive," and it was too easy for Germany to continue to believe, with Mackinder, that whoever ruled the heartland would come to rule the world. Britain added radar and effective defensive fighters to the navy's traditional responsibility in protecting the "moat defensive." Bomber Command's strategic offensive destroyed the rest, homes, and workplaces of its urban population and helped disrupt supplies of raw materials for factories and food for the people. The results of strategic bombing were thus akin to the results of the traditional naval blockade. In British geostrategy the air force took on the same character as the navy, both protecting the "moat defensive" and acting offensively to disrupt enemy civilian life and war-making capacity.

All the belligerents in World War II understood quite clearly that air warfare was one of the main keys to avoiding trench warfare as well as to victory or loss. The problem was how to achieve the right mix of air doctrine and technology. The Japanese did the poorest, tying their air offensive primarily to their navy and failing to develop adequate strategic bombers or an adequate defensive strategy. The best strategic bomber of the war, Boeing's B-29, was deployed against Japanese cities because of its range; its speed and altitude capabilities were rarely needed against an opponent with only rudimentary night defenses. The immolation of Japan's cities by wholesale incendiary bombing was accomplished from very low altitudes, where air defense should have been easy.

The Americans were never really tested on the defensive, except in a small way by the initial Japanese expansion of early 1942. Since it took the British three years to go from the idea of radar-based defense to reality when their backs were against the wall in 1936, it is doubtful that America could have done much better. The American strategic offensive was excellently conducted, once the dangers of the self-defended bomber had been recognized and proper escort fighters had been pressed into service. The optical bombsight and the notion of precision bombing wasted a great deal of research effort and money, if not the lives of aircrew. The American took the right approach to daytime strategic bombing: either bombers would take out important targets or they would act as bait that their escorts would defend. In the latter case, the resulting attrition of German fighters would destroy the Luftwaffe's ability to defend German airspace when the ground invasion began. This attrition was not just of day fighters since the Germans used night fighters by day as well.

The Germans were tested on both the offensive and the defensive, and they lost in both cases. They clung too long to a theory of air war that emphasized the tactical offensive when they needed to switch to the defensive, and they never developed an adequate strategic offensive doctrine or bomber. Whether such a bomber would have turned the tide in the Battle of Britain must remain a moot point. A "Ural bomber" would, however, have been of great use against Russia's withdrawal of its manufacturing capacity east of the Ural Mountains as German troops threatened Moscow. In 1942 the Russians had almost no fighters capable of opposing such bombers and certainly had no night fighters. Compared with the radar efforts of other combatants, the Russians' was by far the weakest. The Soviet Union eventually fought a better tactical air war than did the Germans, but only after massive losses incurred by the German expansions. The Stalinist purges of the late 1930s had removed the enthusiasts for the strategic offensive, which at least ensured that the survivors remained tightly focused on tactical air warfare.

Britain won the first defensive air battles of consequence, two of the key battles of World War II, when it withstood the German offensives in the day Battle of Britain in the fall of 1940 and the night blitz of the win-

*Communications,
Command, and
Control in the War
in the Air*

217

ter of 1941. British strategic bombing by night, in concert with American day bombing, was effective against Germany, though perhaps not to the extent that its proponents hoped. Victory in the Battle of the Ruhr was unequivocal, but the Nuremberg raid and the night Battle of Berlin imposed a cost in bomber aircrew bordering on unacceptable. Psychological problems mounted. The stunning success of the Dresden raid seemed like a pyrrhic victory even at the time. Survivors of Bomber Command now question the morality of attacks on civilian targets more openly than they did at the time, and the wartime justification that this was just retaliation for the firestorms of Coventry and London rings even more hollow today.

But there was a more positive side to both the day and the night air war over Germany. This was the high attrition rate of, in particular, experienced German fighter aircrew from the winter of 1944 on. The loss of high-scoring *experten* was devastating to the Luftwaffe. Inexperienced aircrew were often killed in their first few operational missions, before they had a chance to learn the survival tricks of the old hands; night fighters certainly were no place for such folk. This attrition made the ground invasion of Europe much easier than it might have been. The key to the air war over Germany was therefore not so much that the Allies failed to achieve a clear victory but that the Germans failed to fully exploit the advantages that always accrue to the side defending its homeland. One can only conclude that the Germans lacked an adequate defensive doctrine in the face of a determined strategic offense pressed home even in the face of substantial reversals faced by the Americans by day and the British by night. It must be added that the constant meddling by Hitler and the Nazi Party in matters best left to technical experts did no good.

In the Neotechnic period in general, the Americans were usually more successful than the Germans in organizing research and development, and the Germans usually did better than the British. In the area of radar, however, the order was reversed, with Britain first, Germany second, and America last. The question is, Research for what purpose? Britain, in the teeth of a resurgent Germany, was the first of the three to enter its "warfare state" mode (Edgerton 1991). The threat was recognized as so great that all the normal rules about military research were suspended as early as 1936, so that theoreticians, engineers, manufacturers, and military users all actually began to talk to each other about what was needed and what could be built. Even the normally secretive Naval Signal School shared technology (Callick 1990, 1–7). Although some of this success was personality driven, in particular by Watson Watt, only in Britain did something as revolutionary as TRE emerge in time to affect the war.

America excelled at commercial research but not at military research, not least because of its dangerously pacifist stance between the wars and its excessive reliance on Mahanian geopolitics. Although a very small radar research team did excellent pioneering work at the Naval Research Laboratory (NRL) in the mid to late 1930s, the work was shrouded in almost pathological secrecy. Some of this work was communicated to commercial research laboratories, but not enough to allow, for example, a commercial company like RCA to develop a radar to compete successfully with the products of the NRL (Allison 1981, 105–9). Neither the United States Army nor its Air Corps was informed of this naval radar research. In the aftermath of the Tizard Mission in 1940, American scientists forced the pace by insisting that the Radiation Laboratory at the Massachusetts Institute of Technology be set up along the lines of Britain's TRE because of the obvious failure of NRL's research program (156–59).

The German failure was more complex, and no recovery occurred. In the first place, Hitler's rise to power devastated the technical universities: enrollments were halved because of the anti-intellectualism of the Nazi Party and of Hitler himself (Ulrich 1994, 179). Although good research continued in such pioneering companies as GEMA and Telefunken, as well as in new companies established to work with radar, all lacked competent engineers. The divide between engineers working with well-tested low-frequency radar in the commercial companies and those developing high-frequency radar in a politically driven system did not help matters.

The lack of engineers eventually resulted in large numbers of retired engineers, schoolteachers, and the like being pressed into service, but there simply were never enough technicians to maintain radar systems in the field. The persecution of the Jews in the Third Reich drove those who could get out to do so, thus greatly enriching the research effort of the British and Americans. At the same time, knowledge of Hitler's intentions, spread by both Hitler's *Mein Kampf* and Jewish émigrés, led British and American Jews to become active in areas in which they had technical expertise. Nazism obviously does not explain why Jews were concentrated in the television and radar industries, but it does explain why they worked so hard there to help defeat Nazism.

Britain's pioneering achievements in high-definition, commercial, electronic television go a long way toward explaining its later technical superiority in radar. In part this resulted from the creation of a cadre of competent electronics engineers in the commercial sphere that could apply its skills in radar. In part it was the result of the presence of a reasonably large number of Jews in television who had a particular interest in defeating the Third Reich.

Although America, Germany, and Britain all had the wherewithal to develop high-definition television by the mid- to late 1930s, it is notable that the most rapid strides were made in the technologically weakest polity. EMI claimed that its chair, Isaac Shoenberg, "offered a 405-line service to the BBC as early as February 25, 1935. This was a tremendous gamble. . . . But Shoenberg had confidence that Alan Blumlein could and would build the system in time for competition with Baird Television [the electromechanical system]. By the week of April 10–15, 1935, EMI was able to demonstrate a 405-line picture" (Abramson 1987, 217). The payoffs from this gamble were huge. Whether the British Cabinet played a political role in the BBC decision to go with EMI's electronic rather than Baird's electromechanical television remains a question, but the claim is reasonable. The BBC chose the system that provided the better image, although at the time movies could not be transmitted. From 1936 on,

EMI forced the pace in television development, not least because a large and profitable domestic market was opening up.

Germany's slower development of television seems to have been the result of a slower movement to commercial exploitation of the new technology. Fernseh, which was involved mostly in research, and Telefunken, which saw itself as primarily a producer of commercial receivers, had the same technical capability as Baird, RCA, and EMI in 1935. There is no adequate explanation of Germany's slower commercialization in the current literature. The lack of political interference to force the pace is notable.

American sluggishness was clearly a product of two factors. The Great Depression lingered more severely in America than in rearming Europe, and David Sarnoff, the head of RCA, wanted to develop the best possible system before committing his company to the sizable investment involved. That meant waiting for the report of the National Television Standards Committee, which began its deliberations on competing technologies in early 1940 and reported in early 1941, favoring a 441-line system like Germany's. After the war the NTSC reconvened and issued a second report favoring the current 525-line system. The lack of political interference in this decision is laudable if one subscribes to the theory that such decisions are best left to the market.

I end this chapter with the end of World War II because my main interest here is the failed transition of hegemonic power from Britain to Germany and the eventual hegemonic success of America. After World War II, America, in part through the massive industrial and military research establishments that the war brought on, of which the Radiation Laboratory at MIT was the first, in part through such private research laboratories as Bell Laboratories, and in part through private firms, has come to dominate all fields of radar development.

Compared with the situation in 1940, the time of the Tizard Mission, this was a massive turnaround. America's application of electronics to warfare in 1940 seems shockingly slight in retrospect and certainly requires more explanation than the German and Japanese failures. Only in naval applications did America approach its European competitors. Also slow was the development of high-definition electronic commercial television, in which America should have had a considerable lead by the late 1930s given the work of Zworykin and Sarnoff at RCA and of Farnsworth independently.

It is hard to understand the comparative American failure of the late 1930s except as a geopolitical and political one: a geopolitical failure to understand that a doctrine favoring a strategic air offensive required, by implication, considerable research into air defense and a political failure to force an earlier commercial commitment to high-definition television. The secrecy surrounding naval radar must certainly carry some blame. The USAAC, which was responsible for the strategic offensive, had little understanding of radar's potential before the 1940 Tizard Mission. This may have been in part the result of penny-pinching by the American government, although spending on the growing offensive air fleet was not low in the late 1930s. The implications for America of the fall of France and the near fall of Britain were clear enough. By April 1941 the USAAC had requested designs for strategic bombers able to bomb Germany from North America. The initial specification was for a plane with a range of 12,000 mi at 275 mph carrying a bombload of 4,000 lb (Wegg 1990, 93). Two designs were completed after the war's end to a modified specification calling for a 10,000 mi range and a bombload of 10,000 lb: Northrop's B-35 flying wing and Consolidated's gigantic B-36.

But overall the USAAC seems simply to have fallen victim to an offensive doctrine even more firmly held than Britain's through the mid-1930s. The American airplane industry produced three of the four or five greatest four-engine strategic bombers of World War II but only, in the P-51, one of the greatest fighters, and that only when it was refitted with a British engine. Boeing's four-engine B-17, commissioned in 1934, was a remarkable airplane several years ahead of European thinking. Its European counterparts were the Heinkel He 111 and the Vickers Wellington, both twin-engine planes. Consolidated's B-24, commissioned in 1939, was roughly equal to its British counterparts, the Avro Lancaster and the Handley Page Halifax, both converted from big twin-engine planes. They were, of course, designed for very different purposes, the B-24 as a proper strategic bomber and the British pair as medium and torpedo bombers. However good its performance on paper, the B-24 was not reliable or safe enough to be a truly great bomber, although it was considerably in advance of the one British four-engine bomber designed for the strategic offensive, Short's Stirling. Although the Lancaster was indisputably a great bomber, the Halifax has a more elusive claim. Boeing's B-29, commissioned in early 1940, was as far ahead of European practice in 1940 as the B-17 had been seven years earlier. With such airplanes, the USAAC clearly believed that the self-defended bomber would, indeed, always get through. The USAAC persisted in this belief even in the face of German and British data from 1939 to 1942 showing that unescorted day bombers were easy targets in skies watched by radar and defended by committed pilots in decent fighters. Only the disasters of the unescorted daylight raids of 1943 persuaded the USAAC that it was wrong. Clearly, America's offensive doctrine was based in part on its belief in its "moat defensive," but that was no excuse for continuing an inappropriate doctrine in the light of four years' combat experience by others, one a close ally who had delivered all of its extensive knowledge of radar and its uses on a plate three years previously.

The Japanese failure is easy to dismiss. The Japanese simply lacked the intensive research and development effort needed to produce radar systems outside the laboratory. They suffered even more than the Germans from being a command economy, and the Japanese Naval Air Force relied on an offensive doctrine even more than the Luftwaffe or the U.S. Navy. It compounded such folly by believing that carriers, dive bombers, and light land-based bombers with almost no ability to survive battle damage would allow them to carry it out. Their main bomber, the Mitsubishi G4M Type 1, was called the "Type 1 lighter" by its own crews and the "one shot lighter" by the American fighter pilots who demolished it (Green 1969, 52).

The German failure is most interesting, in particular given the technical edge of the German electronics industry before the war. A substantial technical lead was turned to an increasing gap after 1940. Ideology had too much impact on electronics design, although the German inability to catch up in high-frequency radar resulted in part from a justifiable unwillingness to change horses in midstream. A more severe ideological failure was excessive centralization of decision making and an obsession with military secrecy rivaling that of the American navy. Scientists, engineers, and users simply did not communicate with one another in a creative way. The High Command issued specifications on what it thought front-line users needed and the scientists were expected to deliver. This is the classic failure of command technology. But other factors were also at work. The Third Reich believed it could not afford both "guns and butter." Hitler was convinced that one of the reasons for the loss of World War I was the collapse of civilian morale, so he kept "butter" production up at the expense of guns. Unable to afford a strategic air force and needing a tactical air force to engage in continental expansion, the Luftwaffe built a tactical air force that pretended to be a strategic one. Such an air force did well against Germany's Continental neighbors but collapsed against an opponent who had developed its defensive doctrine using radar to direct its limited supply of fighters.

But this was not so much a technical as a geopolitical failure. As I indicated in chapter 1, Nazi Germany subscribed to the notion that it needed to control the heartland. It updated its offensive geostrategy to support its Mackinderian geopolitics and developed at least part of the technology to implement it. As a consequence of this obsession Germany neglected its defenses and failed to develop the technology needed to overcome enemies who had not neglected their own defenses. The success of Britain's renewed "moat defensive" in the Battle of Britain and the night blitz was the first clear illustration of this. Even more spectacular was the failure of the German offensive against the Soviet Union. Without a long-range strategic bomber to strike east of the Urals the Germans could not prevent the withdrawal and recovery there of Soviet industry. Without control of the Atlantic and the British Isles the Germans could not prevent the Americans from moving large numbers of men and huge amounts of matériel across the ocean. The two great defensive victories in Britain and the Soviet Union plus the ever-increasing strategic air offensive staged by the British and the Americans from 1943 on set the stage for the successful advance of Allied ground troops after mid-1944 and the end of Nazi geopolitical aspirations.

CHAPTER EIGHT

Telecommunications and World-System Theory

Large-scale political organization implies a solution of problems of space in terms of administrative efficiency and of problems of time in terms of continuity. Elasticity of knowledge involves a persistent interest in the search for ability and persistent attacks on monopolies of knowledge. Stability involves a concern with the limitations of instruments of government as well as with their possibilities.

—Harold Adams Innis, *Empire and Communications*

Chapter 7 ends in 1945, with the clear, if delayed, ascendance of America to economic and military hegemony within the capitalist world-system. In this chapter I explore America's attainment of telecommunications hegemony between 1945 and 1971, the key technology for hegemonic power in this period of the world-economy, the response of the other major actors in the capitalist world-economy, and the implications of both the attainment and the response for the future.

HEGEMONIC TRANSITIONS IN THE WORLD-ECONOMY

In the ongoing discussion about world-system theory it is clear that the most interesting question is how hegemonic transitions occur and, in particular, the nature of wars of hegemonic transition. With regard to the last clear transfer, that between Britain and America, to my mind a key question is why it was delayed well beyond the end of the first phase of that war of transition, World War I. Despite disputes over Wallerstein's notion of logistics and the nature of Kondratieff waves, even fairly extreme followers of Wallerstein, such as Peter Taylor (1993, 75), recognize this problem (table 8.1). As Taylor notes, American hegemonic victory in the economic and military struggle we call World War I was "not taken up," so that America was not a genuinely hegemonic power until 1945 and the end of what turned out to be a two-stage war of transition.

The usual suspect paraded before us to account for this failure is American isolationism. In the standard historical interpretation the will to power of America, a state founded in the revolutionary overthrow of an imperial power, avowedly democratic and republican, was not yet great enough to overcome American distaste for European power politics. In this reading American involvement in World War I was an aberration and America's geopolitical destiny was as a Mahanian fortress America, safe behind its Atlantic (and Pacific) walls. Obviously, such an idealist reading does not sit well with a highly materialist world-system theory, but the reading is constantly reiterated, has a huge number of adherents, and contains elements of truth.

Yet this isolationist reading is also clearly flawed on at least two grounds. First, and most straightforwardly, American spending on the military did not drop precipitously after 1918. At the Washington Naval

Table 8.1. Long Cycles and Geopolitical World Orders since 1790

Kondratieff Cycles	Hegemonic Cycles	Geopolitical and Technical World Orders
B Phase 1790–98	British hegemonic cycle begins.	Britain's Paleotechnic world-system matures (first industrial revolution). The French Revolution signals French inability to industrialize to compete with Britain.
A Phase 1815–25	Ascending hegemony. Grand Alliance.	Napoleonic Wars signal French resistance to British ascent.
	Hegemonic victory. Concert of Europe creates balance of power. Mercantilism.	Transition to world order of hegemony and concert underlain by Pax Britannica, 1813–15. British economic and military hegemony of maritime world-system.
B Phase 1844–51	Hegemonic maturity. Switch to free trade.	Balance-of-power system leaves Britain free to dominate world-economy.
A Phase 1870–75	Declining hegemony. Rise of protectionism (neo-mercantilism) and neo-imperialism. German-led "grab for Africa," last colonial frontier.	Transition to world order of rivalry and concert, 1862–71. America adopts protectionism. Germany comes to dominate Europe militarily. Britain still the greatest world power. Transport technology begins to favor land-based polities.
B Phase 1890–96	American hegemonic cycle begins. Protectionism.	American (continental) version of Neotechnic world-system matures (second industrial revolution). Germany lags.
A Phase 1913–20	Hegemonic victory not taken up by America, creating global-power vacuum.	Transition to world order of the British succession, 1904–7. American economic hegemony not matched by comparable American military hegemony.
B Phase 1940–45	Germany redux, Japan attempts to move up in the world-system.	German (high-technology) version of Neotechnic world-system matures in late 1930s.
	Hegemonic maturity. Bretton Woods agreement. Free trade.	Transition to world order of the Cold War, 1944–46. Pax Americana.
A Phase 1967–73	Declining hegemony. "Hidden protectionism" of Japan, EU trade policy.	America misreads military competition from USSR as hegemonic challenge. Bomber gap of late 1950s, missile gap of 1960s. Economic challenge from EU and Japan.

Conference in 1922 America negotiated for its navy to be at least the equal of that of the then hegemonic power, Britain, a significant part of the British fleet being broken up as a result of the conference. In the 1920s the American navy put much time, effort, and money into the development of its advanced aircraft carriers and the planes to fly from them. This would serve America in good stead in World War II. Similarly, the Army Air Corps developed a strategic bombing doctrine in the late 1920s and made itself increasingly independent of the Army. In the late 1930s it procured the bombers, particularly the Boeing B-17, needed to carry out that doctrine. These were emphatically not the acts of an isolationist government. Second, as Taylor points out, both Republicans and Democrats before the New Deal were cadre, not mass, parties (Taylor 1993, 262). The failures in America of both agrarian populism and industrial socialism meant that no mass political party arose to represent a presumed democratic will to peace and isolation. The two cadre parties represented different combinations of elites, but both favored American involvement in the building hegemonic struggle, however much they differed in their approach to winning it. New Deal Democrats may have practiced mass politics at home, but their leadership, exemplified in the patrician Franklin Delano Roosevelt, practiced cadre politics abroad.

Certainly some observers writing in the immediate aftermath of World War I believed that the transition had occurred. Ludwell Denny's *America Conquers Britain* (1930), for example, includes sections with titles such as "The Dear Cousin Myth," "Decline of the British Empire," "Rise of the American Empire," "Britain Lacks Capital," and "The Federal Reserve Rules London." Why, then, was hegemony not "taken up" if America seemed to some observers to have it firmly within its grasp? We must look well beyond domestic politics and isolationism for a reason. We must ask, as most world-system analysts do not, what drives Kondratieff waves, and we must take account of the more naked forms of power struggle, in particular the ability to wage a military struggle.

I argue elsewhere (Hugill 1993), following Berry (1991) and Mensch (1979), that the key to Kondratieff upswings is investment in new technologies in the downswings. Following Hall and Preston (1988), I also argue that new technologies have important precursors in earlier investment cycles and that these technologies are first clearly evident in the great metropolitan areas of the planet. A "carrier wave" precedes the full Kondratieff wave. Although we refer to the age we seem to be currently entering as the "information economy," the B phase of the second Kondratieff wave of the British hegemonic cycle saw that carrier wave begin with massive British investment in a global network of submarine telegraph cables (Bright [1898] 1974; Headrick 1991; Kennedy 1971. See also fig. 2.4), leading to near total British control of global telecommunications by the late 1800s. Cables under nominal American ownership were added, but they attracted substantial British investment, as did the American domestic cable network, Western Union. A joint-purse system prevented competition in the Atlantic market, and the fact that nearly all cables—British, American, and German—ran through Britain, mostly through London, ensured that Britain had access to almost all tele-graphic communication between America and Europe. The current global telecommunications system is thus a product of a hegemonic struggle begun a century ago, although it usually has not been discussed in terms of world-system theory (Boettinger 1989).

The most vocal American opponent of British telecommunications dominance in the first decade of the twentieth century was the U.S. Navy, which strongly supported the development of more sophisticated tech-nology in the form of continuous wave, low-frequency radio telegraphy using Poulsen arc transmitters and the Alexanderson alternator devel-oped at General Electric to break British telecommunications hegemo-ny. Before the success of Poulsen's arc system in 1911 the American navy pointedly preferred German to British radio equipment. The navy also experimented with even more radical telecommunications technolo-gies, including De Forest's vacuum tube amplifiers, as early as 1915. Al-though the navy successfully transmitted speech across the Atlantic in that year, it could not achieve the needed reliability with the soft, low-power tubes of the time (Howeth 1963).

In the immediate aftermath of World War I the navy continued to dominate American geostrategic thinking, especially about the need to control global radio as an alternative to Britain's well-established global cable system. In 1919, at the Versailles peace talks, President Wilson out-lined three arenas of competition between America and Britain that he foresaw in the postwar world. In world-system terms, Wilson was outlin-ing one version of the hegemonic struggle within the core that had be-gun with the rise of America and Imperial Germany in the late 1800s. Wil-son's arenas were oil production, in which he considered America held a clear lead; merchant shipping, led by Britain; and global telecommu-nications, which was up for grabs. Admiral Bullard was sent back to Amer-ica by Wilson to force all American commercial interests into a holding company set up by General Electric, namely, RCA. RCA was intended to freeze Marconi's out of the lucrative American market, to which it had acquired near monopoly rights in 1911. American court decisions that year had assigned primacy to Lodge's crucial tuning patent, bought by Marconi in 1911, thus forcing United Wireless to accept a buyout and be-come American Marconi. RCA initially did precisely what it was intend-ed to do. At Bullard's prompting, GE bought out American Marconi and turned the crucial Marconi patents, as well as its own rights to the Alexan-derson alternator, over to RCA. Other companies with interests in con-tinuous wave, low-frequency, long-range radio followed suit, apparently setting the stage for RCA to successfully challenge a British telecommu-nications hegemony in submarine cables exacerbated by the acquisition of the German cables as war prizes.

But American engineers bet on the wrong technology. In the early 1920s Marconi demonstrated that switching to the much higher fre-quencies that could be generated using simple and cheap vacuum tube transmitters achieved global range at far lower cost. Poulsen's arc and Alexanderson's mechanical alternator generated only low-frequency continuous waves and only with very high energy inputs. To take just one example, when the Kenyan government first considered installing a low-frequency wireless station as part of the British Imperial Chain immediately after World War I, it indicated that if the falls on the River Thika were unable to generate enough energy, another nearby river could also be tapped. The switch to vacuum tubes and beam antennas on much higher frequencies meant that two diesel generators could do the job instead. Exacerbating this technical shift was the mistake made by the FCC under the direction of Secretary of Commerce Herbert Hoover. Just before Marconi clearly demonstrated the technical superiority of high-frequency transmission over very long distances the FCC allocated the low-frequency long waves to the military, the medium-frequency medium waves to the commercial broadcasters (the modern AM band), and the supposedly useless high-frequency short waves to whoever wanted them. It took a decade of international conferences to correct this mistake.

The combination of the technical shift to the higher frequencies and open access to those frequencies meant that the growing American challenge to British telecommunications hegemony faltered at a crucial juncture. In some ways British hegemony actually increased, since Marconi, then the British Post Office, and then Cable and Wireless moved aggressively into long-distance, high-frequency radio communication using beam antennas in telegraphy and then telephony. The map of radio telephone links in the late 1930s (see fig. 3.2) shows Britain almost as securely at the center of a global web as it had been with regard to submarine cables a generation earlier (see fig. 2.4). This view is not shared by Headrick, who notes that American companies "were not saddled like the British with money losing strategic cables to out-of-the way places" (1994, 30). This was true until 1929, but the Cable and Wireless merger in that year meant that the money-losing cables could be dispensed with in favor of the new beam stations, as the Cable and Wireless archives attest. The critical fact remains that the successful American development of a global low-frequency radio system in the teens was short-circuited by the British high-frequency system using beam antennas in the 1920s.

This slippage in America's control of the carrier technologies was also apparent in America's failure to compete hard enough in the important arena of high-definition electronic television, even though some of the most critical items of the technology that made such television possible in Britain by 1936 were American, and in Britain's remarkable success in applying telecommunications in warfare. Although Britain has been legitimately portrayed as generally weak in Neotechnic technology, it was not weak in developing the telecommunications component of that technology in war. The reasons for this unsuspected strength lie in historical contingency. Britain had a global empire and invested heavily in telecommunications as a result. The first of these investments, in the submarine cable system, proved worthwhile, so Britain decided to continue investing in telecommunications. The successful shift to high-frequency radio communication in the 1920s required considerable exploration of the radio-frequency spectrum and how radio waves were affected by the layers of the earth's atmosphere, and this helped Britain gain a slight but key lead in what would emerge as radar. The interaction between radar and television is clear, and the combination of the social forces of production in Britain, where Jewish interests were strongly represented, and clandestine action by farsighted members of the British Cabinet may have forced the speedier development of electronic television in Britain than in America or Germany. As radar came into being in the late 1930s, Britain returned to its traditional geopolitical position, secure behind its Mahanian "moat defensive," in the second half of the war of hegemonic transition (1939–45) instead of being forced to aggressively and expensively interfere in a land war in Europe, as Mackinderian geopolitics had required in the first half of the war (1914–18). Radar also turned out to have considerable offensive utility in the strategic air war that developed as the logical successor to the Mahanian policy of blockade. This slowed Britain's decline as hegemon and helped ensure that the hegemonic successor was America rather than Germany.

The first successful American challenge to British hegemony was thus deferred until after World War II, when it came from a totally different, commercial rather than geopolitical arena, that of the American telephone company, Bell. A second challenge, building rapidly on the first, was a more traditional geopolitical challenge inasmuch as it came from the sphere of the government-financed space program and the development of the geosynchronous communications satellite.

AMERICA ACHIEVES TELECOMMUNICATIONS HEGEMONY

Submarine Telephony

Although radio telephony as developed in the late 1920s on both low and high frequencies was workable, it suffered from numerous problems, the two primary problems being variable speech quality caused by atmospheric interference and the very low capacity of the system. Each line on the map of the late 1930s (fig. 3.2) represents one possible radio telephone circuit. As early as the 1920s it was clear that both problems could be resolved by switching to a wired system, but the technical problems of a submarine telephone cable were immense and the capital cost would be enormous.

Long-distance, ultimately transcontinental telephony had been the goal of Bell Telephone almost from its inception, but the goal proved technically evasive. Bell's first deliberate research and development venture produced the technology of line loading, which by 1906 allowed a low-capacity telephone network to develop over a radius of about 1,000 mi, basically throughout the American East Coast Megalopolis and the Great Lakes Manufacturing Core regions (see fig. 3.5). This success prompted a greatly increased research and development effort with the stated goal of transcontinental telephony in time for the Pacific Exposition in San Francisco in 1915. Although this was achieved using De Forest's new vacuum tube amplifiers, system capacity was low and costs were high. Only thick copper wires without covers could be used over such a distance. Such wires could not be buried but had to be mounted in the open air a reasonable distance apart so that they would not suffer from crosstalk. Such a technology was clearly impractical for submarine telephony.

Continued efforts in research and development produced far better amplifiers and cable by the mid-1930s, so that it became possible to think of submarine telephone cables. Since cables required amplifiers at much more frequent intervals than did open wires (Espenscheid and Strieby 1984), any long-distance cable would have to have vacuum tube amplifiers built into it every few miles and would have to include the electric power supply for those amplifiers. Designing amplifiers that (1) could be manufactured into the cable, (2) did not increase the cable diameter too much, (3) could pass through existing cable-laying machinery as the cable was unreeled, and (4) could survive the hostile environment of the Atlantic seafloor was a formidable challenge for 1930s technology. Even so, Bell proposed a 12-circuit cable using submerged vacuum tube amplifiers at 50 mi intervals as early as 1942, but World War II prevented its construction (O'Neill 1985, 338). An experimental cable was in service between Florida and Havana by 1950, and the first transatlantic telephone cable, TAT-1, was in service by 1956. TAT-1 used one cable in each direction and provided 36 simultaneous channels at the rather conservative 4 kHz channel spacing (table 8.2). It is, however, a measure of the residual strength of British hegemony that 90 percent of this cable had to be manufactured in London by a British company, Submarine Cables Limited (Barty-King 1979, 335).

Probably only Bell, with its American domestic monopoly, high profits, and phenomenal success record in research and development, could have both developed and afforded this pioneer installation. Nevertheless, the British Post Office proposed a higher-capacity, single-cable, more radical technology during its joint planning of TAT-1 with Bell. Bell rejected this technology for TAT-1 and TAT-2 but adopted it for TAT-3 (O'Neill 1985, 350). But together with the American development of four-engine propeller and then jet airliners in the mid- to late 1950s that were able to fly non-stop from New York to London, TAT-1 and its successors revolutionized the Atlantic economy in America's favor, making possible truly transatlantic management, at least in the English speaking world (Hugill 1993, ch. 6). To cite but one example, English Ford was a virtually independent company from the early 1930s to the early 1960s. Almost all of its stock was held in Britain, and virtually all of its managers were British. After English Ford turned from Model A to Model Y in the Depression, it built almost none of the vehicles that Detroit built for the rest of the 1930s, and none at all after 1951. In a visual sense its designs owed something to Detroit, but by the early 1950s they had diverged considerably in a mechanical sense. Ford's use of unit construction, high-revving overhead-valve engines, four-speed all-synchromesh transmissions, and disc brakes all began in England. Yet in the late 1950s Detroit took control. It bought up English Ford stock, installed American managers, who were flown to and from Detroit on a regular basis, and used TAT-1 and its successors constantly to make day-to-day decisions from Detroit. Whereas Henry Ford had treated the head of English Ford, Percival Perry, as an equal, Henry Ford II regarded English Ford as more of a subsidiary. Dagenham went from being a virtually independent fiefdom to being almost a suburb of Detroit by the early 1960s. In less highly capitalized companies than Ford, or perhaps merely those that were more democratically governed, TAT-1 could also have very positive effects, since it enabled more efficient sharing of information among branches of transatlantic corporations and relatively uniform managerial decisions.

TAT-1 initially cost "roughly one million dollars per two-way channel" (O'Neill 1985, 347). Its capacity was raised to 48 channels in 1959 by lowering the bandwidth from 4 kHz to 3 kHz and then doubled by installing time-assignment speech interpolation (TASI), which took advantage of the normal pauses in human speech exchanges to switch lines rapidly between conversations. TAT-1's success was such that demand for transatlantic service grew at 20 percent per year and additional cables were needed. Bell came to own seven TAT systems, although these were technically shared with the country in which they made landfall on the European shore (fig. 8.1). Two British and one French cable were eventually added. This growth in transatlantic telephone traffic is a useful measure of America's attainment of telecommunications hegemony.

After 1956 transatlantic cables were added regularly. TAT-2, a direct copy of TAT-1, was added in 1959 with 96 circuits. The first all-British system, Canada transatlantic telephone (Cantat) 1 with 60 circuits, opened in 1961. TAT-3 opened in 1963 with 138 circuits, and TAT-4 in 1965 with 138. In 1970 TAT-5, the first solid-state system, opened with 845 circuits. The second all-British system, Cantat-2, entered service with 1,800 channels in 1974. By 1974, therefore, 3,173 transatlantic circuits were available. The number would more than double, to 7,173, with the addition of TAT-6 in 1975, and it would increase by another 4,000 with TAT-7 in 1983 (O'Neill 1985, 361, 370; see also table 8.2). TAT-7 marked the effective capacity limit of an analog system. Although a 16,000-channel analog successor to TAT-7 was considered, the future clearly lay with the much cleaner and higher capacity PCM digital transmission through either satellites or fiberoptic cables. American interests were also reflected in the increase in the number of cables through the Caribbean into Latin America and across the Pacific to Asia through Hawaii (fig. 8.2). Such connections were, however, influenced by military as well as commercial interests, in particular during the Vietnam War.

Britain did not totally abandon submarine telephony to America. The design Bell rejected for TAT-1 was far lighter than the final design, had only two-thirds the attenuation of that line, needed fewer amplifiers, and had 70 percent greater capacity. The use of this design for Cantat-1 made Cable and Wireless once again a major company by moving it into telephony, and it commissioned three new cable ships. The 1958 Commonwealth Conference in London planned to include Cantat-1 in a Commonwealth global telephone system—Compac—at a cost of some £88 million to be shared by the Commonwealth countries. By 1963 Compac telephone service had been extended across Canada by microwave relay and across the Pacific by cable from Vancouver to Fiji, New Zealand, and Australia (Barty-King 1979, 336–38, 341). This was an ambitious scheme, but the future lay with satellite technology.

Satellites

Bell Telephone initially attempted to extend the commercial quasi monopoly of international telephone communications it had achieved through the TAT cables into satellite communications with its sponsorship of Telstar. This was resisted by the American government, which preferred to back Intelsat, the International Telecommunications Satellite Organisation. Since the American government, through the procurement of military hardware, controlled the supply of launch vehicles, this attitude is understandable. The 80 kg *Telstar I* was launched in 1962 by a Thor-Delta rocket, which did not have enough power to place it in geosynchronous orbit. This limited *Telstar*'s worth because its footprint was small and it moved constantly, so that ground stations had to move constantly to track it. A large number of such satellites would have been needed for a fairly minimal global service, and the costs of ground stations would have been high. *Telstar I* was followed in 1963 by *Telstar II*, designed be more resistant to the radiation that had damaged its predecessor. As with *Telstar I*, *Telstar II*'s low elliptical orbit made it hard to use. *Telstar* could carry one black-and-white television channel or 600 voice telephone links. Project Relay was a similar low-orbit satellite system developed by RCA under contract to the National Aeronautics and Space Administration (NASA) (Pritchard 1977, 295). Clearly needed, however, were satellites launched into geosynchronous orbits, high enough up

Table 8.2. Submarine Telephone Cable Capacity, 1956–1983

Transatlantic Telephone Cable	Year	Number of Circuits Added	Total Number of Circuits
TAT-1 SB, 4 kHz channels	1956	36	36
TAT-1 SB, 3 kHz channels	1959	12	48
TAT-1 SB, 3 kHz + TASI	1959	48	96
TAT-2 SB	1959	96	192
Cantat-1	1961	60	252
TAT-3 SD	1963	138	390
TAT-4 SD	1965	138	528
TAT-5 SF	1970	845	1,373
Cantat-2	1974	1,800	3,173
TAT-6 SG	1975	4,000	7,173
TAT-7 SG	1983	4,000	11,173

Sources: O'Neill 1985, 341–70; Schenck 1975.
Note: maximum capacity is reported; real capacity varies slightly (see fig. 8.1).
TAT = Transatlantic telephone.
Cantat = Canada transatlantic telephone.
Reducing the spacing between channels from 4 kHz to 3 per channel increased capacity.
TASI = Time-assignment speech-interpolation system.
SB was a first-generation TAT system; SD, SF, and SG were later-generation systems.

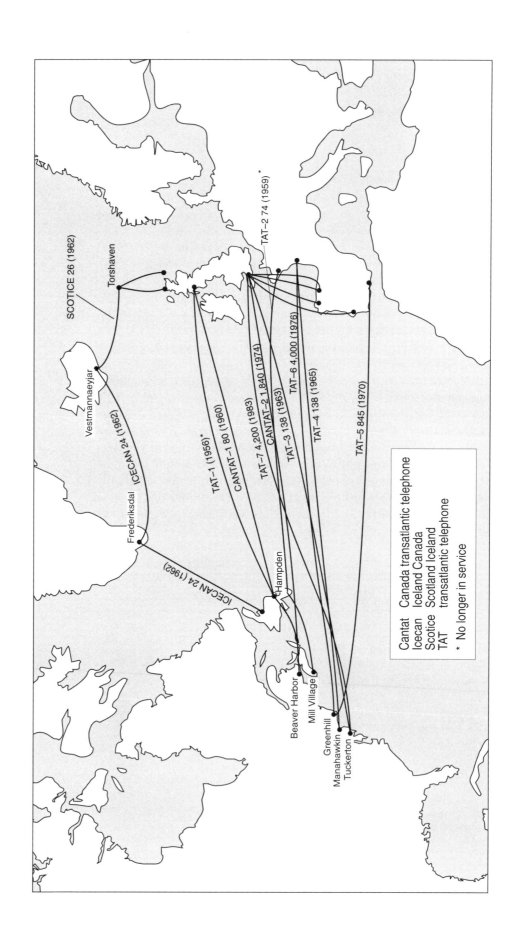

SCOTICE 26 (1962)

Torshaven

TAT-2 74 (1959) *

ICECAN 24 (1962)

Vestmannaeyjar

Frederiksdal

TAT-1 (1956) *

CANTAT-1 80 (1960)

TAT-7 4,200 (1983)

CANTAT-2 1,840 (1974)

TAT-3 138 (1963)

TAT-6 4,000 (1976)

TAT-4 138 (1965)

TAT-5 845 (1970)

ICECAN 24 (1962)

Hampden

Beaver Harbor

Mill Village

Greenhill

Manahawkin

Tuckerton

Cantat   Canada transatlantic telephone
Icecan   Iceland Canada
Scotice   Scotland Iceland
TAT   transatlantic telephone
*   No longer in service

and moving fast enough so that they seemed stationary when viewed from earth. These made for much cheaper ground stations because the ground stations could be fixed or at least movable only in easily controlled and predictable ways to focus on one stationary satellite or another.

The relative success of the *Telstars* and Project Relay as technology provers led to the passage in 1962 of the Communications Satellite Act, which led to the formation in 1963 of the Communications Satellite Corporation (Comsat) to take and hold "leadership for the United States in the field of international global commercial satellite services." A year later, in 1964, 19 nations—Australia, Austria, Belgium, Canada, Denmark, France, Germany, Ireland, Italy, Japan, the Netherlands, Norway, Portugal, Spain, Sweden, Switzerland, the United Kingdom, the United States, and Vatican City—signed on to Intelsat (Hudson 1990, 26, 31). In other words, Intelsat's original members comprised only the developed industrial regions of Europe and North America, plus Japan and Australia. Initial ownership was divided as follows: the United States, 61 percent; Western Europe, 30.5 percent; and Canada, Australia, and Japan, 8.5 percent. As Hudson notes, "These quotas could be adjusted to enable developing countries to participate in up to 17 percent of the total, although the U.S. share would never fall below 50.6 percent" (31). In world-system terms, this was a club for members of the core and their close allies in the semiperiphery, but one that would always be dominated by America. A British member of Parliament stated the obvious when he said that "we shall finally end by starving the transatlantic cable of telegraphic communications from America and . . . Britain will merely end up by renting a line from the Americans" (31). On the other hand, the success of the American space program and the threat of Intelsat persuaded 146 European aerospace companies to form Eurospace in 1961 and the European Space Research Organization and the European Space Vehicle Launcher Organization in 1964. The last two were consolidated into the European Space Agency in 1971 as the Europeans sought to break free of Intelsat and Comsat (32).

Although the geosynchronous *Intelsat I*, also known as *Early Bird*, worked well, as did *Intelsat II* and *III*, the "fully mature phase of satellite communications" opened with the 730 kg *Intelsat IV* in 1971. As launched, *Intelsat IV* could carry up to 12 color television channels or 4,000 to 6,000 voice channels (Pritchard 1977, 295). Almost immediately concerns over capacity encouraged the development and application of multiple-access and multiplexing technologies. By 1977 the application of digital technology through PCM and the use of time-division multiple access and digital speech interpolation allowed capacity to be quadrupled, albeit only with the considerable added expense of much more sophisticated computer equipment at ground stations (300). Such equipment was still substantially cheaper than three more satellites and their launch vehicles.

Of the 20 civilian communications satellites in service in 1977, 19 were American and 1 was Soviet. The 19 American satellites were launched by three families of American military rockets—12 by various marks of

Fig. 8.1. The Transatlantic Submarine Telephone Cables, TAT-1 through TAT-7 *(opposite)*

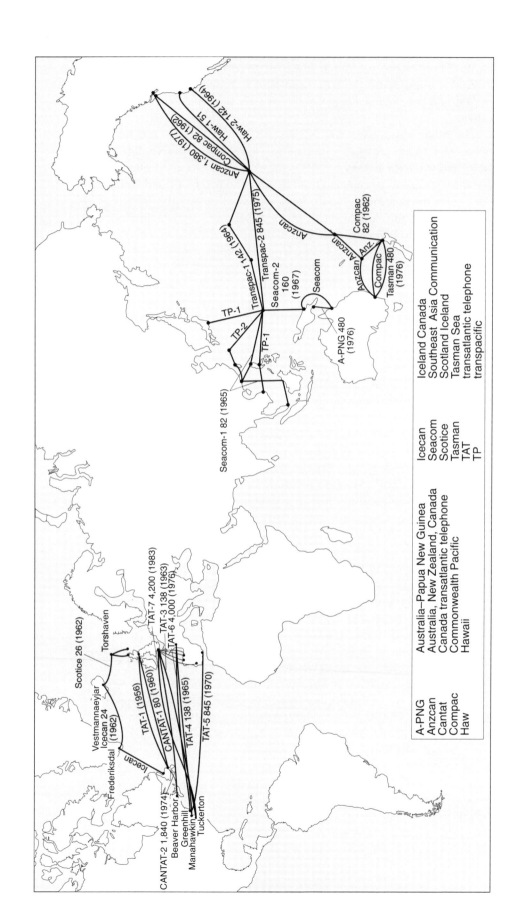

Scotice 26 (1962)

Vestmannaeyjar
Icecan 24 (1962)

Frederiksdal

Icecan

Torshaven

TAT-7 4,200 (1983)
TAT-3 138 (1963)
TAT-6 4,000 (1976)

TAT-1 (1956)

CANTAT-1 80 (1960)

TAT-4 138 (1965)

TAT-5 845 (1970)

CANTAT-2 1,840 (1974)
Beaver Harbor
Greenhill
Manahawkin
Tuckerton

Seacom-1 82 (1965)

TP-1

TP-2

TP-1

Transpac-1 42 (1964)

Transpac-2 845 (1975)

Seacom-2 160 (1967)

Seacom

A-PNG 480 (1976)

Anzcan 1,380 (1977)
Compac 82 (1962)
Haw-1 51
Haw-2 142 (1964)

Anzcan

Anzcan Anz.

Compac

Anzcan Anz.

Compac 82 (1962)

Tasman 480 (1976)

A-PNG	Australia–Papua New Guinea	Icecan	Iceland Canada
Anzcan	Australia, New Zealand, Canada	Seacom	Southeast Asia Communication
Cantat	Canada transatlantic telephone	Scotice	Scotland Iceland
Compac	Commonwealth Pacific	Tasman	Tasman Sea
Haw	Hawaii	TAT	transatlantic telephone
		TP	transpacific

Delta rockets, 1 by an early Thor-Delta, 3 by Atlas-Centaurs, and 3 by Titan IIIc's. Three of these launches were for the European Space Agency, and one was for Indonesia. The later Deltas could handle launches of up to 470 kg, the Atlas-Centaurs as high as 860 kg, and the Titans as high as 1,350 kg (Pritchard 1977, 298–99).

When *Intelsat I* was launched in 1965 it carried only 240 voice circuits, far fewer than the capacity of the submarine cables (Inglis 1990, 399; see also tables 8.2 and 8.3). TAT-4, also opened in 1965, added 588 circuits (Schenck 1975). *Intelsat IV*, however, added at least 4,000 circuits in 1971, far more than the then combined capacity of all the submarine cables. The last major submarine cable, TAT-7, was opened in 1983, and thereafter satellite capacity grew far faster than submarine cable capacity. In fact there is now substantial overcapacity. By the mid-1980s Intelsat was operating 16 geosynchronous satellites, had become a virtual world monopoly, and could be interpreted as "a triumph for U.S. Cold War diplomacy." Although the company, based in Washington, D.C., is avowedly international in outlook, it is claimed to "actively [facilitate] the work of the U.S. electronic intelligence agency" (Asker 1991). In this regard the functions of the American satellite and telephone cable systems by the late 1970s seem to have differed very little from those of the British submarine cable net of the late nineteenth century, which was similarly attacked, mostly by Americans, for both its economic monopoly and its close links to British intelligence. The wheel seems to have come full circle.

There are, however, clear problems with the satellite system. One is that so far the life of satellites has been much more finite than that of the wired systems, although usually longer than planned. *Intelsat I* had a design life of 18 months and lasted 4 years (Pritchard 1977, 295). More recent satellites have been designed to last 10 years (see table 8.2). Another problem is the limited number of parking slots available in geosynchronous orbit, some of which are more desirable than others. The FCC assigns orbital locations for satellites in that part of the orbital arc assigned to the United States by the International Telecommunications Union. The international picture is not so clear, despite broad agreements between America, Canada, and Mexico on parking orbits covering the continental United States, Alaska, and Hawaii (Inglis 1990, 406–9). The entry of new launchers into the market will continue to complicate this problem, as will the growing demand for satellite television

Table 8.3. Intelsat Telephone Circuits, 1965–1981

Intelsat Series	Year of First Launch	Number of Telephone Circuits	Design Life in Years
I	1965	240	1.5
II	1967	240	3.0
III	1968	1,200	5.0
IV	1971	4,000	7.0
V	1980	12,000	7.0
VI	1981	33,000	10.0

Source: Inglis 1990, 415.

Fig. 8.2. Global Submarine Cables, c. 1970 *(opposite)*

broadcasting, which consumes far more bandwidth than do telephone circuits.

CHALLENGES TO AMERICAN TELECOMMUNICATIONS HEGEMONY: SATELLITES AND FIBER OPTICS

America is now in the position that Britain was in at the end of the nineteenth century, although because of the short life expectancy of satellites America is much more vulnerable. America came to enjoy telecommunications hegemony by the late 1970s through the submarine telephone cables installed by Bell and through Intelsat. This hegemony is, however, being challenged as some countries are looking to cheaper satellites to provide alternatives to Intelsat and others are installing fiberoptic submarine cables.

In the 1970s America decided to develop a reusable crewed orbiter, the space shuttle. The greater size of the shuttle allowed the diameter of spacecraft to increase from some 3 m to about 5 m, although larger disposable launchers could have had a similar effect. Diameter imposes severe size limits; large satellites limited to the 3 m in diameter that Delta rockets could handle had to be elongated, and there was a limit to that elongation. A satellite has to rotate, and a long, thin satellite is "inherently unstable dynamically," so that it requires sophisticated damping (Pritchard 1977, 307). Because of optimistic cost projections at NASA the shuttle was funded by the American government, but the program has not been an economic success, especially compared with much cheaper disposable launchers (Inglis 1990, 396–99). Because the shuttle was designed to be crewed and because the decision was made to keep acceleration low so that not only military test pilots but also civilians could fly in it, it cannot reach geosynchronous orbit. It cannot, therefore, repair satellites in geosynchronous orbits, and the satellites it carries that are intended for such orbits must be boosted into them.

The explosion of the space shuttle *Challenger* in 1986 represented a major setback to the American program that was compounded by the malfunction of a Delta rocket launch by the air force that same year. The door was now open for launch vehicles produced in Japan, the former Soviet Union, China, and Europe. Australia's selection of a Chinese launch vehicle for its next generation of satellites was heavily resisted by the American government, if eventually accepted (Hudson 1990, 274). America has also gone out of its way to help the former Soviet space program so that the Russians do not feel a need to sell launches commercially. But the success of the European space program is convincing evidence that Intelsat's "monopoly" of the global satellite system will be loosened if not broken early in the next century, much as Britain's "monopoly" of international submarine telegraph cables was loosened early in this century.

At the same time, the role of satellites is changing. In 1977 a major article in the *Proceedings of the IEEE* could tacitly proclaim American hegemony, excepting one Soviet satellite the author described as "inchoate" (Pritchard 1977). In a similar article in 1990 no such hegemony is apparent, and the article deals with the need for the American satellite industry to change focus. The difference is partly the result of the changed nature of international telephony. According to the article, satellites forced the shift in long-distance telephony from "a mostly analog medium to an all-digital one. Satellites have carried two-thirds of the world's trunk international telephone communications for over two decades and have led in digitization of international links." Unfortunately, digitizing the signal allows the use of fiberoptic cables, "which provide a virtually inexhaustible supply of digital transmission capacity between the points they connect." Satellite service must thus shift focus onto areas with "thinly distributed, but large aggregate populations," onto television rather than communication, and onto mobile communications for land, sea, and aerial vehicles (Campanella et al. 1990, 1039–41). A study by the European Space Agency shows that of the 12 million trucks in daily use in Europe some 4 million would benefit from satellite-based mobile communications, and 4,000 satellite channels are needed to meet this requirement alone (1043). This switch in focus from satellite-based systems back to terrestrial and submarine fiberoptic cable systems will considerably reduce America's ability to tap into the world's information flow. The most efficient challenge is from fiberoptic submarine cables, in which France and Britain are taking the lead (Runge and Trischitta 1986). Even so, an adequate level of American investment assures that there will be no fiberoptic cable monopoly on the scale of the British submarine cable monopoly developed in the nineteenth century (fig. 8.3).

IMPLICATIONS OF THE HISTORY OF COMMUNICATIONS FOR WORLD-SYSTEM THEORY

British hegemony in the fourth world leadership cycle was, at least in part, unusually long lasting because the British came to control global telecommunications in the late nineteenth century. Britain's primary hegemonic challengers at the time, America and Imperial Germany, recognized this and attempted to redress the situation in the early part of this century using low-frequency, long-wave radio. Such attempts at a technological end run around the British submarine cable monopoly invariably failed. Neither America nor Germany was able to convincingly take power from the failing hegemon at the normal point of transition, in the first two decades of this century and following what looked like a normal war of transition in which one challenger, America, emerged as the dominant economic power. Britain stayed ahead of its hegemonic challengers, albeit by the narrowest of margins, by concentrating on the "carrier wave" technologies—high-frequency radio communication, television, and radar—a lead that it established in the 1920s and 1930s and maintained convincingly to 1945.

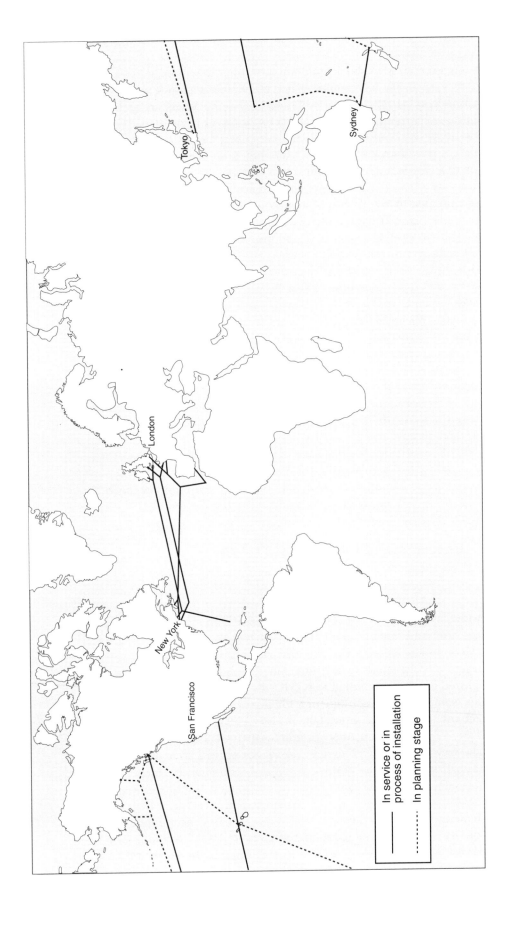

In service or in
process of installation

In planning stage

After World War II America was finally able to control global telecommunications by means of Bell's submarine analog telephone cables after 1956 and the digitized satellite system as developed by Intelsat. By 1971 this apparent monopoly seemed secure in the hands of these two companies. Since then it has been significantly reduced. The impact and implications of the breakup of Bell Telephone are not yet clear. Digitization has allowed the development of fiberoptic cables, which in turn has allowed the development of very high capacity links between major cities around the globe. Satellites, with their limited capacity, cannot compete with fiberoptic cables on such densely traveled routes. The space shuttle has not allowed a cheap satellite repair and replacement service to develop, and the lack of parking places for satellites is also a limiting factor. Competing areas in the world-economy now have viable disposable launchers.

At this juncture in the development of the world-economy it seems that global telecommunications will not be the monopoly property of any one polity or even any current grouping of polities. Two conclusions are therefore valid. First, if the next Kondratieff upswing and the next world leadership cycle are to be driven by efficiencies in information processing and transportation, then we will not see a clear hegemonic transition. No monopoly of global telecommunications currently exists or is likely to exist because of the constant technical change produced by the free interchange of technical information, the relative fluidity now achieved by global capital, and the failure of America both to protect Bell's monopoly and to maintain a dominant space program. Second, contrary to current opinion, the next cycle of investment and growth in the world-economy will be driven not by efficient processing of information but by something else entirely. Since the Kondratieff B phase is currently ending, it is arguable that this something else already exists, at least in some proposed or prototypical form. The lesson of the past hegemonic transitions is that this something else could as easily be a software as a hardware innovation. The lesson of a history informed by social science is that we should not ignore such a possibility and be blinded by the assumption that the future will be merely a continuation of the past. What the past teaches us in this case is that the information economy has been around for some time and is no longer a particularly risky or forward-looking investment. It also teaches us that innovations in the social forces of production—which I take to include, among other things, trade, education, and the social and geographic distribution of wealth—are just as important as innovations in the material forces of production. There has been a regrettable tendency in the capitalist world to focus on the latter at the expense of the former. I read the quotation from Innis that opens this chapter to speak to precisely this problem with regard to "elasticity of knowledge."

The Innis Model

Innis' model suggests two types of communications systems that I have somewhat recast to analyze modern telecommunications. Capacity and

Fig. 8.3. Global Fiberoptic Submarine Cables, 1989 *(opposite)*

cost are the key to my analysis. Low-capacity, high-cost systems approximate Innis' Type 1 systems; high-capacity, low-cost systems approach his Type 2 systems. Low-capacity, low-cost systems (e.g., short-wave radio) are a special case. Innis argued that Type 1 systems tend to produce hierarchical social systems and systems of government. I have extended his argument to include the tendency in a commercial world system for Type 1 systems to embrace commercial monopolies of information flow. Of course there is usually a relationship between that commercial monopoly and the government. Until recently, all the global telecommunications systems (with the exception of short-wave radio) have tended toward Type 1 systems, being low-capacity, high-cost systems. All the low-capacity, high-cost systems have been run by commercial monopolies or government agencies. The British preferred to retain a commercial monopoly in submarine telegraph cables and then wireless, but with clandestine links to government. America used its navy to seek a monopoly in long-distance, low-frequency, continuous wave radio. Marconi's development of commercial high-frequency radio ended the American navy's attempts to control long-distance radio. Marconi's relatively low-cost system, derived as it was from the work of amateurs using mostly war surplus equipment, could be used by anyone with a modicum of training. It was, however, a low-capacity system, and it was quickly monopolized by the British government through first the Post Office and then the quasi-commercial merger of Cable and Wireless. All of this caused high-frequency radio to look like a Type 1 system.

Both low- and high-frequency radio suffered from technical problems, such as fading in and out caused by climatic and diurnal changes, and from lack of security. These problems prompted a search for alternatives. Radio remained a necessity. In the absence of submarine telephone cables and satellites, it was used by commercial monopolies to link national telephone systems across oceanic distances. Governments had no option but to use radio for communications to mobile military units at sea, on land, and in the air. In such cases security was quite literally a matter of life and death. The British and American ability to read German and Japanese codes in the second part of the war of hegemonic transition resulted in the failure of the German U-boat offensive against Britain in the Battle of the Atlantic, the failure of the Japanese naval offensive against America in the Pacific at the Battle of Midway, the death of the greatest architect of Japanese naval power, Admiral Yamamoto, when his transport plane was successfully intercepted by American fighter planes, and the like. The extent to which the reverse is true is less clear. At least some of Britain's stunning losses of merchant ships in the *guerre de course* of the Battle of the Atlantic were caused by insecure commercial telegraphic codes. German submarines knew which ships were in which convoys and what their cargoes were, but convoys were large, slow, and used well-defined traffic lanes in the North Atlantic. Intercepting convoys without advance information was far easier than intercepting U-boats. A simple rule is evident, however. Hegemons and would-be hegemons need secure global communications, and historically such security has been more easily achieved using wired systems than wireless ones, however carefully data are encrypted. Certainly, believing that one's system of telecommunications is secure has usually led to horrendous, indeed fatal errors of judgment.

The need for both quality and security thus encouraged the development of submarine telephone cables, and capital costs once again became a major issue. Since Bell had done the most to develop long-distance, land-based systems, it was well positioned to extend those systems overseas once America assumed military as well as economic hegemony in 1945. Bell was not, however, allowed to extend its monopoly into satellites in the 1960s, and America established the quasi-commercial monopoly Intelsat, which had strong links to the American intelligence system. Doubts about the security of either of these systems, but particularly about the latter, haunt its users today just as they haunted users of the British submarine telegraph cable network in the nineteenth century. It is now America's turn to read everyone's mail. Fiberoptic cables offer an alternative to the Bell cables and to Intelsat and seem particularly well suited to the exchange of PCM digitally encoded data between computers, which need error-free transmission.

Submarine analog telephone cables, digital fiberoptic cables, and satellites all conform to Innis' Type 1 communications systems. Vast amounts of capital are required to install them, although rapidly increasing capacity has meant that costs per message have dropped steadily in real terms. The difference between capital and use costs produces a problem in the continued use of Innis' categories. Type 1 systems can be both state and commercial monopolies. Type 2 systems are commercial but competitive. Currently no true Type 1 systems exist at the international level, despite the attempts by Bell and Intelsat to create them in the 1950s and 1960s, respectively. The rapidly emerging Internet is a classic Type 2 system. I would therefore suggest that one of the major historical prerequisites for hegemonic power, secure Type 2 economic and military communications, cannot exist unless ciphers have been created that cannot be broken. Historically, ciphers have always been broken. Absent a truly secure cipher or some technological breakthrough that is not currently foreseen, it is unlikely that any would-be hegemon will successfully replicate the near monopolies on telecommunications achieved by Britain from the 1860s to the 1930s and by America from the 1950s to the 1990s. The geostrategic implication is clear: since no hegemon can emerge, all would-be hegemons would do better to cooperate in the ongoing globalization of the economy. The alternative is isolationism and withdrawal from the benefits that accrue to specialization and trade.

Tilly's Analysis

Tilly's work suggests that we must further reinterpret our potential futures. His convincing argument that the growth of states is primarily the result of the accumulation and concentration of coercion (Tilly 1990, 27) in the light of my argument above implies that the relatively brief period of powerful territorial states is about to end because the nature of coercion is about to change. Cities, however, will continue to grow on the basis of their accumulation and concentration of capital. Cities may well be located in nation-states, but increasingly they will act in the interest of their citizens in promoting their trading relationships with other cities. Already new cities are growing rapidly in states with particularly beneficial tax structures, and in future cities may even develop outside nation-states so as to avoid their tax laws. Vastly improved telecommunications without clear central control will ensure that whatever their geographic location, such global cities will be linked by a complex web of information. The world seems destined for a period in which coalitions of trading cities (really megalopolitan areas) will dominate wealth production. If history is any guide, they will also come to dominate political discourse, including war.

   It is therefore in the nature of the state and of coercion that we can expect major changes in the next period of development of the world-system. In Tilly's view the history of coercion describes the increasing monopoly of force by the state and the direction of that force against those deemed enemies. To extend Tilly's analysis slightly, enemies could be *(a)* other states, *(b)* non-state-based terrorists, or *(c)* opponents within the state. After recognizably modern states emerged in Europe in the late 1700s and early 1800s, the coercive power of the state eventually trifurcated into strategic power expressed through conventional military air forces to deal with *(a)*, counterinsurgency forces, such as marines, to deal with *(b)*, and domestic police forces to deal with *(c)*.

The Geopolitical Analysis of Meinig and Taylor

Recent work by Meinig (1972, 1986), Wallerstein (1974), and Taylor (1993) has huge implications for modern geopolitical analysis. Meinig foreshadows the essential contribution of Wallerstein in that he takes an implicitly Leninist position on relations between metropolitan cores and their peripheries. Wallerstein codifies those relations into the characteristics of core, semiperiphery, and periphery in an unusually elegant formulation that has attracted more detractors than he deserves. The main thrust in Taylor's writing is that the evolution of mass as opposed to cadre political parties has resulted in the working and even middle classes in any given industrial polity being "bought off" with high transfer payments.

   Meinig's analysis is a subtle one, not least because most of Meinig's work deals with the successful resistance of American colonists to the form of imperialism practiced by Britain and their subsequent adaptation, improvement, and application of that imperialism to the conquest of the central section of the North American continent. Meinig identifies what he calls "the cumulative rigidities of autonomous political territories" (1972, 177). His most recent work (1986, 1993) deals with this problem through the immediately postbellum period of American history. Devolution is a logical result of such accumulated rigidities. As population, wealth, and the like have changed over time, political boundaries have rarely been adjusted to take account of such changes. Exactly the same forces that operate on nations operate on counties, territories, and states in the American West.

Taylor's analysis is particularly applicable to the "golden age" of capitalism following the American achievement of hegemony after 1945. The internal class struggle characteristic of industrial societies at the last hegemonic transition was ended. The development of mass politics and parties meant that nonelites within the core were bought off by high transfer payments, the most obvious being unemployment benefits, social security, and mass higher education (Hobsbawm 1994, 271–72). This led to a decline in class conflict at the price of reduced ability on the part of elites to extract the surplus value of nonelite labor within the core. The class struggle per se did not end, however, but was exported from core to periphery, so that core and peripheral elites combined to repress peripheral nonelites. The accommodation of core nonelites peaked after World War II as decision makers and mass parties strove to ensure that the economic meltdown within the core polities that started in 1929 was not repeated. Cadre parties represent competing elites and sets of elites within a given polity. Mass parties compete on the basis of the different ways they propose to make transfer payments both between classes in their host economies and, more importantly from a Leninist perspective, between the periphery and the core of the world-system.

Taylor also makes explicit Wallerstein's more implicit point that the class struggle so essential to classic Marxism has ceased to exist within the geographic confines of any given nation. After 1945 the struggle came to be between the population of the core allied with the elite of the periphery, on the one hand, and the peasantry and proletariat of the periphery, on the other. The geopolitical problem thus ceased to be a struggle between nations for direct political control over geostrategically important territories—the territorial state—and became a struggle between core and periphery for control of the resources, markets, and cheap labor supplies of the periphery. This struggle found its clearest expression in trade. In such struggles, direct, armed resistance to the power of the core must give way to terrorism. However, in the past decade or so the first part of this equation has begun to change inasmuch as the capitalist polities of the core are returning to a form of internal class struggle, albeit keeping generous social safety nets in place. Cadre parties are probably due for at least a modified reappearance. Core elites are now able to sell their skills and services profitably on a transnational basis, so that their income is no longer restricted by the scale of the national economy. Those whom Robert Reich calls "symbolic analysts" are thus doing very well in comparison with his "routine producers" and "in-person servers," whose incomes are restrained either by the scale of the national economies in which such work is embedded or, worse, by the fact that people in other parts of the transnational economy will do the same work for less (1992, 219). A renewed internal class struggle may be seized upon by demagogues who would return to the territorial nation-state. Such a renewal would break the internal alliance in core states between the elite and the nonelite, which has held so far because of high social transfer payments.

The Multipolar World of the 1990s and Beyond

In *World Trade since 1431* I suggested that the current hegemonic transition would go one of two ways. One way would be to enter more firmly into a multipolar world and continue the hegemonic struggles of the past, with three possible outcomes: 1A, in which America revitalized as NAFTA, the North American Free Trade Association, wins; 1E, in which the European Union wins; and, less likely, 1J, in which Japan succeeds at the head of some currently nonexistent Asian political combine. Combinations are also possible: for example, 1A + 1E against 1J makes sense on cultural and historical grounds and if one views American politics from an Atlantic-centered perspective; 1A + 1J against 1E makes sense if one thinks that America's future lies with its West Coast interests as part of the Pacific Rim. I noted, however, the historical lesson that hegemonic transitions have always included war, whether economic or military.

I then suggested a second, less well-formed alternative future—what I shall call future 2—in which the forces of transnationalism triumph. Several years later I see nothing to suggest that future 2 is less likely. In fact it seems more likely, which is grounds for cautious optimism. Despite poor economic growth in the wealthy world from the early 1970s to the early 1990s, the forces for economic war represented by protectionists and isolationists have not (yet) emerged, although there are the usual reminders that they are still lurking about. The Eurosceptic Tory members of Parliament in Britain and Patrick Buchanan's attempt to capture the Republican Party nomination for the American presidency are obvious examples of neoisolationism. At present both seem more pinpricks than serious problems. No major governmental or opposition party in the wealthy democratic polities has embraced such a position or seems likely to. The naked protectionism of the late nineteenth century was born of the drive by the hegemonic challengers, America and Germany, to isolate themselves from economic domination by the Manchester liberals, with their aggressive free-trade rhetoric and their remarkable ability to flood the domestic markets of other nations with cheap British goods. Although we do not—and probably never will or would never want to—live in a world of totally free trade, no major trading state currently embraces naked protectionism, although there are plenty of covert versions. Japan's Byzantine internal distribution and sales system, America's agricultural price supports, and the European Union's statements that agricultural protection is a necessity to support the family farm are all clear examples. Certainly a major collapse of the stock market might precipitate a neoisolationist response, but the educated elites that run the major trading nations are committed free traders. Delegitimation of those elites could cause severe problems, as it did in the 1920s, but the delegitimation of elites during the last hegemonic transition was caused by the horrors of World War I and the end of confidence in the liberal belief in progress. There seem to be no similarly powerful agencies abroad at present.

Absent such delegitimation and neoisolationism, how can we construe future 2? A transnational world-economy does not necessarily lead to the creation of superstates or nonpolitical supranational organizations. Quite the contrary given modern telecommunications, the lean manufacturing systems pioneered in Japan, and political forces clearly tending toward various forms of political devolution in all major polities. The advantages of economies of manufacturing or bureaucratic scale that came into being in the late 1800s no longer exist. However, the failure of such spatial consolidation and extension of power does not mean the end of geopolitical problems. What it means is the end of the sorts of geopolitical problems that have bedeviled us for roughly a century, problems that were bound up in Mackinder's fears of geographic consolidation and extension and drove British geopolitics early in this century and American geopolitics after 1945. It seems unlikely that we will face another challenge in the heartland of the sort that was posed, in sequence, by Imperial Germany, Nazi Germany, and the Soviet Union. The devolutionary future that Mackinder so desired at the end of World War I, as the only permanent solution he could see to the Eastern question, seems to have arrived. The Soviet Union has devolved into its constituent parts. The emergence of a bureaucratically centralized United States of Europe is being resisted by such polities as Britain, which desires a more Mackinderian future of free association in a free-trade confederacy, and seems open to devolution at home. The United States, after nearly two generations of consolidation and concentration of power in the hands of the federal government, is also showing clear signs of devolution, not into separate polities per se but in the guise of returning power to the states. In the presidency of Bill Clinton, at least, there is remarkable bipartisan agreement on this issue. Early in the twenty-first century China may pose a geopolitical threat at the eastern end of the Eurasian landmass similar to that posed by Germany and the Soviet Union at the western end in this century, but the possibility of desirable devolution in China is by no means absent (McRae 1994, 253).

The success of large, centralized polities from the late nineteenth century on was, in retrospect, historically contingent. It depended on the success of the bureaucratic organization of capital at the national level and the copying of such advantages by government and the military, all of which were made possible by better communications in the form of railroads and telegraphs. These were, in Innis' terms, Type 1 systems subject to considerable monopolization. These bureaucracies emerged as new, often interlocking elites not necessarily committed to liberalism in the same way as the trading and manufacturing elites represented by the Manchester liberals in the first three quarters of the nineteenth century. Mackinder's solution—to break centralized polities into their constituent regional, cultural, and ethnic territories—was technologically impossible when he proposed it. Almost all forms of transportation, whether of human beings, goods, or information, were in the hands of bureaucratically centralized entities, leading to political centralization, rampant nationalism, the new imperialism, and eventually protectionism and isolationism. In our era the much smaller political units that have come into being either by breaking away from former centralized polities or by devolution within existing centralized polities will succeed or fail based on how they recombine through trade. Small political units cannot be economically viable, so economic recombination is necessary. This is now technologically possible using Innis Type 2 noncentralized, relatively nonhierarchical systems of movement—bicycles, automobiles, trucks, airplanes, mobile telecommunications, and the like.

One likely version of future 2 is implied by the work of Hall and Preston and of Tilly. This future would see the increasing economic power of expanded or new megalopolitan areas. Three such megalopolitan areas clearly exist at present: Boston-Washington; London-Paris-Bonn; and Tokyo-Osaka (fig. 8.4). Several more are forming: San Francisco–Los Angeles–San Diego; Dallas/Fort Worth–Houston–San Antonio; Hong Kong–Taipei–Shanghai; Singapore. Given the dominance of North America and Asia in these lists it seems likely that Europe needs at least one more megalopolis. Under the right political circumstances a Berlin-Warsaw-Moscow megalopolis would not be impossible, merely a reinterpretation of German goals of the early 1900s. The need to move high-level managerial personnel around and the current performance of jet transports—subsonic but with hemispheric range—suggests that, absent an efficient, safe, cheap, supersonic jetliner and thus for the foreseeable future, such developments will be limited to the northern hemisphere. A less easily predicted, though much bruited, development would be in the use of very much higher capacity information systems than we currently enjoy to construct virtual realities. It is possible to imagine the movement of holographic images of people rather than people themselves.

A second version of future 2 might develop along social rather than geographic lines, with social policies, social structure, and behavior the prime concerns. At the most basic level is the understanding that unless cheap energy suddenly becomes available and there is a renewed expansion of production, future gains in wealth in advanced capitalist polities will probably have to be gained by improving productivity (Hugill 1993, 320). The crisis in the capitalist polities as a result of the 1973 "oil shock" caused significant productivity gains in such areas as transportation and distribution. The success of such Japanese innovations as lean production and Japan's perfection of the "just in time" system prompted American and, to a lesser extent, European companies to adopt similar managerial technologies, and they experienced similar productivity gains.

Global disparities are now most apparent in services. An alarming share of the gross domestic product (GDP) now goes to medical and legal services in some countries. Much of the debate over American health care costs has been driven by the fact that America spends a far higher proportion of its GDP on health care (13.4 percent in 1991) than any other industrial polity: Japan spends only 6.6 percent of its GDP, and Germany 8.5 percent, to deliver better care (McRae 1994, 38). Equally alarming are the economic costs of crime. The crime rates in Japan and Germany are extremely small compared with America's. From 1970 to

1990 the percentage of the working population of America employed as private security guards doubled, to 2.6 percent, higher than the percentage of police officers (Reich 1992, 269). The murder rate in America was 9.3 per 100,000 people in 1992, compared with 4.1 per 100,000 in Germany, 2.5 per 100,000 in England and Wales, and 1.0 per 100,000 in Japan (Interpol 1995, 73, 6, 175, 114). Between 1960 and 1980 the American prison population doubled, and between 1980 and 1990 it doubled again (Reich 1992, 269n). The population increased 38 percent in the same period. By 1993 America was incarcerating 519 people per 100,000, more than five times as many as any other major industrial polity, at a huge cost to the law-abiding population and at the expense of many other services. In contrast, Japan incarcerated 36 per 100,000, Germany 80, France 84, England and Wales 93, Mexico 97, and Canada 116. Only Russia was worse, at 558 per 100,000 (Mauer 1994, 3).

America has more lawyers per head of population than any other society on the planet, and these lawyers file vastly more product-liability lawsuits. Although action has been taken to reduce the level of punitive damages that can be assessed in a given product-liability suit, the fact remains that such damages cost money, as does the time taken in prosecuting and defending such suits. This does not excuse corporations from engaging in behavior known to be dangerous. In McRae's words, "Put bluntly, if countries wish to carry on becoming richer, their people [individuals and corporations alike] will have to learn to behave better" (1994, 265).

After years of dreams of a classless society and social policies explicitly designed by mass parties to produce such societies by exporting class struggle to the periphery of the world-system, the division of Western polities into haves and have-nots is already becoming evident. As recent economic commentators have noted, "the Third World has moved to parts of Los Angeles and the First World to Singapore" (McRae 1994, 269). This division is largely based on access to and control over information. Even within wealthy megalopolitan or protomegalopolitan centers there are "electronic ghettoes" where "the space of flows comes to a full stop" (Thrift 1995, 31). That the ability to use the information network will be a function of educational attainment has considerable implications for postindustrial societies in general and the educational system in particular. It is unlikely that educational systems conceived to suit late-nineteenth-century agrarian America or mid-twentieth-century industrial Europe and Japan will be well adjusted to the future needs of the information society. Absent broad restructuring of existing educational systems, this opens the possibility for a polity or nascent megalopolis to improve its position within the world-system by devising the right educational system, whatever that may be.

Educational systems embody more than just skills. As the work force shifts to a two-tier structure the lower tier will have to be conditioned to accept high levels of workplace monotony of a different sort than that endured by the previous generations of low-skill workers, those who toiled in fields and factories. There monotony was in manual work.

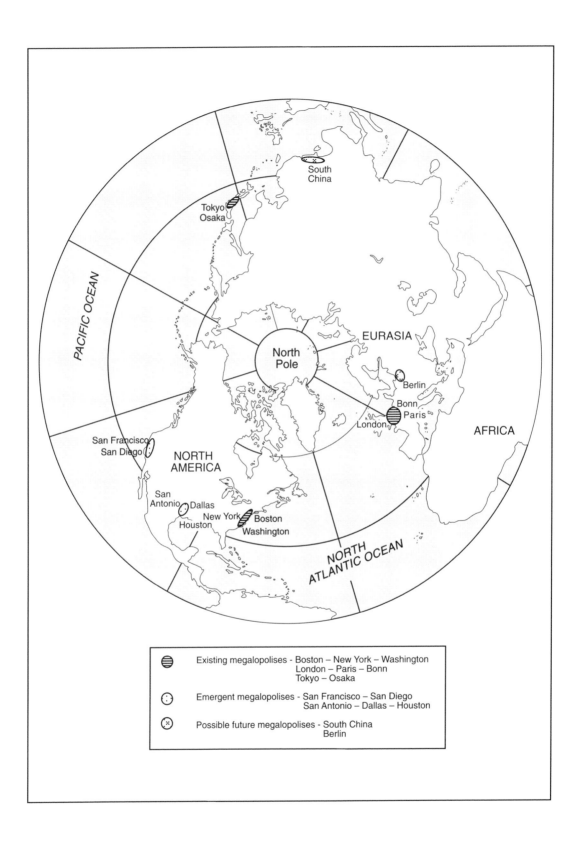

⊜	Existing megalopolises - Boston – New York – Washington London – Paris – Bonn Tokyo – Osaka
⊙	Emergent megalopolises - San Francisco – San Diego San Antonio – Dallas – Houston
⊗	Possible future megalopolises - South China Berlin

Although something of this remains in such occupations as fast food preparation, the information economy generates a different sort of monotony. As the population that operates computer information systems is deskilled, it will have to be conditioned to accept long hours at a computer terminal. Video games, in particular those that stress quick reflexes and mindless response over thought, would seem to do that admirably. Certain types of television shows, in particular easily produced variations on a single theme with no pretense to artistic complexity, such as sports and talk shows, would seem to do the same.

The upper tier will also need conditioning. As political devolution occurs, more and more people will act under much more localized legal systems. And more and more business will be done over the telecommunications links. It will thus become increasingly important to stress personal integrity and responsibility rather than the integrity and responsibility imposed by the state through legal systems. In any case, as the American example shows all too clearly, legal systems can take on a life of their own and impose excessive opportunity costs: money spent on lawyers is money that cannot be spent elsewhere. Formal education too often lacks moral education for any group in the population, not least because of the removal of traditional moral teaching from the curriculum with the removal of religion from schools in many industrialized polities. At least part of the reason for the correlation of capitalism with Calvinism noted long ago by Max Weber in his famous essay on the origins of capitalism lies in Calvinism's inculcation of personal and group responsibility. Such a sense certainly does not have to be produced through religious education, but it has to be produced.

The consequences of the lack of personal and group responsibility are all too evident in such debacles as the collapse of Britain's Barings' Bank brought on by rogue trader Nick Leeson's using London capital to speculate on the Tokyo stock exchange from a base in Singapore. Orange County, California, was brought low by similar rogue trading, as was the Japanese Mitsui Bank out of its New York office. The very term *rogue trader* is obviously used in such circumstances to imply that the problem lies with a single individual rather than the bank or the system within which the bank operates, however much they may also have been at fault. More conservative British bankers use the term *go-go banking* to suggest the obvious problems inherent in high-risk, high-return banking operations. Even the (Marxist) professors of economics in my undergraduate days at Leeds University taught me that interest rates are a measure of risk. To some extent the rapid increase in church membership in America, in particular membership in enthusiastic Protestant sects, is obviously a response to the perceived lack of moral education in formal schooling. The increase in the number of required ethics classes at America's universities not only is a response to the problem but also provides employment for academic philosophers.

Elites face another problem. The transnational elites that have matured in this phase of capitalism owe little to any particular polity or place. They are geographically footloose, and in a devolutionary world in which capital can be moved very freely across national frontiers they

Fig. 8.4. Global Megalopolises: Present, Emergent, and Possible *(opposite)*

are about to become even more so. All polities have a vested interest in attracting such elites, and the competition will result in attractive tax laws, easy residential requirements, and the like. The remarkable boom experienced by the city of Vancouver, British Columbia, as many Hong Kong residents prepared for the takeover of Hong Kong by the People's Republic of China in 1997 is a classic example. However, as Lasch (1995) noted, such placelessness is bound to lead to some form of political backlash from those who are more bound to place and are likely always to be the political majority in most polities. A sense on the part of a majority that elites are using their wealth to retreat to gated suburbs and exclusive residential communities and to escape to tax havens is clearly one route to delegitimation. Devolution will increase both the number of opportunities for tax havens and the backlash against those opportunities. The growing income disparity between symbolic analysts and the rest of the population in core countries will also continue to produce the sort of neoisolationism being peddled in America by Pat Buchanan. The disaffected may form religious or secular communal groups and merely withdraw from society. Alternatively, they may well make common cause with non-state-based terrorist groups against their own elite in an attempt to delegitimate that elite.

The third version of future 2 is implied by the second and is also geographic. Its probable appearance was noted as early as 1964 by Marshall McLuhan when he wrote of the mythic, tribal, and decentralized "global villages" of the future. Ability to use the Internet will be far more important than geographic access to it, since access no longer depends on being in a specific place. Relatively modest infrastructural improvements could already guarantee access to most of the citizens of not only the modern megalopolitan areas but also their host polities. The information society will thus not be as constrained by physical location as were the old agrarian and industrial societies, but it will be more constrained by the degree to which it educates its citizens to use information.

We have returned to a world in which large, centralized, land-based polities seem impossible and competition between their elites seems unthinkable. As transfer and transportation costs drop, a return to a system of smaller, megalopolitan polities bound together by information systems as well as flexible transportation systems thus seems plausible. A return to medieval-style city-states on a larger scale is unlikely. The need for coercion remains because of protection costs. The embedding of medieval city-states within increasingly powerful protonations was not accidental. Although the supranational Hanseatic League was able to police its trade routes from the 1200s to the 1400s, it did so by using self-defending ships. A division of labor in which city-based merchant ships traded and state-based naval ships warred lowered costs in the long run. Given the nature of the state system that evolved in medieval and early modern Europe, such warships fought both other states and non-state terrorists, that is, pirates. Indeed, there was often little to distinguish the two.

We seem to be living in a period in which protection costs are once again higher than transport and transfer costs—the traditional obsession of economic geographers in their modeling of spatial flows of goods and people. When the commodity being transported is information, transport and transfer costs are almost irrelevant and protection costs start to loom very large indeed. There is weekly evidence of this in the European and American media. For example, Japanese laws limit copyrights to 25 years rather than the 50 insisted on in Europe and America, so that cheap, non-royalty-paying copies of songs by such 1960s vocal groups as the Beatles can be sold there. The extent to which, in the absence of some transnational police agency, states will be able to enforce transnational contracts is also problematic. China is almost blatant in what is now called "software piracy," the theft of the intellectual property of others. Problems such as this, if not amenable to negotiation and settlement through government agencies, will eventually call for new forms of counterinsurgency forces, intellectual marines.

As large centralized polities continue their devolutionary meltdown, it is reasonable to predict not the end of geopolitics per se but rather the end of the Mackinderian geopolitics of territorial states and a return to something more like the Mahanian geopolitics of trading states, in which blockade tactics will become even more important. Given the obvious porosity of recent blockades of physical goods (against Iran, Iraq, Serbia, and the like) and the designed porosity of the global telecommunications system, new forms of blockade will have to emerge. Likewise, the polities in which the great megalopolitan areas are embedded will continue to need rapid-response counterinsurgency forces to fight terrorists, whether domestic or external, but it seems sensible to dismantle much of the existing capacity for strategic war making or at least to divert that effort to where it will do more good. The presumed "peace dividend" bruited in America at the end of the Cold War with the Soviet Union will be just that—presumed. Wars will not end, as Tilly has so clearly shown; they will merely change character. As the number of states rapidly increases, the number of small wars will probably also increase. Devolution is likely to continue, driven by telecommunications, the cumulative rigidities of territorial states, and the realization that in the long run wealth is more securely made and retained by trade than by territorial conquest.

Glossary

AI Mark IV. British airborne intercept radar operating on the 1.5 m band and using two CRT displays.

airborne intercept (AI) radar. Radar compact enough to be carried in a night fighter.

air-to-surface vessel (ASV) radar. Radar that could locate ships at sea and was compact enough to be carried in an airplane.

alternator. High-speed generator of alternating current that can carry continuous radio waves.

amplitude modulation (AM). Varying the amplitude of the carrier wave.

bandwidth. The frequency needed to effectively carry the information contained in a signal. More complex signals need greater bandwidth.

Berlin radar. German centimetric radar modeled on British H2S operating at 9.0–9.3 cm (3,250–3,330 MHz) and using two CRT displays.

bombe. British electromechanical device designed to help decode Enigma traffic.

Boozer. British ECM designed to receive Lichtenstein BC and Würzburg emissions.

Cantat. Canadian transatlantic submarine telephone lines from Britain across the North Atlantic to Canada. Two Cantats were laid.

cathode-ray tube (CRT). Tubes used for displaying television and radar images.

Chain Home. British fixed radar operating between 10 m and 13 m that restored the "moat defensive."

Chain Home Low. British scanning mobile radar based on AI Mark IV operating on 1.5 m. It allowed incoming aircraft to be tracked inland.

Combined Cypher Machine (CCM). American Electrical Cypher Machine modified to work with British Typex.

continuous wave radio. Radio transmission on waves that maintain constant amplitude with each cycle.

crosstalk. The undesired transfer of signals between circuits, in particular circuits used for telephony.

damped wave radio. Radio transmission on waves that decrease in amplitude with each cycle.

decibel (dB). A measure of relative levels of current, voltage, or power.

electronic countermeasures (ECM). Active and passive responses to radar.

electromechanical television. Television that used a rotating scanning disc to generate and receive images as opposed to the electronic iconoscope and CRT.

Enigma. German code machine.

flak. German antiaircraft gunfire.

Flensburg. German ECM designed to receive Monica emissions.

frequency-modulation (FM). Varying the frequency of the carrier wave.

Freya. German scanning mobile radar operating at 1.5–4.8 m.

Gee. British electronic navigational system that used three ground transmitters to create a grid of intersecting beams over much of Europe, thus a "map" allowing bombers to locate themselves over enemy territory.

geosynchronous satellite. Satellite in high enough orbit that it appears stationary from the earth's surface.

ground-controlled interception (GCI). A combination of signals from ground radar, AI radar, and IFF sets allowing ground controllers to vector night fighters onto potentially hostile aircraft.

Hertzian waves. Sine waves of electromagnetic emissions. One hertz describes a complete sine wave, or cycle that repeats every second. An early American and British term for Hertzian waves was *cycles per second.*

high-frequency direction finding (HF/DF). Radio direction finding, usually military, using the emissions of high-frequency enemy radio or radar to locate the enemy.

Himmelbett (lit., "bed in the heavens"). German radar software system in which radar-defended "boxes" were extended up into the night skies.

H2S (Home Sweet Home). British ground-imaging radar carried by bombers and operating on the 10 cm band.

iconoscope. Scanning electron beam system at the heart of electronic television. Light from the camera lens falls on a grid of charged photoelectric cells, changing the capacitance of each cell differently, thus creating an electronic image composed of many pixels.

Identification, Friend or Foe (IFF). Electronic device fitted to aircraft that was designed to transmit when electronically challenged on the proper frequency.

Kammhuber line. British name for German system of radar defense in World War II.

klystron. Velocity-modulated vacuum tube that can function as a microwave amplifier.

Knickebein. German navigational system using a radio approach beam followed by a bomber and intersecting beam over a target.

Lichtenstein BC (FuG 202). German radar operating at 490 MHz and using three CRT displays.

Lichtenstein SN (FuG 220). German radar operating at 37.5–118 MHz and using two CRT displays. Designed to resist Window ECM.

Magic. American decrypted Purple traffic.

magnetron. Power-tube oscillator that converts voltage at the cathode directly to microwave energy at 3 gigahertz (GHz) and above.

Mammut. German fixed long-range early-warning radar operating at 1.5–4.8 m.

Mandrel. British ECM designed to jam Freya.

Monica. British radar fitted in the tail of bombers to warn of attacking fighters and operating at 1.5 m.

multiplexing. Carrying more than one electronic signal on the same circuit.

Naxos. German ECM designed to receive H2S emissions.

Neotechnic. The technical era, still current, that began with electric communication in the mid-1800s but is most often associated with electrical energy transmission and the use of the internal combustion engine starting c. 1900.

Oboe. British navigational system using two radar transmitters on the ground and one responding transmitter in a bomber to allow precise marking of targets.

plan position indicator (PPI). CRT display that integrated information from ground and AI radar and from IFF to give the ground controller a clear picture of a fighter intercepting potential enemy aircraft.

Poulsen arc. Continuous wave generator using an electric arc to produce low radio frequencies.

proximity-fuse shell. Antiaircraft gun shell containing miniature radar to make the shell explode when it comes near aircraft.

pulsejet. A simple jet engine fitted to German V-1 cruise missiles.

pulse-code modulation (PCM). Digital pulse modulation that transmits information by allowing the amplitude of the pulses to reach only discrete values.

pulse width. The duration of a burst of electromagnetic energy.

Purple. Japanese Foreign Ministry's version of Enigma.

radar. American abbreviation for "radio direction and range-finding." A continuous or pulsed beam of electrons at the higher radio frequencies is sent out from an antenna. This beam reflects from most objects, and the weak return signal is usually displayed on a CRT.

radio direction finding (RDF). Commercial use of radio beacons spaced at reasonable intervals on the ground to allow a vehicle to locate itself by triangulation. A commercial beacon emitted signals so that a vehicle could follow them to their source. Military RDF located enemy vehicles by homing on their transmissions.

radio-frequency generator. Generator that produces damped or continuous waves, which can be modulated to carry signals. Continuous waves propagate more easily with less power.

SCR 584. American centimetric radar designed to be coupled with antiaircraft guns and proximity-fuse shells to destroy enemy aircraft.

Seetakt. German ship-based fixed radar operating on 375 MHz and used to direct naval gunfire.

Serrate. British ECM designed to receive Lichtenstein SN emissions.

spark-transmission generator. Damped wave generator capable of producing low to medium radio frequencies.

TAT. Transatlantic submarine telephone lines laid for Bell Telephone. Seven TATs were laid.

Typex. British code machine.

Ultra. British decrypted Enigma traffic.

vacuum tube. Continuous wave generator that could produce radio frequencies as high as about 600 MHz.

very high frequency (VHF) band. The radio band between 30 MHz and 300 MHz.

Wassermann. German scanning long-range early-warning radar operating at 1.2–4.0 m.

wavelength. The distance between identical points of two adjacent waves, inversely proportional to the frequency and directly proportional to the speed of propagation, which for electromagnetic phenomena is the speed of light.

Wilde Sau (lit., "wild boar"). German night-fighting technique using day fighters to intercept bombers illuminated from below by the light of burning cities.

Window. British ECM using aluminum foil strips.

world-economy. An economic but not necessarily political entity structured by capitalist principles encouraging relatively open exchange between disparate regions.

world-system. A world-economy within which there is unequal exchange that tends to benefit the core polities. The division of labor is based less on the class structure within a given polity than on differences between geographical regions.

Würzburg. German mobile scanning radar operating at 53 cm, used mostly by the Flak to direct antiaircraft guns.

Würzburg-Riese. Enlarged Würzburg used in fixed locations as long-range early-warning radar.

X-Verfahren. German navigational system using a high-frequency radio approach beam followed by a bomber and three intersecting beams that allowed automatic release of bombs.

References

Unpublished Materials and Research Collections

Cable and Wireless Archives, London.
Imperial War Museum, Aviation Collection, Duxford, England.
Imperial War Museum, London.
Library of the Institution of Electrical Engineers, London.
Public Records Office, London.
Royal Air Force Museum, Hendon, England.

Periodicals and Encyclopedias

Electrician. Weekly trade newspaper, London.
Gibilisco, Stan, and Neil Sclater, eds. 1990. *Encyclopedia of Electronics.* 2d ed. Blue Ridge Summit, Pa.: TAB.
Marconigraph. Monthly house journal of the Marconi Company.
Telefunken Zeitung. Monthly house journal of the Telefunken Company.
Wireless World. Sequel to the *Marconigraph,* published monthly, April 1913–March 1920; fortnightly, April 1920–March 1922; then weekly.
Zodiac. Monthly house journal of the Eastern and Associated Telegraph Companies, later Imperial and International Communications, now Cable and Wireless.

Books and Articles

Abramson, Albert. 1987. *The History of Television, 1880–1941.* Jefferson, N.C.: McFarland.
———. 1995. *Zworykin, Pioneer of Television.* Urbana: University of Illinois Press.
Aders, Gebhard. 1992. *History of the German Night Fighter Force, 1917–1945.* Translated and edited by Alex Vanags-Baginski assisted by Brendan Gallagher. Rev. ed. London: Crécy. Original English edition, London: Jane's, 1979.
Aitken, Hugh G. J. 1985a. *The Continuous Wave: Technology and American Radio, 1900–1932.* Princeton: Princeton University Press.
———. 1985b. Reprint. *Syntony and Spark: The Origins of Radio.* Princeton: Princeton University Press. Original edition, New York: Wiley, 1976.
Albert, Arthur Lemuel. 1934. *Electrical Communication.* New York: Wiley.
Alexanderson, E. F. W. 1984. "Transoceanic Radio Communication." *Proceedings of the IEEE* 72:626–33. Originally published in *Transactions of the American Institute of Electrical Engineers* 38 (1919): 1269–85.
Allison, David Kite. 1981. *New Eye for the Navy: The Origin of Radar at the Naval Research Laboratory.* Washington, D.C.: Naval Research Laboratory.
Andrew, Christopher. 1985. "F. H. Hinsley and the Cambridge Moles." In Langhorne, 22–40.
Andrews, C. F., and E. B. Morgan. 1981. *Supermarine Aircraft since 1914.* Annapolis: Naval Institute Press.
———. 1988. *Vickers Aircraft since 1908.* London: Putnam.
Appleyard, Rollo. 1930. *Pioneers of Electrical Communication.* New York: Macmillan.

Archer, Gleason L. 1938. *History of Radio to 1926.* New York: American Historical Society.

Arnold, H. D., and Lloyd Espenscheid. 1923. "Transatlantic Radio Telephony." *Electrician* 91 (19 October): 424–26, (2 November): 478–82.

Asker, James R. 1991. "Upstart Satellite Companies Press for New Telecommunications World Order." *Aviation Week & Space Technology,* 7 October, 50–63.

Austin, B. A. 1995. "Wireless in the Boer War." In IEE, 44–50.

Baker, W. J. 1972. *A History of the Marconi Company.* New York: St. Martin's.

Barkhausen, Heinrich. 1923. *Elektronen-Röhren.* Leipzig: S. Hirzel.

Barnes, C. H. 1987. *Handley Page Aircraft since 1907.* 2d ed., rev. London: Putnam.

———. 1988. *Bristol Aircraft since 1910.* 3d ed., rev. London: Putnam.

———. 1989. *Shorts Aircraft since 1900.* Revised by Derek James. Annapolis: Naval Institute Press.

Barnett, Corelli. 1988. "Technology, Education, and Industrial and Economic Strength." *Journal of the Royal Society of Arts* 127:118–29.

Barty-King, Hugh. 1979. *Girdle Round the Earth.* London: Heinemann.

Bauer, Arthur O. n.d. *Deckname "Würzburg": Ein Beitrag zur Erhellung der Geschichte des geheimnisumwitterten deutschen Radargeräts, 1937–1945.* Herten: Verlag Historischer Technikerliteratur.

———. 1995. "Receiver and Transmitter Development in Germany, 1920–1945." In IEE, 76–82.

Baugh, Daniel A. 1987. "British Strategy during the First World War in the Context of Four Centuries: Blue-Water versus Continental Commitment." In Masterson, 85–110.

Belfort, Roland. 1928. "The Imperial Cable-Radio Problem." *Electrician* 100 (23 March): 323–24.

Berry, Brian J. L. 1991. *Long-Wave Rhythms in Economic Development and Political Behavior.* Baltimore: Johns Hopkins University Press.

Bewsher, Paul. 1919. *The Bombing of Bruges.* London: Hodder & Stoughton.

Bidwell, Shelford, and Dominick Graham. 1982. *Fire-Power: British Army Weapons and Theories of War, 1904–1940.* London: Allen & Unwin.

Black, R. M. 1983. *The History of Electric Wires and Cables.* London: Peregrinus.

Blackwell, O. B. 1923. "Recent Transatlantic Telephone Tests." *Electrical Communication* 1:12–14.

Blanc, E., G. Dupin, R. Jocteur, A. P. LeClerc, A. Mins, J. Thiennot, and J. P. Trezeguet. 1989. "Optical Fiber Submarine Transmission Systems." *Commutation and Transmission* 11, special issue: 75–94.

Blouet, Brian W. 1987. *Halford Mackinder: A Biography.* College Station: Texas A&M University Press.

Blumtritt, Oskar. 1996. "Electron-Tube Theories: Part of an Emerging Technoscience." Paper presented at the annual meeting of the Society for the History of Technology, London.

Blumtritt, Oskar, Hartmut Petzold, and William Aspray, eds. 1994. *Tracking the History of Radar.* Piscataway, N.J.: IEEE.

Boettinger, Henry M. 1989. "And That Was the Future . . . Telecommunications: From Future-determined to Future-determining." *Futures,* June, 277–90.

Bowen, E. G. 1987. *Radar Days.* Bristol: Hilger.

———. 1988. "The Tizard Mission to the USA and Canada." In Burns, ed., 296–307.

Bowman, Martin W., and Tom Cushing. 1996. *Confounding the Reich: The Operational History of 100 Group (Bomber Support) RAF.* Cambridge: Patrick Stephens.

Bown, Ralph. 1937. "Transoceanic Radio Telephone Development." *Proceedings of the Institute of Radio Engineers* 25:1124–35.

Breit, G., and M. A. Tuve. 1928. "A Test of the Existence of the Conducting Layer." *Physical Review* 28:554–75.

Brereton, Captain F. S. 1911. *The Great Aeroplane.* London: Blackie.

———. 1913. *The Great Airship.* London: Blackie.

Bright, Charles. [1898] 1974. *Submarine Telegraphs; Their History, Construction, and Working.* Reprint. New York: Arno.

British Intelligence Operations Sub-Committee (BIOS). 1950. *Telecommunications and Equipment in Germany during the Period 1939–1945.* London: BIOS, HMSO.

Brittain, James E. 1984. "Commentary on Espenscheid & Strieby." *Proceedings of the IEEE* 72:731.

Brooks, Sydney. 1931. "Electrical Communications in 1930." *Electrical Communication* 9:156–60.

Bruce, J. M. 1957. *British Aeroplanes, 1914–1918.* New York: Funk & Wagnall's.

Bruch, W. 1959. "Stufen der Fernsehempfängerentwicklung in Deutschland" (Stages in TV receiver development in Germany). *Telefunken Zeitung* 32: 229–36, 291.

Buderi, Robert. 1996. *The Invention That Changed the World: How a Small Group of Radar Pioneers Won the Second World War and Launched a Technological Revolution.* New York: Simon & Schuster.

Burch, L. L. R. 1980. *The Flowerdown Link, 1918–1978: A Story of Telecommunications and Radar throughout the Royal Flying Corps and the Royal Air Force.* By the author.

Burns, Russell W. 1986. *British Television: The Formative Years.* London: Peregrinus.

———. 1988. "The Background to the Development of Early Radar, Some Naval Questions." In Burns, ed., 1–28.

———, ed. 1988. *Radar Development to 1945.* London: Peregrinus.

Bussey, Gordon. 1990. *Wireless: The Crucial Decade. History of the British Wireless Industry, 1924–34.* London: Peregrinus.

Buttner, Harold H. 1931. "The Role of Radio in the Growth of International Communication." *Electrical Communication* 9:249–54.

Cable and Wireless. n.d. *A Short History of Cable and Wireless.* London: Cable and Wireless.

Calamia, M., and R. Palandri. 1988. "The History of the Italian Radio Detector Telemetro." In Burns, ed., 97–105.

Callick, E. B. 1990. *Metres to Microwaves: British Development of Active Components for Radar Systems, 1937 to 1945.* London: Peregrinus.

———. 1995. "VHF Communications at RAE, 1937–1942." In IEE, 153–60.

Campanella, S. J., J. V. Evans, T. Muratani, and P. Bartholome. 1990. "Satellite Communications Systems and Technology, circa 2000." *Proceedings of the IEEE* 78:1039–56.

Cantelon, Philip L. 1995. "The Origins of Microwave Telephony—Waves of Change." *Technology and Culture* 36:560–82.

Carver, Field Marshall Lord. 1979. *The Apostles of Mobility: The Theory and Practice of Armoured Warfare.* New York: Holmes & Meier.

Chisholm, Roderick. 1953. *Cover of Darkness.* London: Chatto & Windus.

Clayton, Aileen. 1993. Reprint. *The Enemy Is Listening: The Story of the Y Service.* London: Crécy. Original edition, London: Hutchinson, 1980.

Coales, J. F., and J. D. S. Rawlinson. 1988. "The Development of UK Naval Radar." In Burns, ed., 53–98.

Coates, Vary T., and Bernard Finn. 1979. *A Retrospective Technology Assessment: Submarine Telegraphy. The Transatlantic Cable of 1866.* San Francisco: San Francisco Press.

Cogar, William B., ed. 1988. *Naval History: The Sixth Symposium of the U.S. Naval Academy.* Wilmington, Del.: Scholarly Resources.

Collins, Perry McDonough. 1962. *Siberian Journey Down the Amur to the Pacific, 1956–1857.* Edited by Charles Vevier. Madison: University of Wisconsin Press.

Constable, A. R. 1995. "The Birth Pains of Radio." In IEE, 14–19.

Corum, James S. 1992. *The Roots of Blitzkrieg: Hans von Seeckt and German Military Reform.* Lawrence: University Press of Kansas.

Craemer, A. 1924. "The European Telephone Network." *Electrician* 92 (7 March): 286–89.

Cross, Robin. 1987. *The Bombers: The Illustrated Story of Offensive Strategy and Tactics in the Twentieth Century.* New York: Macmillan.

Crowther, J. G., and R. Whiddington. 1948. *Science at War.* New York: Philosophical Library.

Cusins, A. G. T. 1921. "Development of Army Wireless during the War." *Journal of the Institution of Electrical Engineers* 59:763–70.

Daniels, Josephus. 1963. *The Cabinet Diaries of Josephus Daniels, 1913–1921.* Edited by David Cronon. Lincoln: University of Nebraska Press.

Deloraine, E. M. 1930. "The Use of Short Waves in Radio Communication." *Electrical Communication* 8:213–15.

Denny, Ludwell. 1930. *America Conquers Britain: A Record of Economic War.* New York: Knopf.

Devereux, Tony. 1994. "Strategic Aspects of Radar at Sea." In Blumtritt, Petzold, and Aspray, 157–70.

Donaldson, Frances. 1962. *The Marconi Scandal.* New York: Harcourt, Brace & World.

Douglas, Susan J. 1987. *Inventing American Broadcasting, 1899–1922.* Baltimore: Johns Hopkins University Press.

Drea, Edward J. 1992. *MacArthur's ULTRA: Codebreaking and the War against Japan, 1942–1945.* Lawrence: University Press of Kansas.

Dummer, G. W. A. 1995. "The Design, Development, and Use of Radar Trainers." Paper presented at the conference "The Origins and Development of Radar: Research, Industry, and the Armed Services," Bournemouth University, 8 December.

Duncan, Rudolph L., and Charles E. Drew. 1931. *Radio Telegraphy and Telephony.* New York: Wiley.

Dunlap, Orrin E., Jr. 1941. *Marconi: The Man and His Wireless.* New York: Macmillan.

Eccles, W. M. 1923. "The Amateur's Part in Wireless Development." *Wireless World* 13 (10 October): 50–54.

Eckersley, P. P. 1920. "Duplex Wireless Telephony: Some Experiments on Its Application to Aircraft." *Journal of the Institution of Electrical Engineers* 58:555–71.

Edgerton, David. 1991. *England and the Aeroplane: An Essay on a Militant and Technological Nation.* London: Macmillan.

Erickson, J. 1988. "The Air Defence Problem and the Soviet Radar Programme, 1934/35–1945." In Burns, ed., 229–34.

Erskine-Murray, J. 1921. "Wireless in the Royal Air Force." *Journal of the Institution of Electrical Engineers* 59:693–700.

Espenscheid, Lloyd. 1937. "The Origin and Development of Radiotelephony." *Proceedings of the Institute of Radio Engineers* 25:1101–23.

Espenscheid, Lloyd, C. N. Anderson, and Austin Bailey. 1925. "Atlantic Radio Telephone Transmission." *Electrician* 95 (14 April): 175–77.

Espenscheid, Lloyd, and M. E. Strieby. 1984. "Wide Band Transmission over Coaxial Lines." *Proceedings of the IEEE* 72:732–41. Originally published in *Electrical Engineering* 53 (1934): 1371–80.

Exton, C. H. 1995. "The Secret Factory for the Manufacture of Airborne Radar Equipment at Malmesbury in Wiltshire." Paper presented at the conference "The Origins and Development of Radar: Research, Industry, and the Armed Services," Bournemouth University, 8 December.

Fagen, M. D., ed. 1975. *A History of Engineering and Science in the Bell System: The Early Years (1875–1925)*. New York: Bell Telephone Laboratories.

Fox, Edward Whiting. 1991. *The Emergence of the Modern World: From the Seventeenth to the Twentieth Century*. Cambridge, Mass.: Blackwell.

Franklin, C. S. 1922. "Short-Wave Directional Wireless Telegraphy." *Electrician* 88 (19 May): 593–94.

Freris, L. L. 1995. "A Re-appraisal of the Goldschmidt Generator." In IEE, 71–75.

Friedberg, Aaron L. 1988. *The Weary Titan: Britain and the Experience of Relative Decline, 1895–1905*. Princeton: Princeton University Press.

Fritzsche, Peter. 1992. *A Nation of Fliers: German Aviation and the Popular Imagination*. Cambridge, Mass.: Harvard University Press.

Garlinski, Józef. 1979. *The Enigma War*. New York: Charles Scribner's Sons.

Garnet, Robert W. 1985. *The Telephone Enterprise: The Evolution of the Bell System's Horizontal Structure, 1876–1909*. Baltimore: Johns Hopkins University Press.

Genovese, Eugene D., and Leonard Hochberg, eds. 1989. *Geographic Perspectives in History*. Oxford: Blackwell.

Gibbons, Floyd. [1927] 1959. *The Red Knight of Germany: The Story of Baron von Richthofen, Germany's Great War Bird*. Reprint. New York: Bantam.

Gill, F. 1929. "International Telephony." *Electrical Communication* 7:190–99.

Goldsmith, Alfred N. 1918. *Radio Telephony*. New York:

Goldstein, Joshua S. 1988. *Long Cycles: Prosperity and War in the Modern Age*. New Haven: Yale University Press.

Gollin, Alfred. 1984. *No Longer an Island: Britain and the Wright Brothers, 1902–1909*. London: Heinemann.

———. 1989. *The Impact of Air Power on the British People and Their Government, 1909–14*. London: Macmillan.

Gossling, B. S. 1920. "The Development of Thermionic Valves for Naval Use." *Journal of the Institution of Electrical Engineers* 58:670–703.

Green, William. 1957. *Famous Fighters of the Second World War*. London: Macdonald.

———. 1959. *Famous Bombers of the Second World War*. London: Macdonald.

———. 1969. *Famous Bombers of the Second World War, Second Series*. New York: Doubleday.

———. 1970. *Warplanes of the Third Reich*. New York: Gallahad.

Guerlac, Henry E. 1987. *Radar in World War II*. Vol. 8, *The History of Modern Physics, 1800–1950*. Los Angeles: Tomash for American Institute of Physics.

Gunston, Bill. 1976. *Night Fighters: A Development and Combat History*. New York: Scribner's.

———. 1989. *World Encyclopedia of Aero Engines*. 2d ed., rev. Wellingborough, Northamptonshire: Patrick Stephens.

Haigh, K. R. 1968. *Cableships and Submarine Cables*. Washington, D.C.: U.S. Underseas Cable.

Hall, Peter, and Paschal Preston. 1988. *The Carrier Wave: New Information Technology and the Geography of Innovation, 1846–2003*. London: Unwin Hyman.

Hallion, Richard P. 1984. *Rise of the Fighter Aircraft, 1914–1918*. Baltimore: Nautical & Aviation.

Hanbury Brown, Robert. 1991. *Boffin: A Personal Story of the Early Days of Radar, Radio Astronomy, and Quantum Optics*. Bristol: Hilger.

Hare, Paul R. 1990. *The Royal Aircraft Factory*. London: Putnam.

Headrick, Daniel R. 1988. *The Tentacles of Progress: Technology Transfer in the Age of Imperialism, 1850–1940*. New York: Oxford University Press.

———. 1991. *The Invisible Weapon: Telecommunications and International Politics, 1851–1945*. New York: Oxford University Press.

———. 1994. "Shortwave Radio Communication and Its Impact on International Telecommunications between the Wars." *History and Technology* 11:21–32.

Hinchliffe, Peter. 1996. *The Other Battle: Luftwaffe Night Aces versus Bomber Command.* Osceola, Wis.: Motorbooks International.

Hinsley, F. H., E. E. Thomas, C. F. G. Ransom, and R. C. Knight. 1981. *British Intelligence in the Second World War: Its Influence on Strategy and Operations.* Volume 2. London: HMSO.

Hinsley, F. H., E. E. Thomas, C. A. G. Simkins, and C. F. G. Ransom. 1988. *British Intelligence in the Second World War. Its Influence on Strategy and Operations.* Vol. 3, pt. 2. London: HMSO.

Hitler, Adolf. [1927] 1971. *Mein Kampf.* Translated by Konrad Heiden. Reprint. Boston: Houghton Mifflin.

Hobsbawm, Eric. 1994. *The Age of Extremes: A History of the World, 1914–1991.* New York: Pantheon.

Hobson, J. A. 1938. *Imperialism: A Study.* 3d ed. London: Allen & Unwin.

Homze, Edward L. 1976. *Arming the Luftwaffe: The Reich Air Ministry and the German Aircraft Industry, 1919–1939.* Lincoln: University of Nebraska Press.

Hoth, D. F. 1962. "The T1 Carrier System." *Bell Laboratories Record* 40:358–63.

Hough, Richard. 1983. *The Great War at Sea, 1914–1918.* New York: Oxford University Press.

Howard-Williams, Jeremy. 1992. Reprint. *Night Intruder: A Personal Account of the Radar War between the RAF and Luftwaffe Nightfighter Forces.* Shrewsbury: Airlife. Original edition, Newton Abbot, Devon: David & Charles, 1976.

Howeth, L. S. 1963. *History of Communications-Electronics in the United States Navy.* Washington, D.C.: GPO.

Howse, Derek. 1993. *Radar at Sea: The Royal Navy in World War 2.* Annapolis: Naval Institute Press.

Hudson, Heather E. 1990. *Communication Satellites: Their Development and Impact.* New York: Free Press.

Hugill, Peter J. 1990. "'Miantonomoh,' 'Monadknock,' and Manifest Destiny." Paper presented at the annual meeting of the Southwestern Association of American Geographers, Austin, Tex.

———. 1993. *World Trade since 1431: Geography, Technology, and Capitalism.* Baltimore: Johns Hopkins University Press.

Hugill, Peter J., and D. Bruce Dickson, eds. 1988. *The Transfer and Transformation of Ideas and Material Culture.* College Station: Texas A&M University Press.

Institution of Electrical Engineers (IEE). 1995. *100 Years of Radio.* London: IEE.

Inglis, Andrew F. 1990. *Behind the Tube: A History of Broadcasting Technology and Business.* Boston: Focal.

Innis, Harold Adams. 1950. *Empire and Communications.* Oxford: Clarendon.

———. 1951. *The Bias of Communication.* Toronto: University of Toronto Press.

Interpol. 1995. *International Crime Statistics, 1991–92.* Lyons: International Criminal Police Organization.

Isted, G. A. 1991. "Guglielmo Marconi and the History of Radio—Parts I & II." *GEC Review* 7:45–56, 110–22.

Jackson, A. J. 1987. *De Haviland Aircraft since 1909.* Revised by R. T. Jackson. Annapolis: Naval Institute Press.

———. 1990. *Avro Aircraft since 1908.* Revised by R. T. Jackson. London: Putnam.

James, Derek. 1991. *Westland Aircraft since 1915.* Annapolis: Naval Institute Press.

James, Admiral Sir William. 1956. *The Code Breakers of Room 40.* New York: St. Martin's.

Jewett, Frank B. 1940. "A Quarter Century of Transcontinental Telephony: A Review of Significant Developments in Long-Distance Telephony since the First Transcontinental Conversation in 1914." *Electrical Engineering* 59:3–11.

Johnen, Wilhelm. 1994. *Duel under the Stars: A German Night Fighter Pilot in the Second World War.* London: Crécy.

Johns, W. E. [1935] 1992. *Biggles Learns to Fly.* Reprint. London: Red Fox.

Johnson, George, ed. 1903. *The All Red Line: The Annals and Aims of the Pacific Cable Project.* London: Stanford.

Johnston, R. J., Peter J. Taylor, and Michael Watts, eds. 1995. *Geographies of Global Change: Remapping the World of the Late Twentieth Century.* Cambridge, Mass.: Blackwell.

Jones, F. E. 1988. "OBOE—A Precision Ground Controlled Blind Bombing System." In Burns, ed., 319–29.

Kagan, Donald. 1995. *On the Origins of War and the Preservation of Peace.* New York: Doubleday.

Kasser, J. E. 1995. "Amateur Radio: Past, Present, and Future." In IEE, 120–27.

Kennan, George. 1910. *Tent Life in Siberia.* New York: Arno.

Kennedy, Paul M. 1971. "Imperial Cable Communications and Strategy, 1879–1914." *English Historical Review* 86:728–52.

———. 1987. *The Rise and Fall of the Great Powers: Economic Change and Military Conflict from 1500 to 2000.* New York: Random House.

Kennett, Lee. 1991. *The First Air War, 1914–1918.* New York: Free Press.

Kern, Ulrich. 1994. "Review Concerning the History of German Radar Technology up to 1945." In Blumtritt, Petzold, and Aspray, 171–83.

Kummritz, H. 1988. "German Radar Development up to 1945." In Burns, ed., 209–26.

Langhorne, Richard, ed. 1985. *Diplomacy and Intelligence during the Second World War: Essays in Honour of F. H. Hinsley.* Cambridge: Cambridge University Press.

Lasch, Christopher. 1995. *The Revolt of the Elites and the Betrayal of Democracy.* New York: Norton.

Lenin, Vladimir Ilyich. [1917] 1939. *Imperialism, the Highest Stage of Capitalism.* Reprint. New York: International.

Levine, Alan J. 1992. *The Strategic Bombing of Germany, 1940–1945.* Westport, Conn.: Praeger.

Liddell Hart, Basil. 1925. *Paris, or the Future of War.* New York: Dutton.

———. 1938. *Through the Fog of War.* New York: Random House.

———. 1944. *Thoughts on War.* London: Faber & Faber.

Lipartito, Kenneth. 1989. *The Bell System and Regional Business: The Telephone in the South, 1877–1920.* Baltimore: Johns Hopkins University Press.

Lovell, Sir Bernard. 1991. *Echoes of War: The Story of H2S Radar.* Bristol: Hilger.

Mackinder, Halford J. 1900. "The Great Trade Routes (Their Connection with the Organization of Industry, Commerce, and Finance)." *Journal of the Institute of Bankers* 21:1–6, 137–55, 266–73.

———. 1904. "The Geographical Pivot of History." *Geographical Journal* 23:421–37.

———. [1919] 1942. *Democratic Ideals and Reality: A Study of the Politics of Reconstruction.* Reprint. New York: Henry Holt.

Macksey, Kenneth. 1983. *A History of the Royal Armoured Corps and Its Predecessors, 1914 to 1975.* Beaminster, Dorset: Newtown.

Maclaurin, W. Rupert, and R. Joyce Harman. 1949. *Invention and Innovation in the Radio Industry.* New York: Macmillan.

Mahan, Alfred Thayer. [1890] 1957a. *The Influence of Sea Power upon the French Revolution and Empire, 1793–1812.* 2 vols. Reprint. Boston: Little, Brown.

———. [1892] 1957b. *The Influence of Sea Power upon History, 1660–1783.* Reprint. New York: Hill & Wang.

Mance, Osborne. 1944. *International Telecommunications.* London: Oxford University Press.

Mann, F. J. 1946. "Federal Telephone and Radio Corporation. A Historical Review: 1909–1946." *Electrical Communication* 23:377–405.

Mann, Michael. 1993. *The Sources of Social Power.* Vol. 2, *The Rise of Classes and Nation-States, 1760–1914.* New York: Cambridge University Press.

Mason, Francis K. 1992. *The British Fighter since 1912.* Annapolis: Naval Institute Press.

———. 1994. *The British Bomber since 1914.* Annapolis: Naval Institute Press.

Masterson, Daniel M., ed. 1987. *Naval History: The Seventh Symposium of the U.S. Naval Academy.* Wilmington, Del.: Scholarly Resources.

Mauer, Marc. 1994. *Americans behind Bars: The International Use of Incarceration, 1992–1993.* Washington, D.C.: Sentencing Project.

Maynard, John. 1996. *Bennett and the Pathfinders.* London: Arms and Armour.

McCarthy, Thomas E. 1990. *The History of GTE: The Evolution of One of America's Great Corporations.* Stamford, Conn.: GTE.

McDonald, Philip B. 1937. *A Saga of the Seas: The Story of Cyrus W. Field and the Laying of the First Atlantic Cable.* New York: Wilson-Erickson.

McFarland, Stephen L. 1995. *America's Pursuit of Precision Bombing, 1910–1945.* Washington, D.C.: Smithsonian.

McFarland, Stephen L., and Wesley Phillips Newton. 1991. *To Command the Sky: The Battle for Air Superiority over Germany, 1942–1944.* Washington, D.C.: Smithsonian.

McLuhan, Marshall. 1964. *Understanding Media: The Extensions of Man.* New York: Signet.

McNeill, William H. 1982. *The Pursuit of Power: Technology, Armed Force, & Society since A.D. 1000.* Chicago: University of Chicago Press.

McRae, Hamish. 1994. *The World in 2020: Power, Culture, and Prosperity: A Vision of the Future.* London: Harper Collins.

Mehrtens, Herbert. 1994. "The Social System of Mathematics and National Socialism: A Survey." In Renneberg and Walker, 291–311.

Meinig, Donald W. 1972. "American Wests: Preface to a Geographical Interpretation." *Annals, Association of American Geographers* 62:159–84.

———. 1986. *The Shaping of America: A Geographical Perspective on 500 Years of History.* Vol. 1, *Atlantic America, 1492–1800.* New Haven: Yale University Press.

———. 1993. *The Shaping of America: A Geographical Perspective on 500 Years of History.* Vol. 2, *Continental America, 1800–1867.* New Haven: Yale University Press.

Mensch, Gerhard. 1979. *Stalemate in Technology: Innovations Overcome the Depression.* Cambridge, Mass.: Ballinger.

Meulstee, Louis. 1995. *Wireless for the Warrior: A Technical History of Radio Equipment in the British Army.* Vol. 1, *Wireless Sets No. 1–88.* Broadstone, Dorset: G. C. Arnold.

Modelski, George, and William R. Thompson. 1988. *Seapower in Global Politics, 1494–1993.* Seattle: University of Washington Press.

Molyneux-Berry, R. B. 1988. "Dr. Henri Gutton, French Radar Pioneer." In Burns, ed., 45–52.

Morris, Mandy. 1996. "Home and Garden Go to War: Landscape, Identity, Englishness, 1914–18." Paper presented at the annual meeting of the Association of American Geographers, Charlotte, North Carolina.

Nakajima, S. 1988. "The History of Japanese Radar Development to 1945." In Burns, ed., 243–58.

Nebeker, Frederick. 1996. "World War I and the Birth of Electronics." Paper presented at the annual meeting of the Society for the History of Technology, London.

Nissen, Jack, and A. W. Cockerill. 1987. *Winning the Radar War: A Memoir.* New York: St. Martin's.

Olver, A. D. 1995. "Trends in Antenna Design over 100 Years." In IEE, 83–88.

O'Neill, E. F. 1985. *A History of Engineering and Science in the Bell System: Transmission Technology (1925–1975).* Indianapolis: AT&T Bell Laboratories.

Oslin, George P. 1992. *The Story of Telecommunications.* Macon, Ga.: Mercer University Press.

Oswald, A. A., and E. M. Deloraine. 1926. "Transatlantic Telephony." *Electrician* 96 (4 June): 572–73, (25 June) 666–68.

Overy, R. J. 1980. *The Air War, 1939–1945*. New York: Stein & Day.

Page, R. M. 1988. "Early History of Radar in the US Navy." In Burns, ed., 35–44.

Paris, Michael. 1992. *Winged Warfare: The Literature and Theory of Aerial Warfare in Britain, 1859–1917*. Manchester: Manchester University Press.

Penrose, Harald. 1969. *British Aviation: The Great War and Armistice, 1915–1919*. New York: Funk & Wagnalls.

Petzold, Hartmut. 1996. "Tube Technologies and the Conquest of Frequencies." Paper presented at the annual meeting of the Society for the History of Technology, London.

Pocock, Rowland F. 1967. *German Guided Missiles of the Second World War*. New York: Arco.

———. 1988. *The Early British Radio Industry*. Manchester: Manchester University Press.

Price, Alfred. 1978. *Instruments of Darkness: The History of Electronic Warfare*. New York: Scribner's.

———. 1979. *The Bomber in World War II*. New York: Scribner's.

Prince, C. E. 1920. "Wireless Telephony on Aeroplanes." *Journal of the Institution of Electrical Engineers* 58:377–90.

Pritchard, David. 1989. *The Radar War: Germany's Pioneering Achievement, 1904–45*. Wellingborough, Northamptonshire: Patrick Stephens.

Pritchard, Wilbur L. 1977. "Satellite Communication—An Overview of the Problems and Programs." *Proceedings of the IEEE* 65:294–307.

Putley, E. 1988. "Ground Control Interception." In Burns, ed., 162–76.

Reich, Robert B. 1992. *The Work of Nations: Preparing Ourselves for 21st-Century Capitalism*. New York: Vintage.

Renneberg, Monika, and Mark Walker, eds. 1994. *Science, Technology, and National Socialism*. New York: Cambridge University Press.

Robertson, Scot. 1995. *The Development of RAF Strategic Bombing Doctrine, 1919–1939*. Westport, Conn.: Praeger.

Rosecrance, Richard N. 1986. *The Rise of the Trading State: Commerce and Conquest in the Modern World*. New York: Basic.

Rossi, John P. 1988. "'World Wide Wireless': The U.S. Navy, Big Business, Technology, and Radio Communications, 1919–22." In Cogar, 170–85.

Ruhmer, Ernst. 1908. *Wireless Telephony in Theory and Practice*. New York: Van Nostrand.

Runge, Peter K., and Patrick R. Trischitta, eds. 1986. *Undersea Lightwave Communications*. New York: IEEE Press.

Runge, W. T. 1959. "Vom Funksender zur Elektronik—das Arbeitsgebiet Telefunkens seit 1903" (Telefunken's field of activity since 1903—from spark transmitters to electronics). *Telefunken Zeitung* 32:7–11, 70–71.

Saward, Dudley. 1959. *The Bomber's Eye*. London: Cassell.

Schama, Simon. 1996. *Landscape and Memory*. New York: Knopf.

Schenck, H. H. 1975. *The World's Submarine Telephone Cable Systems*. Contractor Report 75–2. Washington, D.C.: U.S. Department of Commerce, Office of Telecommunications.

Schulz, Reinhard. 1959. "40 Jahre Trägerfrequenz-Technik" (Forty years of carrier frequency engineering). *Telefunken Zeitung* 32:11–21, 71.

Shannon, C. E. [1949] 1984. "Communication in the Presence of Noise." *Proceedings of the IEEE* 72:1192–1201. Originally published in *Proceedings of the Institute of Radio Engineers* 37 (1949): 10–21.

Shiers, George, ed. 1977a. *The Development of Wireless to 1920*. New York: Arno.

———. 1977b. *The Telephone: An Historical Anthology*. New York: Arno.

Siemens, Georg. 1957a. *History of the House of Siemens*. Vol. 1, *The Era of Free Enterprise*. Freiburg: Karl Alber.

————. 1957b. *History of the House of Siemens*. Vol. 2, *The Era of the World Wars*. Freiburg: Karl Alber.

Sinclair, Duncan. 1921. "The Wireless Stations of the British Commercial Airways." *Wireless World* 8 (19 February): 798–810.

Sloan, Norton Q., Jr. 1996. "William Thomson's Inventions for the Submarine Telegraph Industry: A Nineteenth Century Technology Program." Master's thesis, Harvard University.

Smith, Bradley F. 1993. *The Ultra-Magic Deals and the Most Secret Special Relationship, 1940–1946*. Novato, Calif.: Presidio.

Smits, F. M., ed. 1985. *A History of Engineering and Science in the Bell System: Electronics Technology (1925–1975)*. Indianapolis: AT&T Bell Laboratories.

Staal, M., and Weiller, J. L. C. 1988. "Radar Development in the Netherlands before the War." In Burns, ed., 235–42.

Standard Telephone and Cables (STC). 1958. *The Story of S.T.C., 1883–1958*. London: S.T.C.

Strutt, M. J. O., and A. van der Ziel. 1940. "A New Push-Pull Amplifier Valve for Decimetre Waves." *Philips Technical Review* 5:172–81.

Sturmey, S. G. 1958. *The Economic Development of Radio*. London: Duckworth.

Taylor, Peter J. 1993. *Political Geography: World-Economy, Nation-State, and Locality*. 3d. ed. New York: Wiley.

Thompson, Graham M. 1990. "Sandford Fleming and the Pacific Cable: The Institutional Politics of Nineteenth-Century Imperial Telecommunications." *Canadian Journal of Communications* 15:64–75.

Thompson, Robert Luther. 1947. *Wiring a Continent: The History of the Telegraph Industry in the United State, 1832–1866*. Princeton: Princeton University Press.

Thrift, Nigel. 1995. "A Hyperactive World." In Johnston, Taylor, and Watts, 18–35.

Thrower, K. R. 1992. *History of the British Radio Valve to 1940*. Ropley, Hampshire: MMA International.

Tilly, Charles. 1989. "The Geography of European Statemaking and Capitalism since 1500." In Genovese and Hochberg, 158–81.

————. 1990. *Coercion, Capital, and European States, AD 990–1990*. Oxford: Blackwell.

Trischler, Helmuth. 1994. "Self-Mobilization or Resistance? Aeronautical Research and National Socialism." In Renneberg and Walker, 72–87.

Tyne, Gerald F. J. 1977. *Saga of the Vacuum Tube*. Indianapolis: Howard Sams.

Van der Bijl, Henrich. 1920. *The Thermionic Vacuum Tube and Its Applications*. New York: McGraw-Hill.

Vance, James E., Jr. 1992. Reprint. *Capturing the Horizon: The Historical Geography of Transportation since the Transportation Revolution of the Sixteenth Century*. Baltimore: Johns Hopkins University Press. Original edition, New York: Harper & Row, 1986.

————. 1995. *The North American Railroad: Its Origin, Evolution, and Geography*. Baltimore: Johns Hopkins University Press.

Vyvyan, R. N. 1974. *Marconi and Wireless*. Rev. ed. Wakefield, Yorkshire: E. P. Publishing. Originally published as *Wireless over Thirty Years* (London: Routledge & Kegan Paul, 1933).

Wallerstein, Immanuel. 1974. *The Modern World-System: Capitalist Agriculture and the Origins of the European World-Economy in the Sixteenth Century*. New York: Academic.

Wasserman, Neil H. 1985. *From Invention to Innovation: Long-Distance Telephone Transmission at the Turn of the Century*. Baltimore: Johns Hopkins University Press.

Weber, Max. 1983. *Max Weber on Capitalism, Bureaucracy, and Religion: A Selection of Texts*. Edited by Stanislav Andreski. London: Allen & Unwin.

Wegg, John. 1990. *General Dynamics Aircraft and Their Predecessors*. Annapolis: Naval Institute Press.

Wells, H. G. [1908] 1967. *The War in the Air*. Harmondsworth: Penguin.

———. 1933. *The Shape of Things to Come*. New York: Macmillan.

White, R. H. 1925a. "The Theory of the Wireless Beam." *Electrician* 94 (3 April): 392–93.

———. 1925b. "The Wireless Beam." *Electrician* 94 (10 April): 424–25, 430.

Williams, Oscar, ed. 1945. *The War Poets: An Anthology of War Poetry of the Twentieth Century*. New York: John Day.

Wilson, Geoffrey. 1976. *The Old Telegraphs*. London: Phillimore.

Wood, Derek. 1976. *Attack Warning Red: The Royal Observer Corps and the Defence of Britain, 1925 to 1975*. London: Macdonald & Jane's.

Wood, Derek, and Derek Dempster. 1961. *The Narrow Margin: The Battle of Britain and the Rise of Air Power, 1930–40*. New York: McGraw-Hill.

Woodbury, David O. 1931. *Communication*. New York: Dodd, Mead.

Yagi, Hidetsugu. 1984. "Beam Transmission of Ultra Short Waves." *Proceedings of the IEEE* 72:635–45. Originally published in *Proceedings of the Institute of Radio Engineers* 16 (1928): 715–41.

Young, Peter. 1983. *Power of Speech: A History of Standard Telephones and Cables, 1883–1983*. London: George Allen & Unwin.

Name Index

Aders, G., 195–196, 202, 207–209, 214
Aisenstein, Sergei, 188–189
Aitken, Hugh, 84, 88–89, 95
Alexanderson, E. F. W., 114, 116
Arco, Wilhelm von, 89
Armstrong, Edwin, 121, 124, 184–185

Baden-Powell, Robert, 154
Baird, John Logie, 186, 218
Baldwin, Stanley, 170, 175
Balfour, Arthur J., 12, 48
Barkhausen, Heinrich, 154
Bauer, Arthur O., 204
Belfort, Roland, 49, 131
Bell, Alexander Graham, 55, 71
Bennett, Don, 213
Bernstorff, Johann H. von, 47
Berry, Brian J. L., 19, 225
Bewsher, Paul, 159
"Biggles," 152
Bismarck, Otto von, 6
Blumlein, Alan, 189, 218
Boelcke, Oswald, 147–148, 166
Bonaparte, Napoleon, 5, 7, 10, 27
Boot, H. A., 191
Bowen, E. G., 22, 74, 183, 185, 189–190,
    192
Brancker, Major W. S., 148
Branly, Edouard, 88
Breit, G., 178
Brereton, F. S., 171
Bright, Charles, 30, 41
Brooke, J. M., 31
Brunel, Isambard Kingdom, 29
Buchanan, Pat, 244, 250
Buderi, Robert, 176
Bullard, Admiral, 120–121, 226

Callick, E. B., 185, 188
Campbell, George, 64–66
Carranza, Venustiano, 48
Cass, Lewis, 36
Chamberlain, Joseph, 45
Chamberlain, Neville, 22, 163, 175
Chappe, Claude, 25, 27
Chesterton, Cecil, 103, 105
Churchill, Winston, 170
Clark, Andrew Hill, 3
Clausewitz, Karl von, 3, 204
Clinton, Bill, 245
Cockroft, John, 192

Collingwood, Harry, 171
Collins, Perry, 36
Cornell, Thomas Ezra, 36
Corum, James, 168
Cross, Robin, 166

Daniels, Josephus, 121–122
De Forest, Lee, 19, 69, 115, 121, 152,
    226, 228
Dellinger, J. H., 118
Denny, Ludwell, 83–84, 225
Dickson, D. Bruce, 1
Dole, Sanford B., 45
Douglas, Susan, 84
Douhet, Guilio, 15, 168–169
Dowding, Sir Hugh, 23, 178–179, 190
Dummer, G. H., 187, 190

Eccles, W. H., 158
Edgerton, David, 164
Edison, Thomas Alva, 62
Elwell, C. F., 110–111, 113
Erskine-Murray, J., 153
Esau, Abraham, 203
Ewing, Alfred, 47–48

Faraday, Michael, 54, 85
Farnsworth, Philo, 186
Fessenden, Reginald, 115
Field, Cyrus, 31, 37
Fleming, Alexander, 69, 115
Fleming, Sandford, 45
Ford, Henry, 230
Fox, Edward Whiting, 5, 10, 18
Fox, Gustavus Vasa, 37
Friedberg, Aaron, 106–107, 164
Fuller, J. F. C., 157, 168

Galland, Adolf, 175, 182
Gargan, Dick, 188
Gaunt, John of, 9, 159
Gerard, James Watson, 48
Gill, Frank, 77
Gollin, Alfred, 10
Göring, Hermann, 164, 167, 181–182,
    194, 203–205
Green, William, 209

Hall, Peter, 19, 225, 246
Harris, Arthur, 213
Headrick, Daniel, 48, 227

Heaviside, Oliver, 30, 65–66, 87–88
Heinkel, Ernst, 209
Heisenberg, Werner, 203
Hertz, Heinrich, 85–86, 88
Hindenburg, Paul von, 76, 141
Hinsley, F. H., 145
Hitler, Adolf, 6, 22, 164, 166, 173, 182, 184, 203, 217–218, 221
Hobson, J. A., 2
Homze, Edward, 164
Hooper, Stanford C., 121–122
Hoover, Herbert, 123, 125, 227
Howard-Williams, Jeremy, 212
Howeth, L. S., 95, 121
Hülsmeyer, Christian, 176–177

Innis, Harold, 2–4, 17–18, 223, 239–241, 245–246
Isaacs, Godfrey, 101–105
Isaacs, Harry, 104
Isaacs, Rufus, 101–104

James, William, 48
Jameson, Annie, 92
Jameson-Davis, Henry, 92
Jarrell, Randall, 211
Johnen, Wilhelm, 207–208, 214
Johns, W. E., 152
Jones, E. E., 213

Kammhuber, Josef, 23, 193, 196, 204–205, 209
Karman, Theodore von, 156
Kellaway, F. G., 132
Kelvin, Lord (William Thomson), 30, 33, 65
Kennedy, Paul, 106–107
Kennelly, A. E., 87–88
Kesselring, Albert, 182
Kettering, Charles, 174
Kühnhold, R., 196

Lakanal, Joseph, 25, 27, 36
Lasch, Christopher, 250
Leeson, Nick, 249
Lefrance, Jean-Abel, 153–154
Lenin, Vladimir Ilyich, 2
Levine, Alan, 208
Liddell Hart, Basil, 1, 3, 139, 141, 157, 168
Lieben, Robert von, 76
Lindbergh, Charles, 163
Lloyd George, David, 101, 104
Lodge, Oliver, 86, 88, 97–98
Lovell, Bernard, 187

Mackay, Clarence, 43
Mackinder, Halford, 1–3, 5–7, 12–14, 88, 91, 159–160, 215, 221, 245, 251
Mahan, Louis Thayer, 9–10, 14–16, 159, 215, 217, 228, 251

Mann, Michael, 11
Marconi, Guglielmo, 20, 74, 86–90, 92–96, 117, 131, 177, 226–227, 240
Martel, Gifford, 168
Maury, Matthew F., 30–31
Maxwell, James Clerk, 30, 54, 65, 85
Maynard, John, 214
McLuhan, Marshall, 3, 250
McNeill, William H., 155
McRae, Hamish, 247
Meinig, Donald, 242
Meissner, Alexander, 76
Mensch, Gerhard, 225
Messerschmitt, Willy, 180
Midleton, earl of, 132
Milch, Erhard, 196
Mitchell, William, 15, 161, 168–169, 171
Moltke, H. J. L. von, 76, 140
Morse, Samuel F. B., 2
Mumford, Lewis, 1
Murray, Lord, 104
Mussolini, Benito, 169

Nally, E. J., 121
Nissen, Jack, 187, 200
Norman, Henry, 101–102, 105, 124

O'Etzl, August, 27
Overy, R. J., 210

Pemberton Billing, Noel, 148
Pender, John, 32, 39, 48
Perry, Percival, 230
Pickens, Francis W., 38
Plendl, Hans, 203
Pocock, Rowland F., 90–91, 158
Poulsen, Valdemar, 110
Prandtl, Ludwig, 17
Preece, William, 92, 96
Preston, Paschal, 19, 225, 246
Prince, C. E., 117
Prittwitz, K. L. von, 76, 141
Pupin, Michael, 66

Randall, John, 191
Reed, E. J., 37
Reeves, A. H., 213
Reich, Robert, 243
Rhode, Hans, 185
Ricardo, David, 8
Richtofen, Manfred von, 148, 168
Righi, Augusto, 86, 88
Rommel, Erwin, 145, 185
Roosevelt, Franklin D., 225
Roper, A., 43
Rosecrance, Richard, 5
Round, J. M., 153
Rowe, A. P., 23
Ruhmer, Ernst, 109, 114

Runge, Wilhelm, 197–198, 201, 203–204
Russell, Bertrand, 12

Samuel, Herbert, 101–103
Sarnoff, David, 186, 188, 219
Saward, Dudley, 214
Saward, George, 31
Sayn-Wittgenstein, Heinrich *Prinz* zu, 201
Schama, Simon, 13
Schmidt, "Beppo," 196, 212
Shakespeare, William, 159
Shoenberg, Isaac, 186, 189, 218
Sibley, Hiram, 36
Slaby, Adolph, 89, 96
Sloan, Norton, 30
Smith, Adam, 8
Spruance, Raymond, 144
Squier, George Owen, 43
Stone, John Stone, 66, 69
Strowger, Almon Brown, 61

Tank, Kurt, 196
Taylor, A. Hoyt, 177
Taylor, Peter, 223–224, 242
Tilly, Charles, 5–6, 241–242, 246, 251
Tirpitz, Alfred von, 164
Tizard, Henry, 162, 177, 183, 187
Trenchard, Hugh, 169, 171–172
Trenouth, John, 187

Turing, Alan, 145
Tuve, M. A., 178
Tweeddale, Lord, 45

Vail, Theodore, 55–56, 66
Van der Bijl, Heinrich, 153
von Eckhardt, Baron, 48
Vyvyan, R. N., 129, 153

Wallerstein, Immanuel, 223, 242–243
Wasserman, Neil H., 65–66
Watson Watt, Robert, 22, 178, 183, 187, 217
Webb, Sidney and Beatrice, 12
Weber, Max, 11, 249
Wellington, Arthur, 160
Wells, H. G., 12, 14, 160, 170–171
Wever, Walther, 182
White, Anthony C., 62
Whitehouse, Edward O. W., 30
Wilson, Woodrow, 47, 119–120, 226
Wright, Charles, 22

Yagi, Hidetsagu, 118
Yamamoto, Isoroku, 145, 240
Young, Bill, 188
Young, L. C., 177
Young, Owen, 121

Zimmerman, Arthur, 47
Zworykin, Vladimir, 186, 188–189, 219

Subject Index

AI Mark IV radar, 146
airborne intercept radar (AI), 23, 75, 146, 183, 185–186, 192, 206, 214
airplanes, World War I: fighter, 147; reconnaissance, 147–149, 156
air power: and "control without occupation," 171; evolution of, 165; and geostrategy, 10; origins, 160
Air-to-Surface Vessel (ASV) radar, 146, 214
Alabama claim, 37–38
Alaska Purchase, 37
"All Red [cable] Route," 45
alternator, 226; in continuous wave wireless, 113–115, 120
American Telegraph Co., 37
American Telephone and Telegraph Co. *See* AT&T
amplifier, vacuum tube, 62, 69
Amur River, 36, 38
Anglo-American Telegraph Co., 32, 34
antenna: beam, 115, 125, 128–129; as tuner, 89
anti-semitism: and German research effort, 203–204, 218; and Marconi scandal, 101, 103, 105
"apostles of mobility," 15
appeasement, 22, 163, 182
Associated Press, New York, 37
AT&T, 17, 53–57, 65, 75–77, 79–80, 121; and Atlantic telephone, 127–128
Atlantic, Battle of the, 240; HF/DF in, 135
Atlantic Telegraph Co., 30–31
attenuation, 61, 67
Automatic Electric Co., 61

bandwidth, 74
B-Dienst (German cryptanalytic organization), 143–144
beam radio, 49, 56, 115
Bell Telephone Co., 61–63, 65–67, 69, 76–77, 79–82, 115, 127, 239; and "grand system," 55; and *Telstar* program, 231; and transatlantic telephone, 228–230
Bell Telephone Laboratories, 53, 71, 75–76
Bellini-Tossi RDF, 134
Berlin, Battle of, 21, 161, 208, 212, 217

Berlin AI radar, 204, 209
Berlin Conference (1884–85), 12
Biggin Hill Experiment, 179, 190, 199
Bletchley Park (British decryption center), 145
blockade, 9, 15–16, 159, 170, 251
Boer War, wireless in, 141, 154
"bombe" (British decryption machine), 144–145
bomber: as bait, 164; crew attrition, 211; detection before radar, 174; development of strategic, 172–174, 182; self-defending, 169, 171–172, 211
Bomber Command, RAF, 169, 206, 208, 213, 215
bombing, strategic, 164–165; "area bombing," 168, 172; as blockade, 15, 160; by British in World War I, 166–167; cost of, 169, 172; German rejection of, 167–168; by Germans in World War I, 166–167; "terror bombing," 168, 172; by USAAC, 171; by Zeppelins, 170
Boozer ECM, 206
Britain, Battle of, 23, 163, 169, 170–172, 177–183, 217
British Broadcasting Corporation (BBC), 124, 218
British National Museum of Photography, Film, and Television, 186–187
British navy, and new technology, 91
broadcast radio, 122–124, 158
Bruneval raid, 199

C^2 (centralized command and control), 139–141
C^3 (centralized communications, command, and control), 91, 141–143, 146, 154–156, 177, 190, 192–193, 195, 199, 204, 213
Cable and Wireless, 49, 131, 227, 230. *See also* Imperial and International Communications
Cable Landing License Hearings (U.S. Congress), 43
cable ships: *Great Eastern*, 29, 33–34, 39; *Hooper*, 33
Canadian Independent Force, 154
Canadian Pacific RR, completion of, 45
carrier, 71–72, 80. *See also* duplex telegraphy; multiplex transmission

"carrier wave," 225, 237
Chain Home radar, 22–23, 184, 190, 194, 200, 213; and Battle of Britain, 182; origins, 178–179
coaxial cable transmission, telephone, 70, 75, 80
Cocos-Keeling Island repeater station, 46
code: Allied Naval Cipher 3, 143–144; German 0075, 48; German 13040, 47–48; Japanese naval JN-25, 144
code breaking, 17, 47–48, 141, 143; and Bletchley Park, 145
code machines. *See* Combined Cypher Machine; Enigma; Purple; Typex
Co-efficients (London dining club), 12
coherer, Branly, 88
Collins Line, 36
colonial conference, Ottawa (1903), 45
"Colossus" (British non-programmable decryption computer), 144
Combined Cypher Machine, 145
Comité Consultatif International des Communications Téléphoniques (CCI), 77, 79
commercial ciphers, 16, 21
Commercial Pacific Cable Co., 43, 46
Committee for Imperial Defence, 22, 43, 46, 162, 177, 193
Committee for the Scientific Study of Air Defence (Tizard Committee), 22, 162, 177
Commonwealth Conference (1958), and global telephone system (Compac), 230
Communications Satellite Corporation, (Comsat), 233
conferences, international radiotelegraph, 92–97, 123–124, 130, 154
continuous wave technology. *See* alternator; Poulsen arc; and vacuum tube
continuous wave wireless, 16, 84, 89, 102, 109–110
Coventry raid, 184, 217
crosstalk, 67, 229
cruise missiles, German VI, 192–193
cryptanalysis. *See* code breaking

decryption. *See* code breaking
devolution, 242–243, 250; and telecommunications, 245
digital signal, electric telegraph as, 28
direct distance dialing (DDD), 55
diurnal signal variation, 128
Dogger Bank, Battle of, 143
Dresden raid, 217
duplex telegraphy, 33–35

Eastern, 41, 45, 48–49, 131–132; and Commercial Pacific Cable Co., 43
Eastern and Associated Telegraph Companies, 32

Eastern Extension Australasia and China Telegraph Co. (Eastern Extension). *See* Eastern
ECM, 23, 184, 194, 196, 199, 200, 202, 206, 213
electromagnetic spectrum, 85; and first American national radio conference (1922), 123, 125
electronic countermeasures. *See* ECM; *and specific types*
EMI, 185–186, 189, 218; and merger with Marconi, 186
encryption, 21, 143. *See also* code
engineers, electronic, as "war surplus," 158
English Channel, 28, 82; microwave telephony across, 74
Enigma (code machine), 51, 135, 142, 145–146; commercial code machine, as basis for, 143
Eurospace, formation of, 233
experten, attrition of, 207–208, 217

Fanning Island repeater station, 46
Federal Telegraph Co., 102–103, 107, 142; as counter to RCA, 122; and De Forest, 115–116; formation, 110; and U.S. Navy, 107, 111, 113, 120
Fernseh Co., 219
Fighter Command, RAF, 169, 179
fighters: development after World War I, 174–175, 180–182; night, 190–191, 195–196, 206, 212; origins in World War I, 147; strategic, 172, 182
force multiplier, new technology as, 91, 144
Ford Motor Co., transatlantic telephone in management of, 229–230
Four Power Pact, 122, 125–126, 137; as patent pool, 130
frequency-division multiplex carrier, 35
Freya radar, 185, 193–194, 199–200

Gee radar guidance, 212–214
General Electric (GE), 114, 120–121, 226
General Telephone and Electronics Corporation (GTE), 55, 57
geopolitics, 163, 177, 215, 219; America, and Mahanian, 160–161, 217, 223; and communication, 4, 119; and devolution, 245; "Grab for Africa," 12; "Imperial overstretch," 106–107; and Japan, 161; Mackinderian, 2, 91, 160, 221; in multipolar world, 244; return to Mahanian, 159, 228, 251; trading *vs.* territorial states, 7–10. *See also* "moat defensive"
geostrategy, 9, 14–16, 21, 91–92, 169
geosynchronous satellites, 17–18, 231, 233, 235

German missiles, 192–193, 215
Germany, and European telecommuni-
    cations, 80
Gesellschaft für Elektroakustische und
    Mechanische Apparate (GEMA),
    197–198, 218
global wireless: Marconi's plans, 100–
    101, 103, 106; U.S. Navy's plans,
    106–107
ground-controlled interception (GCI),
    189, 191
ground waves, 89
Guernica, bombing of, 163
gutta-percha, 29

H2S radar, 196, 204, 206, 211–214
H2X radar, 214–215
Hamburg, Battle of, 23, 205, 212
hardware technologies, 2
Heartland, 3, 5–6, 12, 91, 159, 215
hegemonic struggle, 8, 16, 84, 107, 159,
    225–226, 228, 244
hegemony, 20, 28, 79, 138, 223, 237, 241
Heligoland Bight, Battle of, 15
Hertzian waves, 85–86, 88
heterodyne receiver, 121
high-frequency direction finding (HF/
    DF), 135, 146
high-frequency wireless transmission,
    15, 21, 125, 128. *See also* short-wave
    wireless transmission
Himmelbett, 193–194, 204, 209, 213
"Homebrew Club," 20, 113
hydrographic surveys, 30, 45

iconoscope, purchase by EMI, 186
Identification, Friend or Foe (IFF), 189,
    204
ideology, and research, 22, 203–204
Imperial and International Communica-
    tions (formerly Eastern), 49, 131–134
Imperial Cable Committee (British), 45
imperial chain: British, 83, 97–98, 101–
    102, 105, 118–119, 129; German, 46,
    104–105
Imperial Wireless and Cable Confer-
    ence (British), 49, 131
imperialism, 2, 242; cultural, 5; new, 12
Indo-European Telegraph Co., 39
information hegemony, 28, 223
International Chamber of Commerce,
    77
international radiotelegraph confer-
    ences, 92–97, 123–124, 130, 154
International Telecommunications
    Satellite Organization (Intelsat), 17–
    18, 231, 233, 235–236
International Telecommunications
    Union, 235
International Telephone and Telegraph
    Co. *See* IT&T

Internet, 20, 250
ionized reflecting layers in atmosphere,
    87, 89, 176, 178
isolationism, American, 223
IT&T, 57, 74, 185

Johnston Island repeater station, 45
Joint Purse agreement, 34
Jutland, Battle of, 143

Kammhuber line, 193, 209
Kennelly-Heaviside layer, 87, 133
klystron, 74–75
Kondratieff waves, 19–20, 223–225, 239

Lichtenstein AI radar, 193, 195, 201–
    202, 204, 206, 209, 213
line loading, 66–67
Lloyd's (insurance co.), and wireless,
    93–94, 134, 142
Lodge patents, 86, 92, 97, 99, 102
Long Lines Department, AT&T, 54
long-wave radio telephony, 133
long-wave theory, 18–20, 224. *See also*
    Kondratieff waves
Lorenz Co., 185, 212
Luftwaffe, 160–161, 167, 193, 195, 199,
    217; electronics technicians shortage
    (1943), 161; as "risk fleet," 164–165;
    social structure of, 201

Magic (decrypted Japanese diplomatic
    messages), 144
magnetron, 74–75, 198, 204, 206; devel-
    opment, 191–192
Mahanian doctrine, 161
Mahanian war of blockade. *See* blockade
Mammut radar, 199
Mandrel ECM, 200
Manhattan Project, 171
Manufacturing Core, 55, 65, 69, 79–80,
    228
Marconi Co., 16, 47, 49, 93; and Alexan-
    derson alternator, 114, 120; beam
    antenna development, 118; British
    company formed, 96; co-operation
    with RCA, 122; EMI merger, to devel-
    op television, 186; and geostrategy,
    88; and military wireless, 141–142;
    "non-intercommunication" policy,
    93; patents, 86, 92–93, 96–97, 102;
    Poulsen patent purchased by, 113;
    range obsession, 88; RDF develop-
    ment, 134; in Russian Empire, 98–
    99; United Wireless purchased by,
    96, 99, 103; and vacuum tube tech-
    nology, 115
Marconi Co., American, 96, 99, 103; sale
    to GE, 121, 226; and Western Union,
    100
Marconi scandal, 92, 97, 101–106; Com-

mons Select Committee hearings, 104–105; and Poulsen syndicate, 106; and U.S. Navy, 105–106
Massachusetts Institute of Technology, 66, 163
Maxwell's theory of electrodynamics, 30, 85–86, 88
mechanical telegraph, 25–27
Megalopolis, 55, 65, 69, 79–80, 228
megalopolitan regions, increasing power of, 246, 251
Miantonomoh, U.S.S., cruise of, 37
microwave: telephone relay across Canada, 230; transmission, 74–75
Midway, Battle of, 144, 240
mirror galvanometer, 33
missiles, as satellite launch vehicles, 231, 233, 235
"moat defensive," 170; air power, and loss of, 13–14, 160–161; American, 161, 215, 220, 224; British, 9, 159, 170; Japanese, 215; radar, and renewal of, 16, 21, 23, 175, 200, 221, 228; Soviet, 23. *See also* geopolitics
Monica radar, 206
Moonshine ECM, 200
multiple-unit steerable antenna, 133
multiplex transmission, 36, 71
Munich Agreement, 163, 175

Nachrichtenmittelversuchsanstalt, 197
Napoleonic Wars, 9–10, 27
NASA, 231
National Television Standards Committee (NTSC; American), 219
Naval Research Laboratory (NRL; American), 217–218
Naval Signal School (British), 217
Naxos ECM, 207, 214–215
Nazis, 161, 163, 164, 177, 182, 197, 202–203, 218, 221; and creation of European telephone system, 80–81; Lebensraum, 165; Oslo Report (on Nazi radar), 199
Necker Island repeater station, 45–46
Neotechnic, 1–2
Neuve Chappelle, Battle of, 139
New York, Newfoundland, and London Telegraph Co., 37
night blitz, 23, 183–184, 217
night fighters: improved, 206, 212; pioneering, 190–191, 195–196
Norden bombsight, 169, 171, 210, 215
Nuremberg raid, 206, 208, 217

Oboe radar guidance, 212–214
outer empire (Canada, New Zealand, Australia, and South Africa), 45

Paris Telegraph and Telephone Conference (1910), 77

patent pool: RCA as, 121; and wireless in World War I, 152–153
plan position indicator (PPI), 187, 189, 199, 204
Post Office (British), 49, 86, 89, 92–93, 96–97, 101, 124, 142; and American equipment, 132; beam wireless stations, 131, 138; Rugby wireless station, 129; and transatlantic telephone cable, 227, 229
Poulsen arc, 17, 95, 102, 106, 142, 226; patent, 110; syndicate role in Marconi scandal, 106; and U.S. Navy, 111
power demands, of wireless transmission, 115, 119, 126, 129–130, 137, 227
precision strategic bombing, as American doctrine, 171, 210
Project Relay satellite, RCA, 231, 233
protection *vs.* transfer costs, 250–251
proximity fuse radar anti-aircraft shell, 192
Pulse Code Modulation (PCM), 71, 75, 213, 230, 233
Purple (code machine), 144

radar, 19–23, 74, 158; Anglo-German origins, 162–163; centimetric, 191; and "death ray," 178; diffusion to America, 162; first patent, 176; and geostrategy, 15–16; ground work for, 87; and "moat defensive," 159; naval development of, 176–177; quality control problems, 201–202; and Radio Research Station, 178; Seeburg plotting table, 204; and television, 163, 227–228
radar countermeasures. *See* ECM
radar guidance (Gee, Oboe, X-Verfahren), 184, 212–214
radar type: AI Mark IV, 146; Berlin, 204, 209; British 1.5 m, 146, 184, 189–192, 196; Chain Home, 22–23, 178–179, 182, 184, 190, 194, 200, 213; Chain Home Low, 190; Freya, 185, 193–194, 199–200; H2S, 189, 196, 204, 206, 211–214; H2X, 214–215; Lichtenstein AI, 193, 195, 201–202, 204, 206, 209, 213; Monica tail warning, 196; other German ground (Mammut, Wasserman, Würzburg-Reise), 199; SCR 584, 192; Seetakt, 185, 194, 199–200; Würzburg, 193–194, 199–200
Radiation Laboratory, modeled on TRE, 162, 218–219
radio direction finding. *See* RDF
Radio Research Station (British), 178
Radio Society of Great Britain, 158
Radio Trust Congressional Inquiry, 124

RAF: Bomber Command, 169, 206, 208, 213, 215; Fighter Command, 169, 179; formation, 165–166

RCA, 17, 84–85, 107, 119–127, 137, 186, 226; and Atlantic telephone, 127; Project Relay satellite, 231, 233; and radar, 217, 219; and Radio Trust, 124

RDF, 134–136, 143, 146, 176, 183

Reichswehr, 167

repeater, electromechanical, 62, 69

repeater stations, 45–46

research, social organization of, 163, 210, 217; in Germany, 202–203; role of Jews, 188–189, 198

Resistenz, 17, 210

Reuters news agency, 39

Room 40, British Admiralty, 47–48, 143, 145–146

Royal Air Force. *See* RAF

Royal Flying Corps (RFC) and Royal Naval Air Service (RNAS), 160, 165–166

Royal Observer Corps (ROC), 179

Ruhr, Battle of, 212–213, 217

satellite telecommunications, 231, 233, 235–236

schrage musik armament, 206

science-based research, in Wilhelmine Germany, 90

science-based technology, at Bell, 66, 82

Seeburg table, 204

Seetakt radar, 185, 194, 199–200

self-defending bomber, 169, 171–172, 211

Serrate IV ECM, 206

ship-to-shore communications, 93, 96, 98–99, 142

short-wave wireless transmission, 125, 128. *See also* high-frequency wireless transmission

Siemens Brothers, 32–33

Siemens und Halske, 33, 61, 76, 89

single sideband transmission, 127

siphon recorder, 33

Slaby-Arco system of wireless telegraphy, 89, 94, 95

software technologies, 2, 8, 21–22, 239

space shuttle, 236

Standard Telephones and Cables (STC) (replaced IT&T), 74–76, 185

Stanford University, 111, 113

strategic bombers, 162, 220, 225

strategic bombing. *See* bombing, strategic

strategic fighters, 208, 211–212

Strowger Automatic Switching Co., 57, 61, 75

submarine cable, 2, 21; and hegemony, 225; and hydrographic surveys, 30,

45; interception of messages by British, 47–48, 140; landing rights, 84; repeater stations, 45–46; subsidies, 41; tariffs, 35, 39; and textiles, 32; transmission speed, 33

submarine cable, Atlantic: financing of, 31–32, 34, 39; Joint Purse agreement, 34

submarine cables and companies: American Telegraph and Cable Co. (1881, 1882), 34; Anglo-American (1873, 1874, 1880), 32, 34; Atlantic (1857, 1858), 29–31, 33, 36, 54, 65; Atlantic (1865, 1866), 29, 34, 54; Atlantic Telegraph Co., 30–31; Bering Straits, 36; Commercial Pacific Cable Co., 43, 46; Direct United States (1875), 34; Eastern and Associated Telegraph Companies, 32; Eastern Extension Australasia and China Telegraph Co., 41, 45, 49, 131–132; English Channel, 28; French Atlantic (1869, 1879), 34; Pacific, 43–46; Red Sea and India Telegraph Co., 39

submarine telecommunications cables, fiber-optic, 237

submarine telephone cable, 54, 228–231

"Sunday Soviets," 23

super-heterodyne receiver, 124

Swedish Roundabout, 47

syntony, concept of, 84

Tannenberg, Battle of, and telephony, 76, 140, 154

telecommunications hegemony, 223

Telecommunications Research Establishment. *See* TRE

Telefunken, 142, 197, 203, 218–219; carrier technology, 80; formation of, 89, 94; and Marconi, 97–98; rejection of first radar patent, 176; and U.S. navy, 95

telegrams, misunderstanding and war, 140

telegraph, 18–20; financing, 36; and geopolitics, 35–36; mechanical, 25–27; revenues, 37; and state bureaucracy, 11, 25; wireless, 16, 17

Telegraph Acts (1896; British), 93

Telegraph Construction and Maintenance Co. (TCM), 32, 39, 46

Telegraph Plateau (on Atlantic floor), 31

Telegraph Purchase Bill (British), 32

telephone, 19–20; automatic exchange, 57, 61; battery power for, 63; and Battle of Tannenberg, 76, 140, 154; "butterstamp" receiver, 62; cable transmission, 67; capacity, 69–70;

carbon transmitter, 19; dial, 57, 61; dual system of, 67; improvement in efficiency of, 62–63; manual exchange, 61; microphone transmitter, 62; and satellites, 231, 233, 235–236; short-wave wireless, 56; tariffs, 55–56, 69, 79; in Third Reich, 80–81; toll, 77; transcontinental, in America, 69–70, 228; transmission, 64; wired transatlantic, 229–230; wireless Atlantic, 116, 127–129, 133, 137; wireless in World War I, 151–152

television, 16, 19, 70–71, 176, 218–219, 227; British development of electronic, 184–189; electromechanical, 186–187; German, 197–198; and radar, 163, 227–228; as source of technicians for radar, 187–188, 198

Television Advisory Committee (British), 186

Telstar, 17–18, 231, 233

territorial state, 5, 7–8, 11–12, 18

Thames, as womb of British Empire, 13

thermionic rectifier, 69

Thomson-Houston Co., 61

Tizard Committee, 22, 162, 177

Tizard Mission, 144, 162, 192, 218–219

trading state, 5, 7–8, 11–12, 18

transcontinental telephone, in America, 69–70

transistor, 82

transmission: digital, 71; microwave, 74–75; theory of, 54, 65–66

TRE, 23, 162, 190, 217–218

Treaty of Six Nations, 36–37

triode vacuum tube, development of, 115

tuning: concept of, 84; patents, 82, 86, 89, 97

type 1 and type 2 communications media, 3–4, 7, 17, 20, 240–241, 245–246

Typex (code machine), 145

Ultra decryption, 51, 135, 144–146

United Fruit Co., 121

United States Army Air Corps (USAAC), 218, 225; development of strategic bombers, 162, 220; failure to develop radar, 161–162

United Wireless Co., 96, 99, 103, 119, 226

U.S. Navy: and hegemonic struggle, 226; and Marconi scandal, 105–106; and Radio Trust, 124; and telecommunications, 2, 16, 46, 94–95, 103, 105, 120

V1 missile, 215

vacuum tube, 69–70, 74, 76, 82; audion, 69; as continuous wave generator, 115; derived from light bulb, 152, 179; Lieben, 76; and radar, 176; "soft" and "hard," 116, 153; war surplus, 157–158

vacuum tube types: British R, 153; French TM, 153; German RE-16, 154; Phillips EF 50 and derivatives, 189–190

very high frequency (VHF) radio, 184–185, 188

"warfare state," Britain as, 164

The War in the Air (Wells), 15, 170–171

Washington Naval Conference (1922), 224–225

Wasserman radar, 199

"weary titan," Britain as, 106–107

Western Electric Co., 57, 77

Western Union, 27, 34–38, 53–54, 62; and Marconi Co., 100

Western Union Extension Co., 36

Westinghouse, 121, 124

Wilhelmine Germany, 23, 142

Window ECM, 194, 202, 206

wired telecommunications, and static warfare, 140

Wireless Experimental Establishment (British), 153

wireless sets: American Type A, 152; British Model 52A, 150; British Model 54A, 150; British Sterling, 149–150; British tank sets, 155; British Type 10, 158

Wireless Telegraph and Signal Co. (later Marconi Co.), 93

wireless telegraphy, 16, 17; in airplanes, 117; in artillery spotting, 147–151; across Atlantic, 87, 117; and the Boy Scouts, 154–155; "quenched spark" system, 89–90; spark, 16, 88–89; and *Titanic* disaster, 95; in war, 140–141

wireless telephony tariffs, 130–131

world leadership cycles, 20

world-system, 7, 18

X-Verfahren radar guidance, 184

Y Service (British), listening to German radio traffic, 200, 204

Yagi beam antenna, 118

Zeppelins: in electronic surveillance, 194; as phallic symbols, 13, 15

Zimmerman telegram, 47–48, 140

ABOUT THE AUTHOR

Peter J. Hugill was born in York, England, in 1945. He received a B.A. in geography and a graduate certificate in education from the University of Leeds, an M.A. in geography from Simon Fraser University, and a Ph.D. in geography from Syracuse University. He is the author of *World Trade since 1431: Geography, Technology, and Capitalism* (Johns Hopkins, 1993) and *Upstate Arcadia: Landscape, Aesthetics, and the Triumph of Social Differentiation in America* (Rowman & Littlefield, 1995) and co-editor, with D. Bruce Dickson, of *The Transfer and Transformation of Ideas and Material Culture* (Texas A&M, 1988) and, with Kenneth E. Foote, Kent Mathewson, and Jonathan M. Smith, of *Re-Reading Cultural Geography* (Texas, 1994). Peter Hugill is professor of geography at Texas A&M University.

Library of Congress Cataloging-in-Publication Data

Hugill, Peter J.
    Global communications since 1844 : geopolitics and technology /
Peter J. Hugill.
        p.   cm.
    "Published in cooperation with the Center for American Places, Santa Fe,
New Mexico, and Harrisonburg, Virginia"—
    Includes bibliographical references and index.
    ISBN 0-8018-6039-3 (alk. paper). — ISBN 0-8018-6074-1 (pbk. : alk. paper)
    1. Telecommunication—History.   2. Geopolitics—History.   I. Title.
TK5102.2.H84   1999
384'.09—dc21   98-42971
                                                                        CIP